T0235336

Direction of Time

Sergio Albeverio · Philippe Blanchard

Editors

Direction of Time

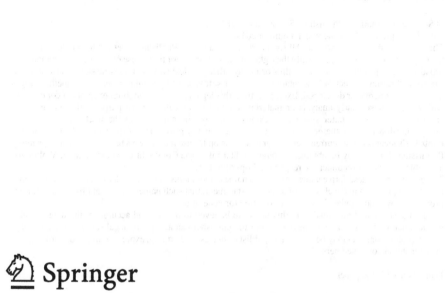

Springer

Editors
Sergio Albeverio
IAM und HCM
Universität Bonn
Bonn, Germany

Philippe Blanchard
Fakultät für Physik
Universität Bielefeld
Bielefeld, Germany

ISBN 978-3-319-38024-7   ISBN 978-3-319-02798-2 (eBook)
DOI 10.1007/978-3-319-02798-2
Springer Cham Heidelberg New York Dordrecht London

Mathematics Subject Classification: 00A06, 00A30, 03B44, 35Qxx, 35Q20, 37D35, 37N05, 37N25, 58Zxx, 70Fxx, 70F16, 70F45, 74A15, 74A25, 78A25, 81S22, 81S40, 81T28, 83Axx, 83Cxx, 83Dxx, 83Fxx, 92D15, 92D25, 92D40

Preface

The present volume collects the Proceedings of the Conference "Direction of Time" that took place at the Center for Interdisciplinary Research (ZiF) of Bielefeld University, January 14 to 19, 2002. The organizing and scientific committee included, in addition to the present editors, Michael Drieschner (Bochum) and Sylvie Paycha (Clermond-Ferrand/Potsdam). We are most grateful to them for a very inspiring and fruitful collaboration. We are also most grateful to Annidita Balslev (Aarhus) for inspiring, highly motivating observations at an early stage of the preparation of the Conference.

The success of the Meeting was due, first of all, to the speakers. Thanks to their efforts, it was possible to take into account recent developments in various directions as well as open problems and to make the Conference an exciting event. We hope that participants and readers will find the articles collected in these Proceedings both interesting and useful. We apologize for the delay in publishing, due to circumstances partly independent of our will and efforts. We hope that the permanent actuality of the topic of this book might attenuate the damages caused by the delay.

We are very grateful to Jean-Claude Zambrini for competent and inspiring advice in the course of the preparation of this volume.

It is a pleasure to express our special gratitude to Marion Kämper for the critical reading of the manuscript and to Hanne Litschewsky for invaluable help in preparing the Conference and in collecting and editing the manuscripts for publication. Without her generous help the book would have never have appeared.

We gratefully acknowledge the financial and logistic support of ZiF.

Bonn, Germany   Sergio Albeverio
Bielefeld, Germany   Philippe Blanchard

Introduction to Time, Its Arrow and the Present Book

The central theme of the meeting was the discussion of fundamental questions like

- What is time?
- Is time an illusion?
- In which sense are past, present, and future to be understood?
- Is time directed?
- What makes time different from space?
- Where does the time arrow come from?

Proposing answers to such fundamental questions requires joint efforts from disparate areas, each bringing in its original point of view. The meeting and this publication which emanates from it will focus only on some aspects of the problematic nature of time (see, e.g. [D, Fraser 1975] for complements). (The references are grouped into four sections, A, B, C, D. Reference [D, Fraser 1975] stands for section D, author Fraser, published 1975.)

Time has been a preoccupation of humanity as a whole from the very beginning of culture and it still remains a basic source of open problems. In fact, nothing is probably more close and familiar to us than time but also, at the same time, so full of mystery.

Time is deeply related to natural phenomena (like periodicities in the sky, cycles in biological processes and evolution), and as such has been of concern in all cultures and religions, which have tried each in its own way to find answers to the basic questions it raises. Also various sciences have been struggling with the concept of time and its multiform aspects. But time is, in addition, an experience which has also been steadily of concern for writers, artists, and philosophers, and humanity at large.

Already from the Presocratic tradition one can observe steady oscillations between a static view of time (like Parmenides) and an all pervading dynamical view of time (like in Heraclitus). Since the illuministic epoch, time has rather been viewed as composed of three components, two very large ones, the past and the future. The third one is tiny, almost non-existing. In some sense, however, we could maintain that none of these three parts has a veritable existence: past, since it no longer exists, future, since it does not yet exist and present, since at each instant it disappears.

In the meeting and in the present publication the main attention is drawn to the question of the "arrow of time", since it is the question which better puts in evidence the multiple visions of time both in different cultures and different sciences ("irreversibility" of life experience contrasted with "reversibility" of time evolution, as described, e.g., in Newton's laws of nature).

In the present book we are examining particularly three main approaches to this problem, namely "time and physics", "time, philosophy, and psychology", and finally "time, mathematics, and information theory".

Let us discuss these three aspects of time, starting always with some general considerations of ours and proceeding then to a short discussion of the specific contributions, in this volume presented.

Time and Physics

Contemporary physics has its origin in classical mechanics as formulated by Galilei and Newton. It involves equations of motion describing the history of an idealized system. In these equations, time is nothing else but a real parameter and the basic Newton equations are invariant under time reflection. On the other hand, as was soon realized, classical mechanics can also be formulated in a way emphasizing the role of invariant quantities: the integrals of motion. The absolute time of Newtonian physics was replaced in the description of classical dynamical systems by the "relative time" introduced 1905 by Einstein (and Poincaré). In special relativity time and space form a whole and depend on the dynamics (e.g., one has the well-known phenomenon of "slowing down of clocks in motion"). The notion of simultaneity is no longer absolute but past and future maintain for each observer their absolute character.

Since Einstein's (and Hilbert's) general relativity (1917), gravity is described as a geometric property of space-time, and in particular time is affected by the presence of matter and energy (although space-time itself, before a metric is put on it, is just conceived as a manifold). The matter distribution determines the space-time geometry and the solutions of the corresponding Einstein's equations. Both in special and in (local) general relativity time does not play a special role, except for being one of the components of a four-dimensional space-time, and making the local Minkowski metric not positive definite, of signature, say, $(+,-,-,-)$.

Cosmology enters, however, into the stages, when general relativity is applied to the whole cosmos. Different cosmological models imply different histories of the universe. The presently most accepted models take into account Hubble's law and lead to a cosmic arrow of time, from an initial "big bang" to the present state of an expanding universe.

Statistical mechanics is concerned with the description of very large systems and constitutes the theoretical foundations of thermodynamics. Although the microscopic equations of motion are reversible, the emerging observed macroscopic evolution is for all practical purposes irreversible, introducing by this a "thermodynamic arrow of time". The question of deriving irreversibility starting ab initio has generated a large amount of investigations, starting on the physical side with Carnot, Clausius, Maxwell, Boltzmann, and Gibbs and on the mathematical side with the works of Poincaré, Zermelo, and others, expanding in the second half of the last century into the general theory of (classical) dynamical systems.

Entropy has emerged as a central notion in the area of thermodynamics and statistical mechanics. In physics entropy is a measure of disorder. It is also a fundamental quantity in information theory and the relations between these two concepts and areas of research are currently under intensive development. Our perception of the direction of time is connected to the fact that entropy in a closed system can only stay constant or increase. The origin of the time arrow as we perceive it depends on two essential properties. The first is a low-entropy initial state. The second is mixing, which is necessary for explaining why a given system evolves and rearranges from a low-entropy (small probability) to a higher-entropy (high-probability) state. The same properties hold true in cosmology. The cosmological arrow of time requires the universe to have started in a low entropy state and the matter to have mixed ever since. Among the possible cosmological models the big-bang ones are still prevailing, but there are also models which are extremely different from these and yet are still being discussed to clarify some basic issues (see e.g. [C, Gödel 1949], [C, Segal 1976, 1996], [B, Stölzner 1996], [B, Yourgrau 1991]).

Although ordinary (non-relativistic) quantum mechanics has dramatically changed our conception of the world, time in it is still a parameter like in Newtonian physics and the fundamental dynamical equation, the Schrödinger equation, stays reversible. But quantum mechanics, on the other hand, is intrinsically probabilistic. Decoherence can serve to relate some quantum probabilities to classical probabilities but it does not make them into classical probabilities. Measurements are a chain of correlated consequences and decoherence does not explain why a particular event is realized in a particular measuring process. Measurement is the irreversible registration of a macroscopic signal, and as such incorporates an intrinsic arrow of time.

The natural extension of quantum theory to include special relativity leads to quantum field theory, which is at the basis of the entire present-day particle physics theory.

A further extension would lead to quantum gravity, a quantization of general relativity. Quantum gravity is concerned with gaining quantitative knowledge of physical phenomena at very high energies. The characteristic scale is the Planck scale of 10^{-35} m, which is very far removed from our every day experience and physical intuition, and cannot be probed directly by experiment. This latter extension has generated very intensive activities using different approaches and points of view (among them string theory, loop quantum gravity, non-commutative geometry, . . .), and showing intriguing connections between particle physics and cosmology. The study of phenomena like black holes needs considerations both from classical and quantum thermodynamics leading to new points of view on the problematic of the arrow of time. It is fair to say, however, that the "theories" trying to cope with such phenomena are still rather just research programs and far from conclusive. In all of them, however, a causality principle seems to prevail, see, e.g., [A, Klein 2007].

Let us also mention the thermal type hypothesis of Connes and Rovelli [C, Connes, Rovelli], which describes time as an emerging statistical property similar to temperature in statistical mechanics. Time appears in this theory as the result of our missing information about reality and it is our approximate knowledge of reality that exhibits a time arrow.

In a recent paper C. Rovelli [A, Rovelli 2000, 2005] proposes a way to describe multiple quantum events in time by introducing a single event that can be formulated without reference to time. In this approach evolution in time is replaced by correlations between events that can be observed in space.

There are other views of space-time like theories based both on the causality principle and the theory of twistors, by R. Penrose [C, Penrose 1994], and the one which postulates microscopically discontinuous (e.g. p-adic) space-times [C, Khrennikov 1999].

Let us also mention in passing that an extreme position of "eliminating time from physics" has been taken by J. Barbour [A, Barbour 1999] (who continued in a tradition, e.g. E. Meyerson [A, Meyerson 1908]), it remains to see whether this is really more than just an interesting play with formalisms.

A calculational tool, "causal dynamical calculation" [A, J. Ambjørn et al., 2008] has been developed for coping with "quantum gravity" over the last few years. The absence of ad hoc extra dimensions and its use of fundamental quantum mechanical principles make this approach conceptually simple. Among the intriguing results of this approach let us mention the heuristic emergence of a classical space-time geometry from quantum fluctuations and of a space-time dimension at very short distance which is not equal to four! It remains, of course, to see whether this approach could be made into a theory which is both satisfactory in a mathematical and physical sense. For further references to the topic "Time and physics" see A.

Let us now summarize briefly the present volume contributions to the subject "Time and physics".

M. Cini's contribution is concerned with topics relating to the different dichotomy representations of change: Parmenides versus Heraclitus, eternity versus timelessness, objective time versus subjective time, time's arrow versus time's cycle, reversibility versus irreversibility. Among these are a discussion of the Earth's history and of biological evolution, and of the radical difference of the role of chance in statistical mechanics and quantum mechanics. Topics covered include the irreversibility of the Second Law, the transition from order to disorder, the irreversibility of quantum measurement, ontic and epistemic uncertainties, and a representation of Quantum Mechanics in phase space by using Wigner's functions.

A. Teta proposes a two-particle model of decoherence in one dimension. It consists of a heavy and a light particle with coordinates R resp. r interacting via a point interaction $\alpha_0 \delta(r - R)$, with α_0 in the real field R and δ standing for Dirac's function. The initial state is chosen to be a product state and the initial wave function of the heavy particle of mass M is a sum of two spatially separated wave packets with opposite momentum. As for the light particle of mass m, it is localized between the two wave packets. The author discusses the asymptotic dynamics of the system in the limit $\varepsilon = m/M \to 0$, and gives an explicit estimation of the error. Moreover, he introduces the reduced density for the heavy particle and gives an explicit estimation of the decoherence effect.

Th. Görnitz confesses that he is convinced that time exists and that nothing is so un-influential as the direction of time. To the question: What is time? he proposes different viewpoints namely those of classical physics (time without importance), of special relativity (an own time for everybody), of general relativity (the flow of time depending on the situation), of quantum theory (the time disappearance and the occurrence of facts at the border between classical and quantum world) and finally of cosmology (time becoming universal and having a beginning). For Görnitz physics as any other human activity can only be an approximation of truth, in many cases a very good one, and it reflects more or less all possible time experiences we have been able to make.

R. Haag starts by recalling a remark by Wolfgang Pauli, who distinguished the "real part of physics" (the phenomena) from the "imaginary axis" (the theory). Theoretical concepts are mental constructs whose "reality value" is debatable. Haag introduces the notion of events as individual facts. He presents the view that the quantum state describes the probabilities for the occurrence of various possible events.

Reality consists of past facts. The future is open. In this way the physicist Haag develops an evolutionary picture of reality similar to that described long before by the philosopher A.N. Whitehead.

A fruitful interaction between Physics and Neuroscience is explored in *von der Malsburg's* contribution. After a short description of our outlook on time, he starts with the description of Cramer's system of communication between potential absorbers in the course of the transmission of a photon. According to this picture along the lines sketched by Tetrode and Fokker the dynamics is formulated in the Eternal Universe by an action principle implying that events at all space-time points stick together by a tangle of advanced and retarded signals from the beginning to the end of time. Nevertheless, the relation between past and future is not symmetric, energy and information propagating always into the future. The adopted viewpoint seems to be in sharp contrast with our traditional intuitive outlook on time according to which the future has no reality yet and is open to the decisions of our free will. What our brain uses permanently is practical predictability allowing our survival. The free will in the Eternal Universe does not change the real future but the potential future of our imaginations and perceptions. The author concludes that there is no contradiction between free will and determinism. On the contrary, free will is impossible without determinism. Reserving the status of reality only to the present is nothing else but another proof of the extreme egocentric perspective of our civilization.

The contribution of *R. Omnès* discusses how decoherence is the most efficient cause of irreversibility in quantum physics, relating the privileged direction of time with the one usually associated with thermodynamics. Moreover, he also relates this arrow of time with a "logical one" as part of the approach to quantum mechanics through "consistent histories". Decoherence implies that the classical world is no more in opposition to the quantum one, but on the contrary is inherently required for its proper formulation.

R. Tumulka's contribution is devoted to the description of a "toy model" inspired by Bohmian mechanics. Introducing two opposite arrows of time, a thermodynamic one related to boundary conditions and another one connected to asymmetry of the microscopic equations of evolving particles, he shows the possibility of having a Lorentz invariant nonlocality in the equations of motion.

J.C. Zambrini: Starting from the fact that the laws of nature are invariant with respect to time symmetry, J.C. Zambrini discusses the operational meaning of the time-reversal operator T. After time reversal, initial conditions become final ones. What is the physical meaning of such a transfer of information from the future in the past? Is a conflict with causality possible? As an example J.C. Zambrini considers first a classical particle moving freely in one dimension as a two boundary conditions problem and shows how the time symmetric Newton equation emerges from an irreversible Hamilton–Jacobi equation. The second part of this contribution is dedicated to the quantum version of this problem by using Feynman's approach to Quantum Mechanics in both real and imaginary time. He considers in this perspective the famous two-slits experiment and shows that this conditioning introduces an irreversibility in an otherwise completely time-symmetric framework. Using the fact that martingales are the probabilistic analogues of the constants of motion he then shows how to guess new quantum symmetries.

Time, Philosophy, and Psychology

The problem of time as a topic of philosophical investigation has very ancient roots. In the eastern traditions, particularly in India, time is discussed within a spiritual and religious "world system", and has many aspects, some of them similar and other very different from the western philosophical tradition. We refer to [B, Balslev 1999], who presents an extensive discussion of time in the Indian tradition and, e.g., [B, Needham 1957], [B, Rawson et al. 1973] resp. [B, Massignon 1951] as a source of information on time in the Chinese resp. Islamic culture.

In the Greek mythology time (Kronos) plays a rather dreadful role. The Presocratic philosophy takes up the mythological themes in the realm of reason.

Two big Presocratic philosophical schools present complementary views of time in their relation to being.

Parmenides (ca. 515 BCE) (and his Eleatic School) puts the being into a central position, time becoming a kind of illusion. Zeno (495–445 BCE) underlies the paradoxes connected with the continuum concept, hence also with time as a continuum.

For Heraclitus (ca. 500 BCE), on the opposite side, time is central and being plays in contrast a rather secondary role. Let us also mention Anaximander (ca. 610–546 BCE), for whom the problems of being and time are not separated, hence an evolution principle takes over a central role.

These philosophical points of view continue their influence through the philosophy of Plato (ca. 428–348 BCE) resp. Aristotle (384–322 BCE). Whereas for Plato events are realizations of ideas, for Aristotle's nature is characterized by the development of events. For him time is not the fullness of being but rather a non-being, a void, a missing being, "it has been and is no more"—"it is coming and is not yet".

The Platonic resp. Aristotelian traditions have influenced the whole course of western philosophy and culture, up to contemporary positions. Let us mention only a few of them, keeping above all in mind that our main goal in the present book is to analyze the directionality of time.

For both Plotinus (ca. 205–270) and Gassendi (1592–1655), e.g., time is already there before the creation of the world. For Lucretius (ca. 97–55 BCE), instead, time does not exist per se, it is indistinguishable from movement.

St. Augustin (354–430) underlines, on the other hand, the personal experience of time (and he is by this a precursor of a tradition which stresses "lived time" as opposite to "measured time", a tradition which in the last two centuries had a strong development in work of philosophers like Kierkegaard, Bergson, Husserl, Heidegger, Sartre and the phenomenological and existentialist "continental" philosophical direction).

For Kant (1744–1804), following Newton (1642–1727), time and space are homogeneous. But above all they are conditions "a priori" for all our experience and knowledge.

Cause and effect, e.g., are not "in the time", rather time is a substratum for the causality relation. The latter is objective, independent of observations.

G.W. Leibniz (1646–1716), and later E. Mach (1838–1916), take a position very different from Newton's one, namely against the substantiality of time (and space), looking at these as beings having no reality outside their relations with objects. Leibniz based this on his metaphysical principle of sufficient reason and the identification of indiscernibles.

For Hegel (1770–1830) time is the possibility of the spirit (and in particular, humanity) to realize itself in history, as an "unfolding" of its very possibilities.

An active, quite well known psychological view of time as "durée" (in contrast with "measured time") was presented by H. Bergson (1898). As we already mentioned, this can be put in relation with St. Augustine's position, but it can also be connected with the romantic school of, e.g., Schelling (1775–1854) and Schopenhauer (1788–1860).

An important philosophical idealistic analysis of time was performed by McTaggart (1908), who denied the very existence of time. This analysis has had a strong impact on successive developments (in the philosophical analytic tradition, but also in other traditions, in connection with the analysis of the concept of causality; McTaggart's position has been reinforced by the construction of Gödel's cosmological model, see [C, Gödel 1949] and, e.g., [B, Dummett et al. 1978, 2006]).

McTaggart in particular distinguishes two basic theories of time, the A type theory (with past, present, future, also called "modal type theory") and the B type theory (which is based on the distinction before–after and is directly related to the causal principle), challenging then the relative validity of both types of theories.

Time seen in this analytic way is in strong contrast with time as conceived in the phenomenological or existential philosophy (Husserl resp. Heidegger, G. Marcel, Merleau-Ponty, Sartre ...), who put time at the very basis of the "Erlebniswelt" (Husserl) or philosophy (Heidegger), or stress it as a succession of present moments, interwoven with projects of the "conscience" into the future (Sartre).

For further references for the topic "Time, philosophy, and psychology" see B.

Let us now summarize the present book's contributions on the topic "Time, philosophy, and psychology".

A.N. Balslev discusses the time metaphors in Indian and Western philosophy especially in relation to the question of the "direction of time". In particular she focuses on such metaphorical designations as "cyclic" and "linear" time, in order to demystify them and make clear how they oversimplify matters and how deeper conceptualizations of time already exist both in Indian and Western traditions.

In particular, she recalls the notion of absolute time (mahakala) in the old Vaisesika school of India, a time which "does not rest in anything else", does not change or flow, but rather is the support of all changes. This notion of absolute time was then challenged by the Buddhist, Jaina, and Yoga schools, as well as by other schools belonging to the Upanisadic tradition. The complexity of the conceptual scenarios of views on time in the Indian tradition is then described, in particular showing how futile it is to try to upheld the reductive "cyclic time" as characteristic of Indian philosophy on time. In particular, A. Balslev points out that one should keep the notions of "cyclic time" well separated to the one of "cosmological cycles", the latter being more proper to Indian philosophy.

H. Barreau: Barreau's starting point is the contrast between irreversibility for physical processes and the need of assigning a direction in time, the latter being conceived as a global one directional development. For this author, cosmology offers two ways for considering the direction of time. There is first an overall development of physical reality if the evolution of the Universe is a function of the cosmic time associated to the age of the Universe. Moreover, there is a second overall development if the cosmic time can also explain the emergence of the complex structures present in the Universe. This second aspect presupposes the first and corresponds to a cooling process which induced a cascade of symmetry breaking associated not only to the different fundamental physical interactions but also to the different degrees of physical reality. He also looks at time as an irreversible process in our lifetime experience. Moreover, he stresses the difference between time in relativistic cosmology and time in a Newtonian Universe. In conclusion he sees in the history of contemporary cosmology a source for a "cosmic arrow of time" and adapts the point of view of some contemporary cosmologists about the "anthropic principle", giving a special role to the Universe in which we live among all universes which are possible at least in principle.

L. Boi discusses symmetry and symmetry breaking in relation to the dynamical action of time. He particularly considers natural and living systems, and morphogenesis. He starts out from the consideration of the development of the thermodynamics. He also stresses R. Penrose's view on a low entropy universe to start with. He then goes over to discuss symmetry and symmetry breaking in nature, in particular in relation to molecular biology and morphogenesis, as well as to wave propagation in neural networks. He points out an interesting observation about relating hallucination phenomena with symmetry breaking phenomena.

He also discusses phase transitions in equilibrium and non-equilibrium statistical mechanics. Moreover, he describes mathematical aspects of bifurcations, singularities, and universality. He then goes over to examine spontaneous symmetry breakdown in the theory of gauge fields and particle physics. Moreover, he examines topics like the topological structure of phase space and the dynamical action of time, stressing also dissipative systems. He closes by presenting some remarks on the geometry of psychological time, borrowing ideas from string theory, and ending with some speculations concerning psychological time conceived as multidimensional and polycyclic.

M. Drieschner starts with a cartoon of Calvin & Hobbes entitled "The time warp" demonstrating that time is something entirely different from space. Drieschner's aim is to discuss the interplay between the structure of time and fundamental questions in the foundations of physics and classical vs. quantum probability theory. Direction is a spatial concept. Neither is the present directed nor the past nor the future. By talking about the "arrow of time" we mean the difference between past and future. In physics as well as in everyday life the past is actual and the future is "potential". To Augustine's question "Quid est ergo tempus?" Drieschner answers: The "direction of time" is not derived from physics and time is rather presupposed for science. Physics would be impossible if we did not start from this "direction" of time. Probability theory plays a key role and is generally applied only to predictions. For the author the most general empirically testable prediction is a probabilistic statement. On that ground he discusses the irreversibility in thermodynamics, Zermelo's reversibility paradox and the measurement problem in quantum mechanics.

H. Lyre gives a survey of views on time and in particular its "directedness" on the basis of the modern philosophy of physics. He starts out by emphasizing the important distinction between time in the sense of "being" and time in the sense of "temporal becoming". He recalls Mc Taggart's A- and B-series, observing that the time reversal's discussion relates to the B-series. He characterizes the Parmenides versus Heraclitus visions of time as "perdurantism or eternalism" versus "endurationism or presentism". Moreover he points out some analogy between Zeno's criticism of the concept of time continuum and the contemporary discussions on the quantum Zeno effect. He recalls R. Penrose's seven possible arrows of time (weak interaction, quantum mechanics, thermodynamics, electromagnetism, psychology, cosmology, and gravitation), going deeper into special and general relativity issues on time (the hole argument, the possibility of instants of time), as well as the thermodynamical information science aspects, closing with the quantum mechanical issues. His favored position is a post-Kantian view of time and its arrow as preconditions of experience.

K. Mainzer discusses direction of time in dynamical systems, stressing interdisciplinary perspectives from cosmology to brain research. In particular, he stresses Hawking's hypothesis of an "early universe without temporal beginning" as "confirmed by the measurement of COBE in 1992". He points at a "cosmic arrow of time from singularity to complexity". He then analyzes time in evolutionary dynamics, stressing the inner or intrinsic irreversible time of an organism or a socio-economic system, distinguished from the "external and reversible clock time". Furthermore, he points at a striking analogy between natural and computational processes, e.g. the fact that even with initial conditions and locally reversible rules many dynamical systems can produce irreversible complex and random behavior". He also analyzes "global communication networks" as similar to the "self-organizing neural networks of the brain".

F. Minazzi analyses the philosophical significance of the relativistic conception of time. He starts by pointing out that in special relativity the question of establishing the "real" and "true" geometry of time "loses all meaning". He stresses the causal theory of space and time as the "most important and significant "philosophical result" of Einstein's theory of relativity".

M. Saniga: After reviewing the most important examples of psychopathology of time like near-death experiences, drug induced states, mental psychoses, mystical states, the author proposes a speculative algebraic model of the time dimension and claims that this model is able to mimic some of those "peculiar and/or anomalous perceptions of time" (the feeling of timelessness, the experience of the dominating past and time standing still). The mathematical construction is based on a pencil (i.e. a linear, single parameter aggregate) of conics in the real projective plane. All the conics touch each other in two points and the corresponding tangent lines meet at a point. Each conic is considered as representing a single event and the selection of a line in the plane and the consideration of its possible intersection properties with the conics is proposed as an interpretation of past, future, and present events.

M. Stöltzner addresses the problem of implementing the unidirectionality of time into two fundamental physical theories which are basically time-reversal invariant, classical mechanics and general relativity. He distinguishes four possible approaches to implement the arrow of time (as a basic axiom, as arising from the laws of an early universe, from microscopic laws or, finally, from "non-lawlike initial conditions"). He then analyses Boltzmann–Exner versus Planck's positions on statistical mechanics. On another issue, he sees Gödel's model of the universe as showing a "semantic incompleteness of general relativity with respect to the concept of time". He then analyses different attempts, in particular by Farman, of putting aside the "time travel malaise" coming from unwanted solutions of the equations of general relativity.

Time, Mathematics and Information Theory

Mathematics and information sciences present several aspects in their relation to the problematic of time, even though a systematic discussion seems to be still lacking. Basic relations are those concerning the process of counting, and the concepts of numbers and time. E.g., Aristotle in discussing the difference between time and motion puts numbers in close relation to time, and stresses the infinite divisibility of the latter. For Plato time was an essential feature of the sensible world, extended, however, from pure geometry, which, in his view, is associated with the eternal world of ideal forms. N. de Oresme (1323–1382) was perhaps the first author who introduced a geometrization of time by applying graphical techniques systematically to represent variations of functions. I. Barrow (1630–1677) regarded time as a mathematical concept having many analogies with a line. He pointed out, however, that time has both "continuum" and "granular" aspects. Newton took time as the "standard independent variable", thinking of it in terms of a one-dimensional continuum.

Kant also stressed numbers as being generated by successive additions "in time". W.R. Hamilton in 1857 influenced by Kant, claimed that just as geometry is the pure mathematical science of space, there must also be a pure mathematical science of time, and that this must be algebra.

L.E.J. Brouwer (1913) in his intuitionistic approach to the foundations of mathematics based his construction on the multiplicity of the intervals of time, "which he regarded as the primary intuition of the human intellect" [B, Whitrow 1980]. He regarded minds as operating by successive acts of attention, reflecting the two terms temporal relation "before/after". G. Cantor (1883), on the other hand, looked at the mathematical continuum as a more fundamental concept than time and space. The neo-Kantian philosopher E. Cassirer reinterpreted Kant's theory of arithmetic as the study of "series", which finds a concrete expression in "time sequences".

Let us mention that discrete models of time ("time consisting of atoms") also have a long tradition, having its origins in certain Buddhism schools and in work of medieval philosophers, like Maimonides (XII century). They also appear with authors like R. Descartes, N. Wiener [C, Wiener 1958], who continues work by Whitehead, on "momentary instants of time", and, more recently, Hamblin [B, Hamblin 1971] and Čapek [B, Čapek 1976]. Whitrow [B, Whitrow 1980] stressed the need to derive the linear continuum of time from an acceptable set of axioms. G.A. Walker (1947) showed how temporal instants can be defined in terms of duration, using a theory of sections in a partially ordered set. This study of mathematical time was continued using only ordinal definitions and postulates, avoiding metrical concepts.

For a thorough discussion of different topological models of time see [B, Newton-Smith 1984]. The same book also discusses metrical aspects of time. An extensive systematic discussion of topological properties versus metrical properties of time is given in the same book. The author argues, in particular, that there are several reasonable possibilities both for the topological and for the metrical structure of time. His starting point is the "standard topological" view of time as an unbounded line segment. He discusses then a description in terms of a first order, respectively, tense logic, following a basic work by Prior [B, Prior 1967, 1968]. He points out that, in analogy with geometry, one can regard "the question of the topological structure of time as an empirical one which is to be decided by reference to investigations of the physical world" [B, Newton-Smith 1984, p. 55]. This is a difficult undertaking, because of the presence of what Quine has called the underdetermination of theory by data. He also argues (contrary e.g. to Swinburne [B, Swinburne 1968]) that time could have any of a set of non-equivalent topologies. He discusses in particular the possibility that time might be closed or cyclical.

Other mathematical discussions of aspects of the time problematic can be found in work concerning the "logic of time", see e.g. [B, Prior 1967, 1968], [B, Rescher 1931, 1971], [B, Pizzi 1974], [B, Burges 1979]. For somewhat unsystematic but interesting aspects of time, numbers and psychoanalysis see [B, von Franz 1978, 1980]. For the foundation of time, numbers and mathematical concepts and their relations in child's development, with relevance to epistemology, see work by J. Piaget and his School (e.g. [B, Piaget 1955, 1965]).

For further references to the topic "Time, mathematics and information theory" see C.

Let us summarize the present book's contributions to the subject of "Time, mathematics, and information theory".

L. Accardi discusses mathematical models of time. He starts out by remarking that the notion of time is very relative to contexts, as illustrated by a provocative question like "how old is an electron?". After recalling some of these contexts and relative discussions, he concentrates on time from the point of view of mathematics. He points out that an axiomatic approach to time should take into account its relation to motion and transformations. He then discusses axiomatic attempts of catching time as a continuum, pointing out the existence of different models (linear time, circular time, "fat time", "fractal time"). In another section, he presents an interesting non-Archimedean model of time, related to a model of an extended real line due to F. Hausdorff. He then goes over to discuss time in classical physics, stressing relations with concepts of the theory of dynamical systems (like recurrence, cyclicity, wandering states, homogeneity). He gives special attention to the formalization of time reversal and related invariants in quantum and classical physics, using in particular the theory of von Neumann algebras.

V. Benci starts from the apparent contradiction between two aspects of the passing of time, namely the fact that time, on one hand, destroys information (second law of thermodynamics) and, on the other hand, creates information (evolution creating more and more complex structures). In this context, one important question is related to the meaning of "information" and its relation to the notions of "entropy". Benci defines Shannon entropy in a new way which highlights its similarity to Boltzmann entropy and gives a definition of algorithmic information content (*AIC*). For Benci the notion of information is primitive and the notion of probability has to be derived from it. He shows that the *AIC*-information content of a finite string σ with respect to a universal computing machine C depends only on σ and then its asymptotic behavior is asymptotically independent of C. Using a compression algorithm he defines a physical quantity measuring the computable information content.

He also introduced another notion of information called *CIC* (computable information content). He shows that *CIC* is a relevant physical quantity and can be split into two components: the entropy associated to the notion of disorder and a macroinformation useful for making predictions. This distinction is similar to the well-known distinction between "free energy" (defined as energy minus temperature times entropy) and "bad energy".

L.S. Schulman discusses the relations between a computer's arrow of time and the psychological or consciousness arrow of time. More precisely, he considers "the extent to which a computer—presumably governed by nothing more than the thermodynamic arrow—can be said to possess a psychological arrow". The author looks at the arrow as a non-primary, rather derived concept, and stresses the subjective elements in the discussion of the second law of thermodynamics.

The contribution by *H.D. Zeh* is related to discussions by L.S. Schulman in previous papers, in which the second law of thermodynamics "is regarded as a "fact" rather than a "dynamical law". In particular this makes it possible to have, in principle, a law with varying directions of time. Zeh argues, both in the classical and in the quantum case, that the examples of such "anomalies" given by Schulman refer to situations which cannot realistically "be assumed to exist during one and the same epoch of the universe". This is illustrated by a discussion of electromagnetic wave propagation, general relativity, molecular systems, classical and quantum dynamical systems, cosmology, and quantum gravity.

In conclusion, let us point out that due to the manifold complexity of the problematic of time and its arrow many aspects have been only touched upon or even had to be left out from the present book and its "introduction". For such aspects we limit ourselves to give some references, e.g. [A, Cerraro 1997], [D, Coveney et al. 1992], [C, Wiener 1958], [D, Ebeling et al. 1990], [A, Hitchcock 2000], [D, Portmann 1951], [D, Vaas et al. 2002], [D, Winfree 1987] (for biology); [D, Gould 1987] (for geology); [D, Aichelburg 1988], [D, Levine 1997], [D, Elias 1987] (for anthropology resp. sociology); [D, Levine 2000], [D, Stadler et al. 2006], [D, Macey 1991] (for history); [B, Brill 1993], [B, Pine 1970], [D, Weis, 1998] (for religion); [A, Paflik 1987], [D, Sandbothe et al. 1994], [D, Architekt 1996], [D, Lelord 2008], [D, Scartezzini 1999], [D, Weis et al. 1994, 1998], [B, Dummett et al. 1978, 2006], [B, Massignon 1951] (for linguistics, arts, culture...). Further references of a general nature are given under D. We hope that these references as well as those under A, B, C might help the interested reader to deepen and extend his insights into the problematic of time. It is, of course, far from being complete and should be conceived as providing a first orientation, to be developed according to the reader's taste and interests.

References

(A) Physics

1. Ambjørn, J., Jurkiewicz, J., Loll, R.: The self-organizing quantum. Sci. Am., 42–49 (2008)
2. Balian, R., et al.: Le Temps: Perception et Mesures. Coll. Interdiscipl. CNRS, Besançon (1994)
3. Barbour, J.: The End of Time. The Next Revolution in Our Understanding of the Universe. Weidenfeld & Nicolson, London (1999)
4. Breuer, R.: Die Pfeile der Zeit, München (1984)
5. Carroll, S.M.: The cosmic origins of time's arrow. Sci. Am. 1, 48 (2008)
6. Cerraro, S., et al.: La freccia del tempo. SISSA, CUEN spl, Napoli (1997)
7. Davies, P.C.W.: The Physics of Time Asymmetry. University of California Press, Berkeley (1977)
8. Davies, P.C.W.: Space and Time in the Modern Universe. Cambridge University press, Cambridge (1977)
9. Davies, P.: About Time. Penguin, London (1995)
10. Deutsch, D.: Fabric of Reality. Penguin Book, Harmondsworth (1997)
11. Earman, J.: Bangs, Crunches, Whimpers, and Shrieks. Oxford University Press, New York (1995)
12. Gardner, M.: The Ambidextrous Universe. Allen Lane, London (1964)
13. Garola, C., Rossi, A. (eds.) The Foundations of Quantum Mechanics – Historical Analysis and Open Questions, Lecce (1993)
14. Genz, H.: Wie die Zeit in die Welt kam. Die Entstehung einer Illusion aus Ordnung und Chaos, RoRoRo (1999)

15. Griffin, D.R.: Physics and the Ultimate Significance of Time: Bohm, Prigogine and Process Philosophy. State University of New York Press, Albany (1986)
16. Guerra, F.: Reversibility/irreversibility. In: Enciclopedia Einaudi, vol. 11, pp. 1067–1106, Turin (1980)
17. Halliwell, J.J., Pérez-Mercader, J., Zurek, W.: Physical Origins of Time Asymmetry. Cambridge University Press, Cambridge (1994)
18. Hawking, S.W.: A Brief History of Time. Bantam Books, New York (1988)
19. Hawking, S.W., Penrose, R.: The nature of space and time. Sci. Am., 44–49 (1996)
20. Hitchcock, S.M.: T-computers and the origin of time in the brain. Neuroquantology 4, 393–403 (2003)
21. Horwich, P.: Symmetries in Time. MIT Press, Cambridge (1987)
22. Horwich, P.: Asymmetries in Time. MIT Press, Cambridge (1987)
23. Janich, P.: Protophysics of Time. Reidel, Dordrecht (1985)
24. Klein, E.: Le facteur ne sonne jamais deux fois. Flammarion, Paris (2007)
25. Kroes, P.: Time: Its Structure and Role in Physical Themes, Dordrecht (1985)
26. Landsberg, P.T.: The Enigma of Time. Adam Hilger, Bristol (1984)
27. Lebowitz, J.L.: Boltzmann's entropy and time's arrow. Phys. Today 46, 32–38 (1993). See also discussions, Phys. Today, Nov. 1994
28. Mainzer, K.: Zeit, Beck (2002)
29. Meyerson, E.: Identité et Realité. Félix Alean, Paris (1908)
30. Paflik, H. (ed.): Das Phänomen Zeit in Kunst und Wissenschaft, Weinheim (1987)
31. Price, H.: Time's Arrow and Archimedes' Point. Oxford University Press, New York (1996)
32. Prigogine, I.: La fin des certitudes. Odile Jacob, Paris (1996)
33. Reichenbach, H.: The Direction of Time. University of California Press, Berkeley (1956)
34. Rovelli, C.: The century of the incomplete revolution: searching for general relativistic quantum field theory. J. Math. Phys. 41(6), 3776–3800 (2000)
35. Rovelli, C.: Che cos'é il Tempo? Che cos'é lo spazio? Di Renzi (ed.) Roma (2005)
36. Rovelli, C.: Forget time (2009). arXiv:0903.3832
37. Saunders, S.: Time, quantum mechanics, and tense. Synthese 107, 19–53 (1996)
38. Schulman, L.S.: Time's Arrows and Quantum Measurement. Cambridge University Press, Cambridge (1997)
39. Sklar, L.: Physics and Chance: Philosophical Issues in the Foundations of Statistical Mechanics. Cambridge University Press, Cambridge (1996)
40. Smolin, L.: The present moment in quantum cosmology: challenges to the arguments for the elimination of time. In: Durie, R. (ed.) Time and the Present Moment. Clinamen Press, Manchester (2000). gr-qc/0104097, PhilSci-archive (2001)
41. Snyder, H.S.: Quantized spaced time. Phys. Rev. 79, 38–41 (1947)
42. Straub, D.: Eine Geschichte des Glasperlenspiels. Irreversibilität in der Physik: Irritation und Folgen. Birkhäuser, Basel (1990)
43. Tiezzi, E.: Steps Towards an Evolutionary Physics. WIT Press, Southampton (2006)
44. von Weizsäcker, C.F.: Aufbau der Physik. DTV, München (1985)
45. Wheeler, J.W.: Frontiers of time. In: Torald Di Francia, G. (ed.) Rend. Scuola Int. Fis. "Enrico Fermi", LXII Corso, Varenna, pp. 395–492. Soc. It. Fis., Bologna, Italy (1979)
46. Zeh, H.D.: Die Physik der Zeitrichtung. Springer, Berlin (1984)
47. Zeh, H.D.: The Physical Basis of the Direction of Time. Springer, Heidelberg (1989). 4th edn. (2001)

(B) Philosophy, Psychology

48. Andureau, E., Enjalbert, P., Farinas del Cerro, L.: Logique temporelle, Etudes et Recherches en Informatique. Masson, Paris (1990)
49. Bachelard, G.: La Dialectique de la durée, Paris (1936)

50. Balslev, A.: A Study of Time in Indian Philosophy, 2nd edn., New Delhi (1999)
51. Baumgartner, H.M. (ed.) Das Rätsel der Zeit, Phil. Analys, Freiburg (1993)
52. Bergson, H.: Essai sur les données immédiates de la conscience. F. Alcan, Paris (1917)
53. Bieri, P.: Zeit und Zeiterfahrung, Frankfurt (1972)
54. Brill, E.J.: Religion and Time, The Netherlands (1993)
55. Burges, J.P.: Logic and time. J. Synth. Log. **44**, 566–582 (1979)
56. Burger, P.: Die Einheit der Zeit und die Vielheit der Zeiten, Würzburg (1993)
57. Chang, W.: Reflections on time and related ideas in the Yijing. Philos. East West **59**, 216–229 (2009)
58. Čapek, M. (ed.): The Concepts of Space and Time: Their Structure and Their Development. D. Reidel, Dordrecht (1976)
59. Denbigh, K.G.: Three Concepts of Time. Springer, Berlin (1981)
60. Dooley, L.: The concept of time in defence of ego integrity. Psychiatry **4**, 13 (1941)
61. Dummett, M.: Truth and Other Enigmas. Duckworth, London (1978)
62. Dummett, M., Bilgranni, A.: Truth and the Past. Columbia University Press, New York (2006)
63. Durie, R. (ed.): Time and the Instant. Clinamen Press, Manchester (2000)
64. Earman, J. (ed.): Notes on the causal theory of time. Synthese **24**, 24–76 (1971)
65. Earman, J.: World and Space-Time: Absolute and Relational Theories of Space and Time. MIT Press, Cambridge (1989)
66. Earman, J., Glymour, C., Stachel, J.: Foundations of space time theories. In: Minnesota Studies in the Philosophy of Science, vol. VIII. University Minnesota Press, Minneapolis (1977)
67. Flood, R., Lockwood, M. (eds.): The Nature of Time. Basil Blackwell, Oxford (1986)
68. Fraisse, P.: The Psychology of Time. Hamper & Ron, New York (1963)
69. Frank, Ph. (ed.): Das Kausalgesetz und seine Grenzen. Suhrkamp, Frankfurt (1988)
70. Freeman, E., Sellars, W. (eds.) Basic Issues in the Philosophy of Time, La Salle (1971)
71. Gale, R.M.: The Language of Time. Routledge & Kegan Paul, London (1968)
72. Gale, R.M. (ed.): The Philosophy of Time. Macmillan, London (1968). Contains an essay by F. Waisman, about topological metric properties of time
73. Gimmler, A., Sandbothe, M., Zimmerli, W.Ch.: Die Wiederentdeckung der Zeit. Wiss. Buchges, Darmstadt (1997)
74. Gold, Th. (ed.): The Nature of Time. Cornell University Press, Ithaca (1967)
75. Grünbaum, A.: Modern Science and Zeno's Paradoxes. Wesleyan University Press, Middletown (1967)
76. Grünbaum, A.: Philosophical Problems of Space and Time, 2nd. enlarged edn. D. Reichel, Dordrecht (1973)
77. Hamblin, C.: Instants and intervals. Stud. Gen. **24**, 127–134 (1971)
78. Heidegger, E.: Sein und Zeit, Tübingen, 17. Aufl. (1993)
79. Honnefelder, G. (ed.): Was also ist die Zeit? Erfahrungen der Zeit. Suhrkamp, Frankfurt am Main (1998)
80. Huggett, N. (ed.): Space from Zeno to Einstein: Classics Readings with a Contemporary Commentary. MIT Press, Cambridge (1999)
81. Husserl, E.: The Crisis of European Sciences and Transcendental Phenomenology. North Western University Press, Evanston (1970)
82. Husserl, E.: On the Phenomenology of the Consciousness of Time (1893–1917). Kluwer Academic, Dordrecht (1991)
83. Landsberg, P.T.: The Enigma of Time. A. Hilger, Bristol (1984)
84. Layzer, D.: Cosmogenesis: The Growth of Order in the Universe. Oxford University Press, London (1990)
85. Le Poidevin, R., McBeath, M.: The Philosophy of Time, Oxford (1993)
86. Libet, B.: Mind Time. The Temporal Factor in Consciousness. Harvard University Press, Cambridge (2004)

87. Lucas, J.R.: A Treatise on Time and Space. Methuen, London (1973)
88. Lyre, H.: Time and information. In: Proc. Workshop "Time, Temporality Now", Ringberg Castle (1996)
89. Massignon, L.: Time in islamic thought, man and time, Bollinger Ser. **30**(3), 108–109 (1951)
90. Mc Lure, R.: The Philosophy of Time, Times Before Time. Routledge, London (2005)
91. Mellor, D.H.: Real Time. Cambridge University Press, Cambridge (1984)
92. Mercier, A.: Petits prolégomènes à une étude sur le temps. Theoria **29**, 277–282 (1963)
93. Minkowski, E.: Le temps vécu. Artrey, Paris (1933)
94. Mittelstaedt, P.: Der Zeitbegriff in der Physik, Mannheim (1980)
95. Needham, J.: Science and Civilisation in China, Cambridge (1959)
96. Newton-Smith, W.H.: The Structure of Time. Routledge & Kegan Paul, London (1984)
97. Novrikov, I.D.: The River of Time. Cambridge University Press, Cambridge (1998)
98. Piaget, J.: Die Bildung des Zeitbegriffes beim Kinde, Zürich (1955)
99. Piaget, J.: Psychologie et épistémologie de la notion du temps. Verh. Schw. Naturf. Ges., Gent (1965)
100. Pine, N., God and Timelessness. Routledge & Kegan Paul, London (1970)
101. Pizzi, C. (ed.): La Logica del tempo. Boringhieri, Torino (1974)
102. Pomian, K.: L'ordre du temps. Gallimard, Paris (1984)
103. Prior, A.N.: Past, Present and Future, Oxford (1967)
104. Prior, A.N.: Papers on Time and Tense. Oxford University Press, London (1968)
105. Rawson, Ph., Legeza, L.: Tao, the Chinese Philosophy of Time and Change (Art & Imagination). Thames and Hudson Ltd. (1973, 1987)
106. Rescher, N., Garson, J.: Temporal Logic. Springer, New York (1931)
107. Rescher, N., Urquhart, A.: Temporal Logic, Wien (1971)
108. Reusch, S. (ed.): Das Rätsel Zeit. Ein philosophischer Streifzug. Wiss. Buchges, Darmstadt (2004)
109. Russell, B.: On order time. Proc. Camb. Philol. Soc. **32**, 216–228 (1936)
110. Russell, B.: Mysticism and Logic, and Other Essays. Unwin Books, London (1963)
111. Salmon, W.C. (ed.): Zeno's Paradoxes. Bobbs-Merrill, Indianapolis (1970). Repr. Hackett, Indianapolis (2001)
112. Sandbothe, M.: Die Verzeitlichung der Zeit. Wiss. Buchges, Darmstadt (1998)
113. Sartre, J.P., Being and Nothingness. Philosophical Library, New York (1956)
114. Savitt, S.F.: Time's Arrows Today. Recent Physical and Philosophical Work on the Direction of Time. Cambridge University Press, Cambridge (1995)
115. Schlegel, R.: Time and the Physical World. Dover, New York (1961)
116. Shallis, M.: On Time. Penguin, London (1981)
117. Sherover, C.M.: The Human Experience of Time, New York (1975)
118. Sklar, L.: Space, Time, and Spacetime. University of California Press, Berkeley (1974)
119. Smart, J.J.C. (ed.): Problems of Space and Time. Macmillan, New York (1964)
120. Stöltzner, M.: Gödel and the theory of everything. In: Háyek, P. (ed.) Gödel. Springer, Berlin (1996)
121. Swinburne, R.: Space and Time. Macmillan, London (1968)
122. Tasić, V.: Mathematics and the Roots of Postmodern Thought. Oxford University Press, London (2001)
123. Tiezzi, E.: The Essence of Time. WIT Press, London (2002)
124. Tiezzi, E.: The End of Time. WIT Press, Southampton (2003)
125. von Franz , M.L.: Time, Rhythm and Repose, Thames and Hudson (1978)
126. von Franz, M.L.: Zahl und Zeit – Psychologische Überlegungen zu einer Annäherung von Tiefenpsychologie und Physik. Suhrkamp, Stuttgart (1980)
127. Weizsäcker, C.F.v.: Zeit und Wissen, München (1992)
128. Whitrow, G.J.: The Natural Philosophy of Time, 2nd edn., Oxford (1980)

129. Winfree, A.T.: When Time Breaks Drown. Princeton University Press, Princeton (1987)

130. Wood, D.: The Deconstruction of Time. Humanitas Press, Atlantic Highlands (1989)

131. Yourgrau, P.: The Disappearance of Time, Kurt Gödel and the Idealistic Tradition in Philosophy. Cambridge University Press, Cambridge (1991)

132. Zimmerli, W.Ch., Sandbothe, M. (eds.): Klassiker der modernen Zeitphilosophie, Darmstadt (1993)

(C) Mathematics, Information Theory

133. Albert, D.: Time and Chance. Harvard University Press, Cambridge (2000)

134. Connes, A., Rovelli, C.: Von Neumann algebras automorphisms and time—thermodynamics relation in general, covariant quantum theories. Class. Quantum Gravity **11**, 2899 (1994)

135. Ekeland, I.: Le calcul, l'imprévu, les figures du temps de Kepler à Thom. Seuil, Paris (1987)

136. Gödel, K.: An example of a new type of cosmological solutions of Einstein field equations of gravitation. Rev. Mod. Phys. **21**, 447–450 (1949)

137. Hausdorff, F.: Brief an Landauer. In: Hausdorff, F., Werke, G., Stegmaier, W. (eds.) Bd. VII (1902)

138. Khrennikov, A.: P-adic information spaces, infinitely small probabilities and anomalous phenomena. J. Sci. Explor. **13**, 665–679 (1999)

139. Levichev, A., Kosheleva, O.: Intervals in space-time: A.D. Alexandrov is 85. Reliab. Comput. **4**, 109–112 (1998)

140. Penrose, R.: Shadows of the Mind. Oxford University Press, London (1994)

141. Penrose, R.: The Road to Reality. Vintage, New York (2005)

142. Schnell, K.: Eine Topologie der Zeit in logistischer Darstellung, Inaugural. Diss., Münster (1938)

143. Segal, I.: Mathematical Cosmology and Extra Galactic Astronomy. Academic Press, New York (1976)

144. Segal, I., Nicoll, J.F.: Statistics of a complete high-redshift quasar survey and predictions of nonevolutionary cosmologies. Astrophys. J. **419**, 496–504 (1996)

145. Shapiro, S.: Thinking About Mathematics. Oxford University Press, London (2000)

146. Tifft, W., Cocke, W. (eds.): Modern Mathematical Models of Time and Their Applications to Physics and Cosmology. Kluwer, Dordrecht (1997)

147. von Franz, M.L.: Zahl und Zeit, Suhrkamp (1980)

148. Weizsäcker, E.U.v.: Offene Systeme. I. Beiträge zur Zeitstruktur von Information, Entropie und Evolution, Stuttgart (1986)

149. Weyl, H.: Raum, Zeit, Materie, Darmstadt (1961)

150. Wiener, N.: Time and the science of organization. Scientia **93**, 199–205 (1958)

(D) Other Areas, General

151. Ageno, M.: Le origini della irreversibilità. Bollati-Boringhieri, Torino (1992)

152. Aichelburg, P.C. (ed.): Zeit im Wandel der Zeit. Vieweg, Braunschweig (1988)

153. Aschoff, J., et al.: Die Zeit – Dauer und Augenblick. Piper, München (1989)

154. Atmanspacher, H., Ruhnau, E. (eds.): Time, Temporality, Now: Experiencing Time and Concepts of Time in an Interdisciplinary Perspective. Springer, Berlin (1997)

155. Baert, P. (ed.): Time in Modern Intellectual Thought, Amsterdam (1998)

156. Butterfield, J. (ed.): The Arguments of Time. Oxford University Press, London (1999)

157. Callender, C. (ed.): Time, Reality, and Experiences. Cambridge University Press, Cambridge (2002)

158. Connes, A., et al.: On Space and Time. Cambridge University Press, Cambridge (2009)

159. Coveney, P., Highfield, R.: Anti-Chaos – der Pfeil der Zeit in der Selbstorganisation des Lebens, Reinbek (1992)

160. Cramer, F., Der Zeitbaum, Frankfurt (1953)

161. Der Architekt 3 (1996)

162. Ebeling, W., Engel, H., Herzel, H.P.: Selbstorganisation in der Zeit, Berlin (1990)

163. Ekeland, I.: Das Vorhersehbare und das Unvorhersehbare. Die Bedeutung der Zeit von der Himmelsmechanik bis zur Katastrophentheorie, München (1985)

164. Elias, N.: Über die Zeit, Arb. zur Wissenssoziologie II, Frankfurt 3. Aufl. (1987)

165. Enzensberger, H.M.: Zickzack. Suhrkamp, Frankfurt (1997)

166. Falk, D.: Search of Time – Journeys Along a Curious Dimension. Nat. Maritime Museum, Greenwich (2009)

167. Fraser, J.T., et al. (eds.): The Study of General Time I, II, III. Springer, New York (1975)

168. Gent, W.: Das Problem der Zeit. Eine historische und systematische Untersuchung, Hildesheim (1965)

169. Genz, H.: Wie die Zeit in die Welt kam – Die Entstehung einer Illusion aus Ordnung und Chaos. RoRoRo, Hamburg (1999)

170. Giorello, G., Sindoni, E., Sinigaglia, C.: I volti del tempo. Bompiani, Milano (2001)

171. Gold, Th. (ed.): The Nature of Time. Cornell University Press, Ithaca (1967)

172. Gould, S.J.: Time's Arrow, Time's Cycle: Myth and Metaphor in the Discovery of Geological Time. Harvard University Press, Cambridge (1987)

173. Hack, M., et al.: L'idea del tempo, UTET (2005)

174. Kamper, D., Wulf, Ch. (eds.): Die sterbende Zeit, Darmstadt (1987)

175. Landsberg, P.T. (ed.): The Enigma of Time. Hilger, Bristol (1982)

176. Lelord, F.: A la poursuite du temps qui passe. Odile Jacob, Paris (2008)

177. Levine, R.: A Geography of Time. Basic Books, New York (1997)

178. Levine, R.: Eine Landkarte der Zeit. Wie Kulturen mit Zeit umgehen. Piper, München 3. Aufl. (2000)

179. Macey, S.L.: Time. A Bibliographic Guide. Garland, New York (1991)

180. Mainzer, K.: Zeit – Von der Urzeit zur Computerzeit. C.H. Beck, München (1995)

181. Maturana, H.R.: Die Natur der Zeit, pp. 114–125 (1997) in [B, A. Gimmler et al.]

182. Novello, M.: Le circle du temps. Atlantic Science (2001)

183. Portman, A.: Die Zeit im Leben der Organismen, Eranos Jahrb. **20** (1951)

184. Poulet, G.: Etudes sur le temps humain I–IV. Plon, Paris (1952)

185. Prigogine, I.: La Nascita del Tempo, Theoria, Roma (1988)

186. Prigogine, I., Stengers, I.: Entre le Temps et l'Eternité. Arthème Fayard, Paris (1988)

187. Prigogine, I., Stengers, I.: Das Paradox der Zeit. Piper, München (1993)

188. Reusch, S. (ed.): Das Rätsel der Zeit. Wiss. Buchges, Darmstadt (2004). With a connected bibliography

189. Sandbothe, M., Zimmerli, W.Ch. (eds.): Zeit – Medien – Wahrnehmung. Wiss. Buchges, Darmstadt (1994)

190. Scartezzini, J.L.: Visions de l'espace-temps, EPFL, Architecture (1999)

191. Schultz, P.E.W.: Zeit – das Abstrakteste des Abstrakten – Neuropsychologische Aspekte subjektiver und objektiver Zeit, Berlin (1998)

192. Sergent, B., Guilland, L.: L'imaginaire du Temps, Herméneutiques Sociales, No. 1

193. Stadler, F., Stölzner, M. (eds.): Time and History. Ontos Verlag, Heusenstamm (2006)

194. Toulmin, S., Goodfield, J.: The Discovery of Time. Penguin, Harmondsworth (1965)

195. Vaas, R.: Zeit und Gehirn. Spektrum, München (2001)

196. Vaas, P., et al.: Wenn die Zeit rückwärts läuft. Bild Wiss. **12**, 44–66 (2002)

197. Vaas, P., et al.: Zeit ist nur eine Illusion. Bild Wiss. **1**, 46–64 (2008)

198. van Fraassen, B.C.: An Introduction of the Philosophy of Time and Space. Random House, New York (1970)

199. Weis, K. (ed.) Was ist Zeit – Zeit und Verantwortung in Wissenschaft, Technik und Religion. DTV, München (1994)

200. Weis, K. (ed.) Was treibt die Zeit – Entwicklung und Herrschaft der Zeit in Wissenschaft, Technik und Religion. DTV, München (1998)

201. Winfree, A.T.: When Time Breaks down, the Three-Dimensional Dynamics of Electrochemical Waves and Cardiac Arrhythmias. Princeton University Press, Princeton (1987)
202. Yoder, J.G.: Unrolling Time: Christiaan Huygens and the Mathematization of Nature, Cambridge (1988)

Bonn, Germany   Sergio Albeverio
Bielefeld, Germany   Philippe Blanchard

Contents

Contributors

Luigi Accardi Centro Vito Volterra, Università di Roma "Tor Vergata", Rome, Italy

Anindita Niyog Balslev Høyberg, Denmark

Hervé Barreau CNRS, Strasbourg, France

Vieri Benci Dipartimento di Matematica Applicata "U. Dini", Università degli Studi di Pisa, Pisa, Italy

Luciano Boi École des Hautes Études en Sciences Sociales, Centre de Mathématiques, Paris, France

Marcello Cini Dipartimento di Fisica, Università La Sapienza, Rome, Italy; INFM, Sez. di Roma, Rome, Italy

Michael Drieschner Institut für Philosophie, Ruhr-Universität Bochum, Bochum, Germany

Thomas Görnitz Institut für Didaktik der Physik, Johann Wolfgang Goethe-Universität, Frankfurt/Main, Germany

Rudolf Haag Schliersee-Neuhaus, Germany

Holger Lyre Philosophy Department, University of Magdeburg, Magdeburg, Germany

Klaus Mainzer Carl von Linde-Akademie, Technische Universität München, München, Germany

Fabio Minazzi Università degli Studi dell'Insubria, Varese, Italy; Accademia di architettura dell'Università della Svizzera italiana, Mendrisio, Switzerland

Roland Omnès Laboratoire de Physique Théorique, Université de Paris-Sud, Orsay, France

Metod Saniga International Solvay Institutes for Physics and Chemistry, Free University of Brussels (ULB), Brussels, Belgium; Astronomical Institute of the Slovak Academy of Sciences, Tatranská Lomnica, Slovak Republic

L.S. Schulman Physics Department, Clarkson University, Potsdam, NY, USA

Michael Stöltzner University of South Carolina, Columbia, USA

Alessandro Teta Dipartimento di Matematica "G. Castelnuovo", Universitá di Roma "La Sapienza", Rome, Italy

Roderich Tumulka Department of Mathematics, Rutgers University, Piscataway, NJ, USA

Christoph von der Malsburg Institut für Neuroinformatik and Department of Physics and Astronomy, Ruhr-Universität Bochum, Bochum, Germany; Dept. of Computer Science, and Program in Neurosciences, University of Southern California, Los Angeles, USA; Frankfurt Institute for Advanced Studies, Frankfurt am Main, Germany

J.C. Zambrini GFMUL, Lisbon, Portugal

H.D. Zeh Waldhilsbach, Germany

List of Participants

Luigi Accardi, Centro Vito Volterra, Roma
Rudolf Ahlswede (deceased), Universität Bielefeld, Bielefeld
David Albert, Columbia University, New York
Sergio Albeverio, Universität Bonn, Bonn
Aninidita Niyogi Balslev, University of Copenhagen, København
Julian B. Barbour, College Farm, Oxon
Hervé Barreau, Strasbourg
Vieri Benci, Università di Pisa, Pisa
Philippe Blanchard, Universität Bielefeld, Bielefeld
Luciano Boi, École des Hautes Études en Sciences Sociales, Paris
Anne Boutet de Monvel, Paris
Egbert Brieskorn (deceased), Eitorf
Erwin Brüning, University of Durban-Westville, Durban
Martin Carrier, Universität Bielefeld, Bielefeld
Marcello Cini (deceased), Università degli studi di Roma "La Sapienza", Roma
Andreas Dress, Universität Bielefeld, Bielefeld
Martin Drieschner, Ruhr-Universität Bochum, Bochum
Detlef Dürr, Universität München, München
Rodolfo Figari, Università di Napoli "Frederico II", Napoli
Thomas Görnitz, Johann Wolfgang Goethe-Universität, Frankfurt am Main
Francesco Guerra, Università degli studi di Roma "La Sapienza", Roma
Rudolf Haag, Schliersse-Neuhaus
Andreas Kamlah, Universität Osnabrück, Osnabrück
Claus Kiefer, Universität zu Köln, Köln
Günter Küppers, Universität Bielefeld, Bielefeld
Holger Lyre, Ruhr-Universität Bochum, Bochum
Klaus Mainzer, Universität Augsburg, Augsburg
Jens Mennicke, Bielefeld
Roland Omnès, Université Paris-Sud, Orsay
Sylvie Paycha, Complexe Universitaire des Cézaux, Aubiere
Metod Saniga, Université Libre de Bruxelles, Brüssel

Walter Schneider, Baden-Rütihof
Lawrence S. Schulman, Clarkson University, Potsdam
Michael Stöltzner, Universität Bielefeld, Bielefeld
Alessandro Teta, Università di L'Aquila, L'Aquila
Roderich Tumulka, Ludwig-Maximilians-Universität, München
Rüdiger Vaas, Stuttgart
Jean-Claude Zambrini, Gruppe de Física-Matemática da Universidade de Lisboa,
  Lisboa
Dieter Zeh, Waldhilsbach

Chapter 1
Is Time Real?

Marcello Cini

Abstract We first give a detailed historical analysis of the different representations of change and time. After that we discuss the question of reversibility and irreversibility in the classical and in the quantum world. We shall follow the good practice of going from the simpler to the more complicated.

Keywords Classical mechanics · Statistical physics · Quantum mechanics · Irreversibility

1 The Representations of Change

1.1 From Change to Time

1.1.1 Nothing Changes/Everything Changes

Two opposing views of the world have marked more than 25 centuries of philosophy since its birth in Ancient Greece. The first one, due to Parmenides, states that nothing changes: change is only appearance. According to Popper, his argument goes as follows [1]: (1) Only being exists; (2) non-being does not exist; (3) non-being would be the absence of being: the vacuum; (4) vacuum cannot exist; (5) if there is no vacuum the world is full: there is no space for movement; (6) movement and change are impossible.

The second one, due to Heraclitus, states that everything changes: "All things are always in movement ... even if this escapes our sensations". Things are not real things, they are processes, they are continuously changing. "Panta rei". They are

The editors mourn in deep sorrows the passing away of Marcello Cini, on Oct. 22, 2012. He was brilliant physicist, epistemologist and author. By his writings and "engagement" in social and political issues he had a strong influence upon us and we badly miss him.

M. Cini
Dipartimento di Fisica, Università La Sapienza, Rome, Italy

M. Cini
INFM, Sez. di Roma, Rome, Italy

S. Albeverio, P. Blanchard (eds.), *Direction of Time*,
DOI 10.1007/978-3-319-02798-2_1,
© Springer International Publishing Switzerland 2014

1

like fire: a flow of matter, or like a river. "Nobody bathes twice in the same river". The apparent stability of things is only a consequence of the laws which constrain the processes of the world [2].

Today we are still at grips with the same conflict. In fact it simply reflects the dispute between those who believe the universe to be ruled by absolute and eternal laws of nature written in mathematical language, and those who see it as a network of interconnected processes and prefer to look for explanations of phenomena based on generalizations of empirical evidence.

I personally share the view of J.R. Oppenheimer who says: "These two ways of thinking, the one which is based on time and history, and the one which is based on eternity and timelessness, are two components of man's effort to understand the world in which he lives. Neither is capable of including the other one, nor can they be reduced one to the other, because they are both insufficient to describe everything".

1.1.2 Objective Time, Subjective Time

For a long time men have discussed the nature of time. It is appropriate, I think, to start this discussion by quoting Augustine's statement: "Time does not exist without a change produced by movement".[1]

The first thing to do, in fact, is to dissipate the belief that the concept of *time* is a necessary premise for describing and interpreting change. The reverse is instead true. Already Aristotle, several centuries before Augustine, said that "time is the number of change, in accordance to what comes before or after", and recognized explicitly that without something that changes there is no time: "Since we have no cognizance of time when we do not detect any change, while on the contrary when we perceive a change we say that time has elapsed, it is clear that there is no time without change and movement" [4].

There is more. The concept of time may arise only from a comparison between two processes of change. A period of stillness may be long or short compared with another one, as well as a change may be quick or slow only compared with the rate of another one. The comparison between two processes outside us leads us to conceive time as an objective entity, while the comparison between the change of an exterior object and our internal, conscious or unconscious, rhythms, leads to the concept of a subjective time.

Let us start from the first one. It should be clear that "objective time" is not a substance flowing at a constant rate, as many colloquial expressions such as "time flows" or "clocks measure the flow of time" imply. It is sufficient to notice that the velocity of this hypothetical fluid would be of one second per second in order to realize that it is a tautological nonsense. A more precise proof that absolute time does not exist comes, as is well known, from Einstein's relativity: time is a form of relationship between succeeding events in different space locations. Even after

---

[1]Sant'Agostino, *La Città di Dio*, quoted in [3].

Einstein, however, the idea that time "contracts" or "dilates" implies a misleading reification of the concept, which hinders its comprehension.

The second notion of time—which defines it, according to Kant, as an innate *a priori* capacity of human perception of synthetizing events in the form of temporal sequences before having access to any kind of experience—should be equally criticized. The knowledge accumulated in two centuries of scientific development has shown how tight are the connections existing between mind and body on the one hand, and between the individual and society on the other one. On factual grounds, in addition, the classic studies of Piaget [5] on the development of the notion of time in the child have shown how scarcely innate it is.

Neither of these concepts, therefore, can be defined without recognizing that both are reciprocally connected within the pattern of the social fabric in a given historical context. It is appropriate at this point to quote Norbert Elias, an author who has investigated in depth this point of view: "Time is not the reproduction of an objectively existing flux, nor a form of common experience of all men, antecedent to any other experience... The word "time" is, so to say, the symbol of a relationship created by a group of human beings, endowed with a given biological capacity of remembering and synthetizing, between two or more series of happenings, one of which is standardized as a frame of reference or a unit of measure of the other one" [6].

1.1.3 Time's Arrow, Time's Cycle

Besides the dichotomy between objective and subjective time another dichotomy, which also goes back to the depth of ages, contrasts two conceptions of change— reversibility or irreversibility—as mutually exclusive, it consequently reflects itself on two conflicting views of time. An example of the latter is the human life, which inexorably flows from birth to death; an example of the former is the motion of celestial bodies, with their eternal going forwards and coming back.

"A crucial dichotomy—writes Stephen J. Gould—covers the most ancient and deep themes of western thought about the central subject of time: two visions, linear and circular, are resumed under the notions of time's arrow and time's cycle. At one end—time's arrow—history is viewed as an irreversible sequence of unrepeatable events. Each moment occupies a distinct position in this sequence, and altogether they tell a story of successively connected events moving in one direction. At the other end—time's cycle—events have no significance as distinct episodes with a causal impact on a contingent history. Apparent motions are part of repeated cycles and differences of the past will become realities in the future. Time has no direction" [7].

These two conflicting ways of conceiving time have alternatively dominated human cultures. At the roots of western culture, according to Gould, we find in the Bible the arrow of time. Not always and not everywhere, however, this vision of time has marked the birth of civilization. According to Mircea Eliade [8] the majority of peoples in the history of mankind have believed in a cyclic time, and considered the arrow of time as inconceivable and even frightening.

Only in recent times, however, the notion of an arrow of time—again according to Gould—has become "the familiar and orthodox conception for the majority of cultured western people". Without it the idea of progress, or the concept of biological or cosmic evolution would be impossible. This phenomenon is in fact very recent, since its origins can be traced back to the introduction in physics by Sadi Carnot in the early decades of the 19th century of a way of looking at the world based on the irreversibility of natural phenomena, alternative to the one on which the galilean revolution was based, which made of the reversibility of any kind of motion the key for explaining everything that happens.

It is, however, only a century after Carnot, in the second half of the 20th century, that the metaphor of the arrow of time has become a component of the "metaphysical core" of many contemporary scientific disciplines. As a striking example I quote from a report of the physicist Jean Pierre Luminet the following list of five different arrows of time actually envisaged in this discipline [9]:

1. The radiative arrow (spherical waves always propagate outwards from a source).
2. The thermodynamic arrow (transformations in an isolated system always proceed in the direction of increasing entropy).
3. The microscopic arrow (weak interactions show an asymmetry between decays and inverse reactions).
4. The quantum arrow (the interaction between a measuring instrument and a microscopic object changes irreversibly the state of the latter).
5. The cosmological arrow (the universe apparently expands irreversibly).

1.1.4 Causality and the Two Forms of Time

A close connection between time's arrow and time's cycle can be found by using the concept of causality. Intuitively, to explain an event one has to find its cause. Of the four aristotelian types of causes (material, formal, final, and efficient) only the last one is still considered a cause in a proper sense, because it considers the occurrence of an event or the accomplishment of an action as a necessary and sufficient condition for the occurrence of a subsequent event. The latter is therefore the effect of the former. This sequence establishes an arrow of time: the effect always comes after its cause. This is what we call linear causality. However, this apparently trivial remark turns out to entail non-trivial consequences when the same cause and the same effect are repeated in a steady sequence of time's cycles.

A simple example of this relation between the two representations of time is given by the connection of the stress applied to an elastic body to its deformation (strain). If the stress is applied suddenly and remains constant thereafter (step function) the strain starts from zero and increases with time reaching asymptotically a final value. This is due to the presence of internal friction which dissipates into heat a part of the work done in deforming the body. The arrow of time goes from the application of the stress towards the inception of the strain. On the other hand, if the applied stress is periodic, the response also is periodic. There is no longer something which comes before (or after) something else: when a steady state is reached the peak of the strain comes after the preceding peak of the stress, but before the next one, and vice versa. Let us see how the relation between the two comes out.

In the periodic case we can express the strain S in terms of the unit stress $e^{i\omega t}$ simply by multiplying it (Hooke's law) by a complex response factor:

$$S = [A(\omega) + i B(\omega)] e^{i\omega t}$$

where $A(\omega)$ and $B(\omega)$ are two apparently independent functions of the frequency ω, characteristic of the material of which the body is made: the elastic modulus and the damping coefficient. In the case of a sudden application of the stress the time evolution of the strain can be obtained in terms of $A(\omega)$ and $B(\omega)$ by means of a Fourier transform of the step function as an integral of periodic exponentials.

The remarkable result is that, by simply imposing the causality condition, namely that the effect must be zero before the onset of the cause, the two functions $A(\omega)$ and $B(\omega)$ turn out not to be independent, but rather to be connected by a relation of the form

$$A(\omega) = \int B(\omega')/(\omega - \omega')\, d\omega'$$

and vice versa with A and B exchanged. Relations of this type are called *dispersion relations* and have played an important role in physics in different fields. I personally discovered it [10] in 1947 in dealing with the field of elasticity, without knowing that it had been discovered 20 years before by Kramers and Kronig [11, 12] in the field of optics (where A and B represent the refractive index and the absorption coefficient of light in a medium). In the mid-1950s similar relations have been found for scattering amplitudes in elementary particle physics [13], by using a generalized form of causality expressing the impossibility of connecting causally two events in space-time connected by a spacelike distance. I have published a brief history of dispersion relations [14] in *Fundamenta Scientiae* many years ago.

1.2 Reversible and Irreversible Changes

1.2.1 Reversibility of Motion: Galileo

A lantern oscillates in Pisa's Cathedral. A young man—we are at the end of the 16th century—does not pay much attention to the religious service. His attention is attracted by the lantern. Slowly the oscillations are damped, the amplitude reduces gradually until the motion comes to an end. In these times anyone would have interpreted the phenomenon as a verification of the Aristotelian doctrine: in its natural motion a body tends to reach its "natural" place, the lowest possible attainable. This is the important "fact". The phases of the motion's rise are only accidental consequences of the initial "artificial" motion impressed on the body by an external impact.

But the young man—Galileo—looks at the phenomenon with a different eye. Oscillations are the important "fact". They disclose that, if one neglects as accidental the gradual damping, the downward and the upward motions are equally "natural", because one is the reverse of the other one. Only from this point of view is it possible to ask oneself how long it takes the pendulum to perform a complete oscillation: if the downward and the upward motion are qualitatively different there is no oscillation. Only by deciding to unify conceptually the two motions is Galileo able to find that the time required is practically constant and independent of the amplitude. The pendulum becomes the symbol of cyclical time.

1.2.2 Reversibility and Irreversibility in the Earth's History: Burnet, Hutton, and Lyell

With Newton's triumph this new way of looking at things penetrated into all the domains of science. Gould's reconstruction of the work of three pioneers of modern geology clearly illustrates this diffusion [15]. The task they had to accomplish, one after the other, was to explain the empirical discovery that the Earth's history had originated many millions of years before the biblical date of Creation.

The first one, Thomas Burnet, a man of the church, bound to the necessity of conciliating this explanation with the Holy Writ, divides the span of time between the Creation of the Earth and its Final End into recurring cycles whose phases change from one to the other, but maintain a substantially similar pattern.

The second one, James Hutton, solves the problem within the boundaries of science by elaborating a conception of the Earth as a "machine" in which disrupting forces and restoring forces balance each other continuously in order to reproduce a cycle of events which rigorously reproduce themselves: the first ones eroding mountains and continents, the second ones rising them and reconstructing them. The influence of Newton's thought is explicitly recognized by Hutton himself: "When we find that there are means cleverly devised in order to make possible the renewal of the parts which necessarily decay ... we are able to connect the Earth's mineral system with the system by means of which celestial bodies are made to move perpetually along their orbits".

The third and most famous of them, Charles Lyell is usually known for upholding the thesis that the same forces acting today have been responsible for all the gradual geological changes of the past (uniformism) against Georges Cuvier, according to whom the history of the Earth is a succession of unrepeatable and unpredictable events (catastrophism). Here again Lyell's view was based on the belief that "many enigmas of both the moral and the physical worlds, rather than being the effect of irregular and external causes, depend on invariable and fixed laws." The metaphor of the time's cycle is therefore for him the conceptual tool for explaining the Earth's history in terms of repeated phases of reversible changes (climatic and morphological) produced by alternating upward and downward movements of lands and seas, leading to the gradual and steady variation of the different forms of life.

1.2.3 Irreversibility of Thermodynamical Transformations: Carnot

Sadi Carnot wrote: "Because by reaching in any way a new caloric equilibrium one may obtain the production of motion power, any new equilibrium reached without production of motion power should be considered as a true loss; in other words, any change in temperature not due to a change in volume of bodies, is nothing else than a useless attainment of a new equilibrium" [16].

A newtonian scientist would never use words as *loss* and *useless*. They are concepts referring to man, not to the phenomenon in itself. The caloric loss is lost for man. The variation of temperature without production of motion power is useless for man.

This new way of looking at nature has two consequences. The first one is that the type of "law" looked for by Carnot is qualitatively different from Newton's laws, which prescribe what has to happen. His aim is to discover the interdictions set by nature to the use of its forces, to determine the constraints which limit their reciprocal transformations. His "laws" establish what is forbidden. The consequence is that the type of abstraction needed to pursue this aim is different in the two cases. For Carnot dissipation is important, while for Newtonians dissipation is negligible.

As is well known, Carnot proved in this way that any transformation involving the transfer of heat from a source to a body in order to produce mechanical power must inevitably involve the irreversible transfer of a part of this heat to another body at a lower temperature.

1.2.4 Irreversibility of Biological Evolution: Darwin

Darwin explains the evolutionary process of life on Earth by the concurrent action of two factors: a mechanism (on the nature of which Darwin does not express himself) producing a variability of the somatic features of the different individuals belonging to a given species, and a filter which selects the individuals with the most convenient features for survival, leading to the formation of species better adapted to the changing environment. Natural selection is the result of the capacity of these individuals to reproduce at a higher rate than the more disadvantaged ones, whose descendants gradually are extinguished.

At the beginning of the XX century, with the rediscovery by Hugo de Vries of Mendel's law, the origin of variability is traced back to the random mutations of a discontinuous genetic material possessed by the individuals. The breaking in of chance produces irreversibility. The model of evolution which has dominated in the community of biologists (New Synthesis) for almost 60 years is therefore characterized by the gradual irreversible change of the population of a species under the action of the two complementary processes of random generation of genetical variability and deterministic selection of the fittest phenotypes.

At the beginning of the 1970s a new model was introduced by N. Eldredge and S.J. Gould. Their theory of punctuated equilibria rejects the gradualism of the standard evolutionary process and replaces it with a discontinuous process in which species remain unchanged for long periods (millions of years) until they disappear abruptly (thousands of years) and are replaced by new ones. Both their birth and their death may often be due to chance.

This intervention of chance at the two levels of individuals and of species leads Gould to conclude his book *Wonderful Life* with the words:

> And so, if you wish to ask *the* question of all the ages—why do humans exist?—a major part of the answer, touching those aspects of the issue that science can treat at all, must be: because *Pikaia* survived the Burgess decimation. This response does not cite a single law of nature; it embodies no statement about predictable evolutionary pathways, no calculation of probabilities based on general rules of anatomy or ecology. The survival of *Pikaia* was a contingency of "just history". [17]

2 From Macro to Micro

2.1 From Reversibility to Irreversibility and Back

2.1.1 From Dichotomy to Statistics

A macroscopic volume V of gas, at normal pressure and temperature, contains a number N of molecules of the order 10^{23}. Suppose that initially the volume is divided by a partition in two non-communicating volumes V_R and V_L, and that all of the N molecules are contained in V_R. The density in V_R will be $d_R = N/V_R$ and in V_L it will be $d_L = 0$. If the partition is removed, very quickly in both volumes the densities will be equal to N/V. At the macroscopic scale the change is irreversible.

On the other hand, if N is of the order of a few molecules, it may happen that d_R and d_L will be different. Perhaps we may even find again $d_R = N/V_R$ and $d_L = 0$. The change may therefore be reversible. Its probability can be easily calculated, and amounts to $1/2^N$. The sharp dichotomy has become a statistical evaluation. Of course, when N is 10^{23} the probability of reversal becomes ridiculously small. The spontaneous expansion of a macroscopic quantity of gas in vacuum is therefore "for all practical purposes", always irreversible.

This simple argument shows that the Second law of thermodynamics, introduced by Clausius and Thomson for macroscopic bodies, has a microscopic justification. However, things are not that simple.

2.1.2 Boltzmann, Loschmidt and Zermelo: Time's Arrow or Time's Cycle?

The central point of the debate which animated the physicist's community at the end of the 19th century is about the nature of the Second law. The reversibility of newtonian motions is in fact incompatible with the Second law of thermodynamics, which excludes the possibility of reversing the direction of the spontaneous trans-formation leading to the equilibrium state of a system from a non-equilibrium one. We all know that this fact was not easily recognized and that Boltzmann in 1872 worked out a theorem (H theorem) which seemed to prove that the irreversible tran-sition from any non-equilibrium state to the equilibrium one was a consequence of Newton's law. We also know that Loschmidt first (1876) and Zermelo later (1896) rejected Boltzmann's result and presented counterexamples showing that his claim was untenable. I am not going to dwell on the details of the dispute (denoted in the following as BLZ), which had been forgotten for almost a century, and was recon-sidered only recently in two interesting books to which I refer [18, 19].

The thing which matters here is that Boltzmann, as a consequence of this dispute, changed his approach to the problem, and introduced a distinction between initial conditions which lead to an evolution towards equilibrium and those which tend to lead the system away from it. Since it turns out that the former are enormously more numerous than the latter ones, the irreversibility of the Second law can be reconciled, according to Boltzmann, with the reversibility of newtonian motion. The Second law loses therefore the character of absolute necessity, which was attributed to it up to that moment by the majority of the physicist's community, to acquire the status of a probabilistic prediction about the properties of a system made of a great number of elementary constituents. The law of increasing entropy expresses therefore a statistical property: the great majority of evolutionary paths lead from less probable to more probable states.

2.1.3 Order, Disorder, and Information

The free expansion of a perfect gas presented in Sect. 2.1.1 can be interpreted as a transition from order to disorder. In fact the initial state (all the molecules are concentrated in V_R) is more ordered than the final state (molecules may be in V_R as well as in V_L). Order is, however, a "subjective" concept. We "know" that all the molecules are in V_R initially, while we do not know at the end where any given molecule is. We can describe the change as a loss of *information* on the position of the molecules. The expansion is, however, an "objective" phenomenon. We can describe it in terms of increase of *entropy*. It turns out, as is well known, that the two quantities are proportional, with a minus sign in front.

All is clear, therefore, at the macroscopic scale. However, as Loschmidt and Zer-melo claimed, at the microscopic scale the motion of any given molecule is "in principle" completely determined, once its initial state is completely given, by its collisions with the other ones and against the walls. Since the forces are conserva-tive, the classical motion of each molecule is reversible and the collective motion of all of them should be equally reversible. Is this claim well founded? What are its implications?

Apart from the consideration that the motion of molecules is not classical, but is ruled by the laws of quantum mechanics [this argument will be discussed in Sect. 2.2.1], it is clear that a physical experiment capable of proving this kind of reversibility will never be possible, even in the case that the number of molecules could be drastically reduced. However, the experiment can be simulated in a computer, and, of course, it works. A whole new field of research has developed along these lines (Molecular Dynamics).

2.2 Reversibility/Irreversibility in the Quantum World

2.2.1 The Role of Chance in Quantum Mechanics

At the level of the dynamics of an individual system, newtonian motion is reversible. Irreversibility is brought in, we have seen, by the necessity of describing, by means of the probability distribution in phase space of classical statistical mechanics, the macroscopic properties of a collection of a great number of particles.

Quantum mechanics, however, had to take into account two new facts. The first one is that different events may follow from apparently equal external and initial conditions; the second one is that it is impossible to fix exactly the value of all the variables of a given system. Position x and momentum p of a particle are the simplest example of incompatible variables. Heisenberg's principle sets the lower limit of $h/4\pi$ to the product of their uncertainties.

The solution of the problem, as we all know, was found by releasing the connection between the state of the system and its variables. The first one, represented by means of a suitable (wavelike) function, was still completely determined by the initial conditions and the laws of motion, and the latter were left free of acquiring at random, with different probabilities given by that function, one of the possible values within their range of variability. The difference with classical statistical mechanics is radical: while in the latter the probabilistic description of a system's state is simply due to our ignorance of the precise value of its variables, which nevertheless do actually have a precise value, the probabilistic nature of the quantum system's properties is considered, by the overwhelming majority of physicists, to be "ontological".

Now the question arises: at what stage does chance come in? The usual answer is: the evolution of the wave is deterministic and reversible, while the measurement brings in randomness and irreversibility. Of course this answer introduces a lot of problems of a fundamental nature: on the role of the "observer", on the power of man's mind to manipulate "reality" and so on. I will come back to these questions at the end.

2.2.2 Irreversibility of Quantum Measurement

The origin of irreversibility is therefore generally ascribed to the so called "wave function collapse" or "reduction of a wave packet" produced by the act of measuring a quantum variable by means of a macroscopic measuring instrument. As is well known, the problem was tackled, in the early days of quantum mechanics, by Bohr, who *postulated* the existence of classical objects in order to explain how the quantum objects could acquire abruptly, in the interaction with them, sharp values of either position or velocity. This dichotomy between classical and quantum worlds was questioned by von Neumann who insisted that, after all, also the macroscopic objects should obey the laws of quantum mechanics.

The problem, in my opinion, should be formulated as follows. On the one hand we have microscopic objects (*quantons*) which have *context dependent properties*. This means that these properties, which have generally blunt values, only occasionally, but not simultaneously, may acquire sharp values. This happens when a quanton interacts with a suitable piece of matter which constrains it to assume, *at random but with a given probability*, a sharp value. On the other hand our everyday experience shows that macroscopic objects have *context independent properties*. *It becomes therefore necessary to prove that the existence of macroscopic pieces of matter with context independent properties is not a postulate* (as Bohr assumed) *but follows from the equations of quantum mechanics themselves.*

2.2.3 Ontic and Epistemic Uncertainties

This question was investigated and answered by my group in Rome 20 years ago in two papers [20, 21], which at the time received some attention (*Nature* dedicated a whole page of comment to the second one [22]). It is, however, fair to give credit to K. Gottfried [23] for having correctly approached the problem many years before.

In these papers we proved that when a quanton P in a given state interacts with a suitable "instrument" S_q made of N quantons, the difference between the probabilistic predictions of quantum mechanics on the possible outcomes of this interaction and the predictions of classical statistical mechanics, for an ideal statistical ensemble in which a classical instrument S_c replaces S_q (with the same values of its macroscopic variables), tends to vanish when N becomes very large ($\gg 1$). This means that, after all, Bohr was right in assuming that classical bodies exist. Needless to say, our result proved also that Schrödinger's cat cannot be at the same time dead and alive, simply because it is a macroscopic "object".

A similar problem—namely whether a single particle, whose wave function is represented by two distant wave packets, materializes instantly in one or the other only when its position is measured—has been investigated, I believe with success, by Maurizio Serva and myself a few years ago [24], and further clarified in collaboration with Philippe Blanchard [25]. In this case we can explicitly calculate the uncertainties Δx and Δp of position and momentum, which appear in the well known general expression of the Heisenberg uncertainty principle. The standard inequality becomes

$$(\Delta x \Delta p)^2 = (h/4\pi)^2 + \left[(\Delta x \Delta p)_{\text{csm}}\right]^2$$

where $(\Delta x \Delta p)_{\text{csm}}$ is the uncertainty product of the corresponding probability distribution of classical statistical mechanics. The second term, therefore, expresses an epistemic uncertainty, while the first one expresses the irreducible nature of chance at the quantum level.

This interpretation of the uncertainty principle solves the paradox of the particle localization in one of the two distant isolated wave packets. In fact we can conclude that the particle actually was in one or the other even before the measurement was performed, because the large Δx has a purely classical epistemic origin.

This clarifies also the different nature of ontic randomness and epistemic randomness. The first one is reversible (no dissipation) the second one is irreversible (the loss of information [entropy] increases with time).

Here again we are faced with the old debate (BLZ) of explaining how a macroscopic system (particle + instrument) made of a great number of microscopic objects can acquire a property (irreversibility of evolution) that his elementary components do not have. The answer is the same: the great majority of evolutionary paths lead from less probable to more probable states.

2.2.4 A Unified Statistical Description of the Quantum World

If randomness has an irreducible origin the fundamental laws should allow for the occurrence of *different* events under *equal* conditions. The language of probability, suitably adapted to take into account all the relevant constraints, seems therefore to be the only language capable of expressing this fundamental role of chance. If the probabilistic nature of the microscopic phenomena is fundamental, and not simply due to our ignorance as in classical statistical mechanics, it should be possible to describe them in probabilistic terms from the very beginning.

The proper framework in which a solution of the conceptual problems discussed above should be looked for is therefore, after all, the birthplace of the quantum of action, namely phase space, where no probability amplitudes exist. It is of course clear that joint probabilities for both position and momentum having sharp given values cannot exist in phase space, because they would contradict the uncertainty principle. Wigner [26] however, introduced the functions called pseudoprobabilities (which may assume also negative values) to represent quantum mechanics in phase space, and showed that by means of them one can compute any physically meaningful statistical property of quantum states. It seems reasonable therefore to consider these functions not only as useful tools for computations, but as a framework for looking at quantum mechanics from a different point of view.

This program has been recently carried on [27] by generalizing the formalism of classical statistical mechanics in phase space with the introduction of a single quantum postulate, which introduces mathematical constraints on the set of variables in terms of which any physical quantity can be expressed (usually denoted as characteristic variables). It turns out, however, that these constraints cannot be fulfilled by ordinary random numbers, but are satisfied by the mathematical objects called by Dirac *q-numbers*. The introduction of these q-numbers in quantum theory is therefore not assumed as a postulate from the beginning, *but is a consequence of a well defined physical requirement*. The whole structure of quantum mechanics in phase space is therefore deduced from a single quantum postulate without ever introducing wave functions or probability amplitudes.

This approach has some advantages. First of all, many paradoxes typical of wave-particle duality disappear. On the one hand in fact, as already shown by Feynman [28], it becomes possible to express the correlations between two distant particles in terms of the product of two pseudoprobabilities independent from each other. All the speculations on the nature of an hypothetical superluminal signal between them becomes therefore meaningless. Similarly, the long debated question of the meaning of the superposition of state vectors for macroscopic objects may also be set aside as equally baseless.

Secondly, this approach eliminates the conventional hybrid procedure of describing the dynamical evolution of a system, which consists of a first stage in which the theory provides a deterministic evolution of the wave function, followed by a hand made construction of the physically meaningful probability distributions. The direct deduction of Wigner functions from first principles solves therefore a puzzling unanswered question which has been worrying all the beginners approaching the study of our fundamental theory of matter, all along the previous 75 years, namely "Why should one take the modulus squared of a wave amplitude in order to obtain the corresponding probability?" We can now say that there is no longer need of an answer, because there is no longer any need to ask the question.

Finally it should be stressed that it is not the practical use of the formalism of quantum mechanics, of course, which is put in question by the approach suggested here. However, from a conceptual point of view, the elimination of the waves from quantum theory is in line with the procedure inaugurated by Einstein with the elimination of the ether in the theory of electromagnetism. Maybe it can provide a new way of musing on the famous statement of Feynman: "It is fair to say that nobody understands quantum mechanics".[2]

References

1. Popper, K.: The World of Parmenides. Routledge & Kegan, London (1998). It. transl. Il Mondo di Parmenide, p. 126. Edizioni Piemme, Casale Monferrato (1998)
2. Popper, K.: ibid. p. 39
3. Luminet, J.P.: Le Temps et sa Flèche, p. 59. Flammarion, Paris (1994)
4. Aristotle: Physica IV, 11 (218)

---

[2]The approach reported above at the Bielefeld Conference in 2001 to non relativistic Quantum Mechanics has been further extended [29] to quantum field theory. In this paper it is shown that a coherent development of the original formulation by its founders [30, 31] leads to a formulation of quantum field theory in terms of ensemble averages of the field's dynamical variables, in which no reference at all is made to the Schrödinger wave functions of "first quantization". In this formulation the wave particle duality is no longer a puzzling phenomenon. The wave particle duality is instead, in this new perspective, only the manifestation of two complementary aspects (continuity vs. discontinuity) of an intrinsically non-local physical entity (the field) which objectively exists in *ordinary three dimensional space*.

5. Piaget, J.: Le developpement de la notion de temps chez l'enfant. Presses Universitaires de France, Paris (1946)
6. Elias, N.: Saggio sul tempo, p. 59. Il Mulino, Bologna (1986)
7. Gould, S.J.: Time's Arrow, Time's Cycle. Harvard University Press, Cambridge (1987). It. transl., p. 24, Feltrinelli, Milano (1989)
8. Eliade, M.: Le mythe de l'eternel retour. Gallimard, Paris (1949). It. transl., p. 70, Borla, Torino (1968)
9. Luminet, J.P.:. ibid. p. 78
10. Cini, M.: J. Appl. Phys. 21, 8 (1950)
11. Kramers, H.A.: Atti Congr. Fis. Como (1927)
12. Kronig, R.: J. Opt. Soc. Am. 12, 547 (1926)
13. Gell-Mann, M., Goldberger, M., Thirring, W.: Phys. Rev. 95, 1612 (1954)
14. Cini, M.: Fundam. Sci. 1, 157 (1980)
15. Gould, S.J.: ibid.
16. Carnot, S.: Reflexions sur la puissance motrice du feu, p. 23. Blanchard, Paris (1953)
17. Gould, S.J., Wonderful Life, p. 323. Norton, New York (1989)
18. Hollinger, H.B., Zenzen, M.J.: The Nature of Irreversibility (1985)
19. Ageno, M.: Le Origini dell'Irreversibilità, Torino (1985)
20. Cini, M., De Maria, M., Mattioli, G., Nicolò, F.: Found. Phys. 9, 479 (1979)
21. Cini, M.: Nuovo Cimento 73B, 27 (1983)
22. Nature 302, 307 (1983)
23. Gottfried, K.: Quantum Mechanics, sect. 20. Wiley, New York (1966)
24. Cini, M., Serva, M.: Found. Phys. Lett. 3, 129 (1990)
25. Blanchard, Ph., Cini, M., Serva, M.: In: Albeverio, S., et al. (eds.) Ideas and Methods in Quantum and Statistical Physics. In Memory of Raphael Hoegh-Krohn. Cambridge University Press, Cambridge (1992)
26. Wigner, E.: Phys. Rev. 40, 749 (1932)
27. Cini, M.: Ann. Phys. 273, 99 (1999)
28. Feynman, R.P.: In: Hiley, B.J., Peats, F.D. (eds.) Quantum Implications, p. 285. Routledge & Kegan, London (1987)
29. Cini, M.: Ann. Phys. 305, 83–95 (2003)
30. Born, M., Heisenberg, W., Jordan, P.: Z. Phys. 35, 557 (1926)
31. Dirac, P.A.M.: Proc. R. Soc. A 114, 243 (1927)

Chapter 2
A Simple Model for Decoherence

Alessandro Teta

Abstract The meaning of decoherence as a (practically) irreversible process in Quantum Mechanics is discussed. Also a simple two-particle model is introduced consisting of a heavy (the system) and a light (the environment) particle and the decoherence effect is explicitly computed on the heavy particle due to the presence of the light one.

It is generally believed that one of the main distinctive character of Quantum Mechanics is the superposition principle.

From the mathematical point of view, it simply means that if one has two possible states for the system then also any their (normalised) linear combination is a possible state, due to the fact that the state space of the system has a linear structure.

The key point is that a superposition state in general describes entirely new physical properties of the system which cannot be argued from the knowledge of the component states separately.

A typical example considered here is the case of a particle in one dimension described, in the position representation, by the superposition state $\psi_t(x) = \frac{1}{\sqrt{2}}(\psi_t^+(x) + \psi_t^-(x))$, where ψ_t^+, ψ_t^- are two normalised and orthogonal states at time $t \geq 0$. If one computes the probability distribution of the position of the particle one obviously has

$$|\psi_t(x)|^2 = \frac{1}{2}|\psi_t^+(x)|^2 + \frac{1}{2}|\psi_t^-(x)|^2 + \mathrm{Re}\big(\psi_t^+(x)\overline{\psi_t^-(x)}\big) \tag{1}$$

Then if the supports of the two states are not disjoint, the interference term $\mathrm{Re}(\psi_t^+(x)\overline{\psi_t^-(x)})$ is relevant and it is responsible for the interference fringes observed in real experiments involving microscopic objects, e.g. the two slits experiment.

A. Teta (✉)
Dipartimento di Matematica "G. Castelnuovo", Universitá di Roma "La Sapienza", P.le Aldo Moro, 5, 00185 Rome, Italy
e-mail: teta@mat.uniroma1.it

S. Albeverio, P. Blanchard (eds.), *Direction of Time*,
DOI 10.1007/978-3-319-02798-2_2,

It is remarkable that, due to the presence of the interference term, one cannot interpret the state ψ_t as a classical statistical mixture of identical particles which are in ψ_t^+ or ψ_t^- with probability one half.

The possibility of producing such interference is one of the most relevant characteristic behaviours of the microscopic world which is accurately described applying the rules of Quantum Mechanics.

On the other hand the Schroedinger equation has universal validity and in particular it can be used to describe systems consisting of a macro object coupled with a micro object. In such a situation it is easily seen that a superposition state of the micro object can be transferred to the macro object as a result of the dynamical evolution.

This means that the theory predicts the existence of superposition states and the highly non-classical interference effects also for macro objects which, of course, are not usually observed in our everyday life.

This apparent paradox can be explained if one realises that superposition states are in fact fragile and then they can be destroyed even by a weak interaction with an environment. Such dynamical and practically irreversible mechanism of suppression is usually called decoherence.

In the last 30 years the phenomenon of decoherence has been described in the physical literature using many different models (see e.g. [4] and references therein).

Nevertheless only few of these results are mathematically proved and then a further analysis in the direction of a rigorous study of simple models in which the approximations used are controlled is required.

Here we shall describe a first attempt of rigorous derivation of the decoherence effect in a two particles system ([2], see also [3] for results in the same direction).

The basic tool for the analysis is the representation of the state by a density matrix, i.e. a positive, trace-class operator ρ_t, with $\mathrm{Tr}\,\rho_t = 1$, acting on the Hilbert space of the system $\mathcal{H}$.

If in particular $\rho_t^2 = \rho_t$, i.e. ρ_t is a projector on some $\xi_t \in \mathcal{H}$, then one recovers the usual description in terms of the wave function ξ_t and ρ_t is called a pure state.

In the general case $\rho_t^2 \neq \rho_t$ the state is called a mixture.

The difference between pure and mixed states can be understood in terms of the entropy $S(\rho_t) = -\mathrm{Tr}\,\rho_t \log \rho_t$; for a pure state the entropy vanishes (corresponding to the maximal information available on the system) while it is strictly positive for a mixture (corresponding to our degree of knowledge on the preparation of the state).

If one considers an isolated particle described by the superposition (pure) state ψ_t introduced above, the corresponding density matrix in the position representation is given by the kernel

$$
\begin{aligned}
\rho_t^P(x, x') &= \overline{\psi_t}(x)\psi_t(x') \\
&= \frac{1}{2}\overline{\psi_t^+}(x)\psi_t^+(x') + \frac{1}{2}\overline{\psi_t^-}(x)\psi_t^-(x') \\
&\quad + \frac{1}{2}\overline{\psi_t^+}(x)\psi_t^-(x') + \frac{1}{2}\overline{\psi_t^-}(x)\psi_t^+(x')
\end{aligned}
\tag{2}
$$

The last two terms in (2) are usually called off-diagonal terms and they are responsible for the interference effects (in fact the probability distribution for the position $\rho_t^p(x, x)$ reduces to (1)).

On the opposite side one can considers the mixed state for the same particle

$$\rho_t^m(x, x') = \frac{1}{2}\overline{\psi_t^+(x)}\psi_t^+(x') + \frac{1}{2}\overline{\psi_t^-(x)}\psi_t^-(x') \qquad (3)$$

obtained from (2) by eliminating the off-diagonal terms. In such a case all the interference effects are cancelled and one can say that the particle is in ψ_t^+ or in ψ_t^- with probability one half, i.e. one has a classical statistical mixture of ψ_t^+ and ψ_t^-, corresponding to our ignorance on the preparation of the state.

In this sense we can say that a quantum particle described by ρ_t^m exhibits a classical behaviour.

Notice that $S(\rho_t^m) = \log 2$, which is the entropy associated to a classical bit with two possible levels of probability one half.

Between the two extreme cases ρ_t^p and ρ_t^m one can have an intermediate situation in which the off-diagonal terms are non-vanishing but reduced with respect to the pure case.

If one considers the more general situation of a particle interacting with an environment it is convenient to introduce the notion of reduced density matrix. Let x, y be the coordinates of the particle and the environment, respectively, and let $\rho_t(x, y, x', y')$ the corresponding density matrix in the position representation.

If the environment is considered practically not observable, we can only be interested in the expectation values of (bounded) observables A^x relative to the particle, i.e. operators acting only on the x variable.

Then, applying the standard rules of Quantum Mechanics, one has

$$\langle A^x \rangle_{\rho_t} = \text{Tr}(A^x \rho_t) = \text{Tr}_x(A^x \hat{\rho}_t), \qquad \hat{\rho}_t(x, x') = \int dy \rho_t(x, y, x', y) \qquad (4)$$

where Tr_x denotes the trace with respect to the coordinates of the particle and $\hat{\rho}_t$ is the reduced density matrix. It is now clear that $\hat{\rho}_t$ is the basic object for the investigation of the dynamics of the particle in presence of the environment.

More precisely, we shall consider an initial state for the particle plus environment in a product form $\rho_0 = \rho_0^p \otimes \rho_0^e$, where ρ_0^p is a superposition (pure) state of the particle of the form (2) and ρ_0^e is a state for the environment.

The reduced density matrix of the system at time zero is $\hat{\rho}_0 = \rho_0^p$ and, clearly, $S(\hat{\rho}_0) = 0$. Due to the interaction between the particle and the environment, at any time $t > 0$ the density matrix ρ_t is no longer a product state and the reduced density matrix $\hat{\rho}_t$ is in general a complicated mixture, with $S(\hat{\rho}_t) > 0$ (i.e. in the transition from $\hat{\rho}_0$ to $\hat{\rho}_t$ there is an obvious loss of information since the degrees of freedom of the environment have been neglected).

We shall say that the environment has produced a decoherence effect on the particle if, after some short time t, $\hat{\rho}_t$ takes a form very close to (3).

Such a kind of result can be proved under suitable condition on the environment.

We shall consider here the extremely simple case of one heavy particle (the system) interacting with a light particle (the environment) via a delta potential in dimension one.

The self-adjoint hamiltonian in $\mathcal{H} = L^2(R^2)$ describing the two particles is

$$H = -\frac{\hbar^2}{2M}\Delta_x - \frac{\hbar^2}{2m}\Delta_y + \alpha_0\delta(x - y), \quad \alpha_0 > 0 \tag{5}$$

and we consider the initial state

$$\rho_0(x, y, x', y') = \rho_0^p(x, x')\rho_0^e(y, y'), \qquad \rho_0^e(y, y') = \overline{\phi_0}(y)\phi_0(y') \tag{6}$$

where $\rho_0^p(x, x')$ is given in (2) and

$$\psi_0^{\pm}(x) = \frac{1}{\sqrt{\sigma}}f\left(\frac{x \pm R_0}{\sigma}\right)e^{\pm i\frac{P_0}{\hbar}x}, \qquad \phi_0(y) = \frac{1}{\sqrt{\delta}}g\left(\frac{y}{\delta}\right),$$

$$\sigma, \delta, R_0, P_0 > 0, \quad f, g \in C_0^{\infty}(-1, 1) \tag{7}$$

According to (6), (7), the heavy particle is initially in a superposition of two wave packets, one localised in $-R_0$ with momentum P_0 and the other localised in R_0 with momentum $-P_0$; the light particle is localised in the region around the origin.

The model hamiltonian (5) has been considered for the sake of simplicity and the solution of the corresponding Schroedinger equation can be explicitly computed (see e.g. [6]).

In fact we are interested in the case in which the mass ratio $\varepsilon = \frac{m}{M}$ is small and in such a regime the evolution becomes particularly simple.

Since the dynamics is linear, we can analyse the evolutions of the wave packets ψ_0^+, ψ_0^- separately.

If we consider the wave packet ψ_0^+ coming from the left, we expect that it propagates almost freely and, after a time of order $\tau = \frac{MR_0}{P_0}$, it reaches the origin.

Then, due to the presence of the δ potential, the wave function of the light particle is partly reflected far away to the right and partly is transmitted, i.e. it remains localised around the origin.

Obviously, the wave packet ψ_0^- coming from the right produces an analogous effect, i.e. part of the wave function of the light particle is reflected far away to the left and the remaining part is transmitted.

This means that, after a time of order τ, only the transmitted parts of the wave function of the light particle have a common support.

The result is that in the reduced density matrix of the heavy particle $\hat{\rho}_t$ the diagonal terms are almost unaffected while the off-diagonal terms are reduced and the reduction is stronger if the transmitted wave is smaller.

The above intuitive picture can be proved in a rigorous way. In fact, assuming $\varepsilon \ll 1$ and, moreover, $\delta \ll R_0$, $\sigma \ll \frac{1}{\alpha} \ll R_0$, where $\alpha \equiv \frac{m\alpha_0}{\hbar^2}$, for $t > \tau$ one has

$$\hat{\rho}_t(x, x') = \frac{1}{2}\overline{U_t^0\psi_0^+}(x)U_t^0\psi_0^+(x') + \frac{1}{2}\overline{U_t^0\psi_0^-}(x)U_t^0\psi_0^-(x')$$

$$+ \frac{\Lambda}{2}\overline{U_t^0\psi_0^+}(x)U_t^0\psi_0^-(x') + \frac{\Lambda}{2}\overline{U_t^0\psi_0^-}(x)U_t^0\psi_0^+(x') + \mathcal{E} \tag{8}$$

$$U_t^0\psi_0^\pm = e^{-i\frac{t}{\hbar}H_0}\psi_0^\pm, \qquad H_0 = -\frac{\hbar^2}{2M}\Delta, \qquad \Lambda = \int dk\left|\tilde{\phi}_0(k)\right|^2\frac{k^2}{\alpha^2 + k^2} \tag{9}$$

and the small error $\mathcal{E}$ can be explicitly estimated, uniformly in $t > \tau$ (see [2] for details).

Notice that the parameter Λ is less than one and it represents the fraction of transmitted wave for a particle initially in ϕ_0 and subject to a point interaction of strength α.

Thus the effect of the light particle is to reduce the off-diagonal terms and this means a (partial) decoherence effect on the heavy particle.

The model considered here is clearly too simple and it can only have the pedagogical meaning to show explicitly a dynamical mechanism producing decoherence.

A more reasonable model of environment would be a gas of N (non-interacting) light particles.

In this more general situation we can expect that the effect of each scattering event is cumulative and then we would get the same expression (8) with Λ replaced by Λ^N which, for N large, means complete decoherence.

A similar argument has been heuristically justified in [5], while a rigorous derivation starting from the Schroedinger equation for the N-particle system is given in [1].

References

1. Adami, R., Figari, R., Finco, D., Teta, A.: On the asymptotic dynamics of a quantum system composed by heavy and light particles. Commun. Math. Phys. **268**(3), 819–852 (2006)
2. Duerr, D., Figari, R., Teta, A.: Decoherence in a two-particle model. J. Math. Phys. **45**(4), 1291–1309 (2004)
3. Duerr, D., Spohn, H.: Decoherence through coupling to the radiation field. In: Blanchard, Ph., Giulini, D., Joos, E., Kiefer, C., Stamatescu, I.-O. (eds.) Decoherence: Theoretical, Experimental and Conceptual Problems. Lect. Notes in Phys., vol. 538, pp. 77–86. Springer, Berlin (2000)
4. Giulini, D., Joos, E., Kiefer, C., Kupsch, J., Stamatescu, I.-O., Zeh, H.D.: Decoherence and the Appearance of a Classical World in Quantum Theory. Springer, Berlin (1996)
5. Joos, E., Zeh, H.D.: The emergence of classical properties through interaction with the environment. Z. Phys. B **59**, 223–243 (1985)
6. Schulman, L.S.: Application of the propagator for the delta function potential. In: Gutzwiller, M.C., Ioumata, A., Klauder, J.K., Streit, L. (eds.) Path Integrals from meV to Mev, pp. 302–311. World Scientific, Singapore (1986)

Chapter 3
On the Different Aspects of Time in the Fundamental Theories of Physics

Thomas Görnitz

Abstract Humans seem to be the only animals who know they have to die some-times. Therefore, the concept of time is fundamental to us and we have to argue from the basis of our transitory existence. This fundamental is only partially represented in the different fundamental physical theories. Newtonian physics, special and general relativity, quantum theory and cosmology have different models for the evolution of a system in time. In no one of these theories alone the fundamental difference between future, now, and past is expressed. The transition from future possibilities to past facts happens at the so-called measuring process. A model is presented to explain this transition without reference to an observer, which is outside of physics, and also without reference to unobservable fictions like many universes. As a conclusion it becomes evident that the formulae in physics we have found by trial and error should not be believed in like a revelation. They have a wide range of applications but they do not reign from eternity to eternity—and not irrespective of the existence of time and space. On the other hand, it becomes evident that all aspects of our possible human time experiences are reflected in one way or another in some of the theoretical concepts of physics.

Keywords Aristotle · Augustine · Big bang · Bohr · Classical physics · Cosmology · Cyclic time · Delayed choice · Deterministic theories · Dirac · Direction of time · Double slit · Drieschner · Einstein · Facts · Feynman · Frame of reference · Galileo · General relativity · Hubble · Illusion of time · Individual quantum process · Lorentz · Maxwell · Measuring process · Michelson · Morley · Newton · Plato · Podolsky · Quantum razor · Quantum theory · Rosen · Schlüter · Schrödinger · Special relativity

In science in general, especially in physics, arguments can be understood more easily if an attempt is made to clarify their motivation. Humans seem to be the only animals who know they have to die. Therefore the concept of time is fundamental

T. Görnitz (✉)
Institut für Didaktik der Physik, Johann Wolfgang Goethe-Universität, Max von Lauer-Str. 1, 60438 Frankfurt/Main, Germany
e-mail: goernitz@em.uni-frankfurt.de

S. Albeverio, P. Blanchard (eds.), *Direction of Time*,
DOI 10.1007/978-3-319-02798-2_3,
© Springer International Publishing Switzerland 2014

to us and we have to argue from the basis of our transitory existence. This conference has conveyed the impression that many of the participants seem to desire that time has no direction, perhaps that time does not really exist. The author confesses that he belongs to a minority of the participants and does not share these ideas. As a former grave digger he is convinced that time does exist and that nothing is so un-influential as the direction of time.

The experience of time may appear from different perspectives. An impression of a never-beginning and never-ending flow may be otherwise contrasted by the perception of our finite life span. Some aspects of time are indeed cyclic—the alarm clock rings every morning—and other aspects are definitively not cyclic—our ageing is not reversible. A common aspect of time for all humankind may be the experience of time not flowing in a uniform way. Additionally in moments of joy or of meditation it seems that time may only be an illusion. Such feelings about time seem to contradict the exact concept of time in physics that we have learned from Newton.

Such different views provoke the question: What is time?

The best answer may have be given by Augustine: "If nobody asks me, then I know it, but if someone asks me, I cannot answer." For physics such an answer is not enough.

The Greek philosophers Plato and Aristotle have spoken about time as a "movable picture of the immortal" and as "a number, measuring the movement with respect to the sooner and the later." Here it becomes clear that time has to do with "measuring" and is connected with "sooner and later".

The measuring of time started with astronomy. Until now our unit for time has rested on a solitary astronomical fact, the day, and is measured in parts of it, namely the second. Even though at present the second is not defined by the apparent rotation of the sphere of stars any more but through an atomic process, the amount of time for a second has always remained the same. Therefore astronomy was the starting point for a scientific concept of time as well as the starting point for science at all.

1 Classical Physics—Time Without Importance

The first great success in mechanics was the description of the solar system using the tools of Newton's mechanics. Newton was the hero in the history of science, who did no longer see the force as a part of metaphysics, which had been the case from Aristotle until Galileo.

The Calculus was the instrument that enabled to define the acceleration as a momentary change of a velocity. This concept of mathematics caused a vision of time as an indefinitely dividable line.

From this epoch onwards all the mechanical problems appear to be solvable. This line of success was followed by the theory of electrodynamics. Although electrodynamics could not be reduced to mechanics, both theories have the same structure concerning time. These two areas, particle mechanics and later on the electromagnetic field, are governed by differential equations in time. This well-known fact is of special interest for the concept of time in these theories. For them time is the most general parameter that is used to describe a system. In both theories we have a fully deterministic regime as far as only the fundamental aspects will be considered and all pragmatic extensions are ignored, such as friction or dampening which could only be explained by quantum theory.

In such deterministic theories, the systems path in time is absolutely fixed such as in a film. Nothing really happens; all the "facts" are fixed on the film roll and only the uninformed observer has an illusion of time.

The great success of both theories has had an impressive force for the definition of time in physics until now. In these theories, time is degraded to a parameter, which loses all its connections to the fundamental difference between earlier and later, between the before of an event and the after. The fundamental equations in these theories allow a free "backward and forward" movement in time.

This is consequent, because in such a theory nothing really new can happen. Only the ignorant observer has the illusion of a flow of time. Einstein spoke of the "illusion of time" in this sense.

The fundamental equations in classical physics are invariant with respect to a reversal of the direction of time. In the same way as a film can be shown backwards, the physical models in classical physics can run forward or backwards in time.

Therefore it is absolutely reasonable that there is indeed a problem with the "direction of time" in these models.

A purely unilateral evolution in time cannot be reached without additional assumptions, surmounting the range of the fundamental concepts.

The normal way in physics is to introduce statistical concepts, like in statistical thermodynamics.

However, classical physics does not know an objective chance.

For its models, chance is a result of incorrect knowledge, thus it is subjective. However, the statistical models work very well. This can be understood if one realises that all these extensions can indeed be based on quantum theory.

Before we engage in this new part of physics, we have to handle an extension of classical physics: At the beginning of the 20th century, the concept of time in classical mechanics would be supplemented by special relativity.

2 Special Relativity—His Own Time for Everybody

The introduction of special relativity resulted from the difference in the structure of time between classical mechanics and electrodynamics. Maxwell's equations do not allow real processes in space and time to go faster than the velocity of light. Einstein's interpretation of Michelson's experiment was that this velocity is the same for any observer in a vacuum. This result has the inevitable consequence that for observers moving in different ways the concept of simultaneity becomes nontrivial.

Two observers, mutually moving in respect to another, should notice a difference of their wristwatch times.

Since the setting up of the large accelerators this is no more a mere theoretical consideration. It is rather one of the most well funded experimental results in physics. It does not only mean that the wristwatches run differently. This effect is only a reflection to the fact that in the different frames of reference every timelike process appears to be different. It is a reciprocal relation. Every observer sees the process in the moving frame more slowly than his own. Therefore some people have spoken of the mere subjective character of time.

3 General Relativity—The Flow of Time Depends on the Situation

General relativity is a theory on the relation between space and time and the amount of matter and energy that is present there. This theory enables to replace the action of a gravitational force by the curvature of space-time. Space and time become dynamic structures.

In other words, there is no more gravitation. All bodies move along straight lines, but the "straightness" of these lines are governed by the masses that curve space-time.

Time flows slower in places where the curvature is large when it is seen from and compared with places of smaller curvature. This is a theoretical concept that is strongly supported by experiments. The experimental facts reach from differences in atomic clocks within the gravitational field of the earth or moving around the earth up to GPS and to astronomical appearances around massive stars.

I my view the fundamental problem of general relativity roots from its mathematical structure. I agree with the assertion that the energy loss of the famous double pulsar is a result of its emitting gravitational waves. But this is a result of the linearisation of General Relativity and if gravitational waves exist in this form then general relativity in its rigid form cannot be true. I will come back to this point in connection with cosmology.

4 Quantum Theory—The Disappearance of Time

Whereas classical physics can be understood as a theory of facts, in essence quantum theory is a theory of possibilities.

But both areas of theories—the classical and the quantum one—are governed by differential equations. In principle the Schrödinger equation is of the same structure as Maxwell's equation. Regarding time, we see that both theories give a deterministic description of a system. This means that for a quantum system a real course of time with its distinction between earlier and later is not granted either.

For classical systems the time structure presents an order of facts although not the direction of this order. The system can be observed every time. Therefore the disappearance of real time does not become so evident in the case of classical physics. Although the film is fixed as a whole, the pictures on it can be seen as real events.

In quantum theory no order of events is given, because there are no events.

In quantum theory only the change in possibilities is fixed. In Feynman's path integral formalism the time structure of quantum systems gets clear expressions.

The system is able to probe all the possible geometrical paths as it were "outside of time". This sounds strange for the layman, but in Bohr's "individual quantum process" a timelike structure does not exist at all. "Delayed choice" is the most spectacular expression for this strange behaviour.

For a scattering process on a double slit it is possible to determine whether the particle was able to go through both holes or only through one of them by experimental arrangements. The decision on this settlement can be made at such a late time—measured on the laboratory clock—that the occurrence should have happened already on the basis of classical pictures.

This behaviour is strongly connected to the so-called Einstein-Podolsky-Rosenstates. Such states describe the possibilities of quantum systems to have correlations that seem not to be confined to the restrictions of special relativity. They express correlations between space as points. If such states can be extended in space then there are other frames of reference in which they appear as extended in time. This opens up possible views of the physics of time on phenomena like "prophecy" that have been banished from science since the darkest middle ages.

5 The Real Structure of Time—The Facts Are Created on the Border Between Classical Physics and Quantum

If there is no place for the real structure of time either in classical physics or in quantum theory, where can we find it? Where is the place for the fundamental difference between before and after?

The occurrence of facts is on the border between quantum theory and classical physics. The so-called measuring process is exactly the place where the facts come into existence.

A measuring process ends an individual quantum process with its unitary time evolution. In this evolution nothing has happened. Everything remains in the state of possibility. Then the projection sets a point! Now no one is able to go back behind it and so a fact is created.

"Measuring process" sounds very anthropocentric. What happens if there is nobody who performs the measurement?

The essence of this process is the loss of information about the phase relation of the quantum system. After it there is again—or formulating this for the purists—there can be a new pure state again. By the information loss it is impossible to reconstruct the former state, the state before the measurement and so a fact is created.

Any measuring process can be traced back to a very simple model. It was proposed by Schlüter [1] and then improved by me [2].[1] For the creation of a fact the main point is that *a quantum particle, for instance a photon, carries the information into the darkness of the universe and will never come back*.

Is there a possibility for an "objective measuring process" in a strong sense? The answer is "it depends".

---

[1]Cf. Fig. 1.

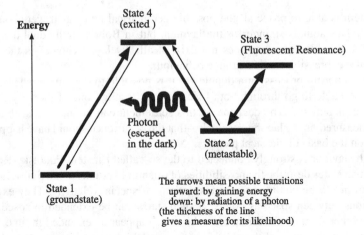

Fig. 1 A model for the measuring process

If one believes in actual infinities, like actual infinite degrees of freedom of the electromagnetic field or an actual infinite time for a scattering process, than the answer is "yes". Infinite time is equivalent to an exact preservation of energy. If one believes—as I do—that infinitely many degrees of freedom are only a metaphor, then the observer has to take its responsibility. This means that he has to decide at which value the decoherence process can be seen as finished. So he has to set, say "10^{-30} is equal to 0" or so.

The observer is responsible that there are no mirrors in space. Then he can decide that an outgoing photon will never come back if it is not back after some time—say some seconds or minutes.

The experimentalists have shown that these are not only academic thoughts. The so-called "quantum razor" has shown that by reflection a mirror can be the cause that an apparent scattering process—which would be a measuring process in the sense of quantum theory—remains a unitary process. But a unitary process can never be a measure.

In my view on quantum theory the dark sky in the night is a central condition for the occurrence of time. This means also that cosmology has to do with the interpretation of quantum theory.

So in the end we cannot avoid looking at cosmology.

6 Cosmology—Time Becomes Universal and Receives a Beginning

The essential aspect of modern cosmology is its self understanding as a part of physics. This means that modern cosmology pretended to be an empirical science. Such an interpretation has consequences.

In physics cosmology is understood normally as a part of general relativity.

It is a correct and important mathematical statement that any exact solution of general relativity is always a complete cosmic space-time.

On the other hand—as a physical statement—it is also correct that almost all the solutions of general relativity are only pure mathematics, because they must be wrong descriptions of nature. All these solutions are different—and in classical physics different solutions on the same object in the same situation cannot be right at the same time.

Therefore it seems useful to look on cosmology not only though the glasses of general relativity—but of course with their support.

The most fundamental empirical data—the red shift of the galaxies and the background radiation—show that space-time seems not have existed for an indefinitely long period. Therefore the idea of a Big Bang is supported by the astronomical data with overwhelming significance.

Due to his philosophical belief Einstein was fixed on the idea of an infinite duration of the cosmos. Einstein's belief was such a strong one that he changed his equations. He did it because in their original form they did not allow a never-beginning and never-ending universe. But Einstein possessed the scientific greatness to acknowledge the empirical facts that were presented by Hubble only few years later.

If one speaks about a finite cosmic time, it is only meaningful if it is possible to define a universal time for the cosmos as whole. In general relativity many of the solutions that do not differ too strongly from the empirical data possess this opportunity.

One interesting question related to time could be that such a universal time is not only universal but also fundamental. In contradiction to this idea the concepts of general relativity seem to state that not one frame of reference is favoured over another one. But this may be a claim that goes somewhat to far.

The essence of an equation is equivalent to the set of all it solutions. Almost all the solutions of general relativity have nothing to do with nature because they are models of the cosmos that are not actualised. Then it might be true that the idea of "the non-existence of a distinguished frame of reference" is also not realised in nature. There is a delightful paper of Dirac who stated that Einstein has maintained much in this connection [3].

The natural microwave radiation is coming in equally from all directions for a suitable observer. If you take another observer who is moving relatively to that first observer, he will see it coming stronger in the direction to which he is proceeding and less strongly from behind him. So it will only be symmetrical with respect to one observer. There is thus one preferred observer for which the microwave radiation is symmetrical. You may say this preferred observer is at rest in some absolute sense, maybe he is at rest with respect to an ether. That is just contradicting the Einstein view. ...

It is possible to observe the velocity of the earth through the ether as defined by the microwave radiation. One finds that the earth and the whole solar system are moving very rapidly, with a speed that can be observed.

The only reason why Michelson and Morley got a null result, why they failed to observe the motion of the earth in an absolute sense, was because their technology was inadequate. ...
With the more modern technology, there is an absolute zero of velocity.

... You might say that, with the microwave radiation showing Einstein was wrong, would destroy relativity. But it has not destroyed the importance of Einstein's work. The importance lies in another respect. ...
The real importance of Einstein's work was that he introduced Lorentz transformations as something fundamental in physics.

Dirac confirmed that the background radiation defines a universal frame of reference in the whole of cosmic space.

Such a universal frame of reference would make it easier to understand and to interpret the nonlocal effects of quantum theory.

But there is need for a remark. The sky was not at all dark in the early time of the universe. In the juvenile universe for any outgoing photon there was a same one that would come back. Photons of the same energy and polarisation are all indistinguishable, so it is as if there were ideal mirrors everywhere.

The occurrence of facts can be traced to the above model, therefore in the early time of the universe there were no facts. This means there was no time, because the essence of time is the distinction between the before and after of an event, before and after the occurrence of a fact. This picture does fit into the ideas that Kiefer presented. Time appears like a classical concept that comes into being "after the Big bang".

The consequence of this is that the Big Bang cannot belong to a scientific cosmology having an empirical pretension. As Drieschner has pointed out empirical science is a concept without value as long as no time can be defined.

Here there is a need for a second remark.

If the occurrence of facts depends—roughly speaking—on the disappearance of information, then the cosmic space has to be spatially infinite or has to expand at least as fast as the velocity of light. The first conjecture would be the end of any empirical pretension on cosmology. In such a horrible case our experience would cover exactly zero per cent of the object under consideration—the cosmos. I am not an experimentalist, but I think such a value is not a safe basis in any part of science. The other possibility, the expansion of a closed cosmic space with velocity of light, as I proposed ten years ago, is supported by the new data.

Nevertheless, one critical problem is left.

If the creation of facts depends on the everlasting disappearance of information, then we have to remind ourselves that nothing can escape from the cosmos. If one applies quantum theory on the whole cosmos then time may disappear in the end.

7 Possible Conclusions

Now we come to a central point for the philosophy of science.

Physics should not and cannot give up it pretension to be an empirical science. Empirical ideas presuppose time, so time was always there before physics. It is like a Physicist who is in any case a child or human first and then becomes a physicist later. On the other hand, cosmology as a theory on the evolving of space and time is an unavoidable part of physics, it must be done or physics will become inconclusive. So what is a possible conclusion?

I think that physics as any other human effort cannot catch the whole truth. It can be only an approximation to truth, and it is already a very good one. It is very successful but with it we are not yet sitting on God's own chair. The formulae we have found by trial and error should not be believed in like a revelation. They have a wide range of applications but they do not reign from eternity to eternity—and not irrespective of the existence of time and space.

Coming back to my starting point, I believe it becomes evident that all aspects of possible time experiences are reflected in one way or another in one of the theoretical concepts of physics.

References

1. Schlüter, A.: Der wachsende Kosmos und die Realität der Quanten. Nova Acta Leopoldina NF **69**(285), 127–135 (1993)
2. Görnitz, T.: Quanten sind anders. Spektrum, Heidelberg (1999)
3. Dirac, P.A.M.: Why we believe in the Einstein theory. In: Gruber, B., Millman, R.S. (eds.) Symmetries in Science. Plenum, New York (1980)

Chapter 4
Quantum Events and Irreversibility

Rudolf Haag

Abstract It is pointed out that the conceptual structure of Quantum Physics already implies irreversibility arising from the bipartition of temporal evolution into the "Quantum State" on the one hand and real events (e.g. observation results) on the other, connected by probability assignments with intrinsic indeterminacy.

Keywords Quantum events · Time direction · Irreversibility

The topic of this conference "Direction of Time, Irreversibility" is a subject which has accompanied me for decades. I considered it again and again, sometimes changing my opinion or at least the emphasis. I had lengthy discussions with quite a number of colleagues and found this often very tedious because in these questions one is likely to meet strong convictions and even firmly entrenched prejudices. So I hold no great expectations that I can present on five pages anything that could change the previously held opinion of anyone. But I shall try to isolate essential issues, to state and partly justify the possible answers I want to suggest.

The major opinion among physicists is that

(1) within physics the appearance of irreversibility is appropriately described by the second law of thermodynamics;
(2) this law is derivable from statistical mechanics;
(3) in this derivation it is immaterial whether we use classical theory or quantum theory.

The main point I want to make here concerns item (3). It is the assertion that Quantum Theory, as we know and use it, contains a basic element of irreversibility whose relation to the second law of thermodynamics is not usually considered. But let me first, very briefly, elaborate on item (2) above.

Statistical Mechanics gives a coarse grained description which endows a macroscopic state with a thermodynamical probability, the number of microscopic states

Contribution to the Conference "The Direction of Time", Bielefeld, January 2002.

R. Haag (✉)
Waldschmidtstr. 4b, 83727 Schliersee-Neuhaus, Germany
e-mail: haag.rudolf@t-online.de

S. Albeverio, P. Blanchard (eds.), *Direction of Time*,
DOI 10.1007/978-3-319-02798-2_4,
© Springer International Publishing Switzerland 2014

which have the same coarse grained appearance. This coarse graining by itself will not produce an asymmetry of the time direction. Starting from a microscopic theory which is invariant under time reflection it is logically impossible to arrive at irreversibility on the macroscopic level without further assumptions. These are well known but often glossed over.

For example, in Boltzmann's collision equation for a dilute gas, starting from classical mechanics of a large number of molecules, it is assumed that before each collision there is no correlation between the molecules engaged in it. After a collision the pair of molecules concerned are correlated. But this correlation is wiped out by subsequent collisions with different molecules and its cumulative effect will not become relevant in time intervals of interest.

More generally we may start from a macroscopic state of relatively low thermodynamic probability (entropy) and ask for its subsequent development. We can expect that it will move towards a state of higher probability. This is fine. But we should not ask then how the macrostate looked in the past. Otherwise we might conclude that also then it was a state with higher probability. In particular, if we consider a process of approach to equilibrium and start arguing from a time somewhere in the middle of this process, having only this information and using it to ask what the situation was at an earlier time then we should conclude that the entropy has been higher then. If instead we use Boltzmann's equation to calculate backward we find that the entropy decreases in the past. Miraculously this agrees with the actual history up to some specific early time t_0 but from then on it deviates from it and ultimately, in the remote past, it leads to a singularity. This special time t_0 can in laboratory experiments be attributed to the willful action of the experimenter, starting his investigation. On the cosmic scale the experimenter might be replaced by God creating the big bang.

Another type of situation is an open system such as the earth. We may distinguish the outgoing radiation (characterized by Sommerfeld's boundary condition) from the incoming radiation, originating from outside sources. The outgoing radiation escapes and the asymmetry between incoming and outgoing radiation leads to some rough flux equilibrium with entropy production and irreversibility on the small scale.

Moving now to Quantum Physics we should first clarify a few general questions concerning the scope and method of physics. This is because the advent of Quantum Mechanics has injected some doubts about what we mean by "reality" and by the conventional picture of an "outside world" as distinguished from the impressions in our consciousness. That this insecurity is serious is exemplified by a title like "Reality or Illusion?" chosen for the description of some recent experiments. I have to be very brief about this question but what I find helpful is a remark by Wolfgang Paul (not Pauli!) who distinguished the "real part of physics" from the "imaginary axis". The former consists of phenomena, but not arbitrary phenomena. To be acceptable for physics there has to be agreement between many people about its occurrence and description. So it is not dependent on the consciousness of any individual person. And it must be reproducible which means that we must be able to classify the phenomena considered and their circumstances of their appearance into equivalence classes. I shall call a phenomenon subject to these conditions an "event" in order to indicate that it may be abstracted from human cognition. All empirical material of Quantum Physics consists of such events. The prototype is the response of some detector. The imaginary axis of physics is the theory. Its concepts are mental constructs whose "reality value" is debatable.

In the formal structure of the theory there appear two basic concepts: Quantum States and Observation Results. The latter correspond to events (if you wish "individual elements of reality"). A quantum state, on the other hand, describes probabilities for the occurrence of various possible events. Whatever mental picture we associate with the term "probability", in the physical applications we have in mind it always amounts to the counting of a relatively frequency within an ensemble combining many individual cases. Thus it depends on the definition of equivalence classes of circumstances and of type of events. Events are individual facts, the quantum state relates to possibilities.

Now we must recognize that an "observation result" is a very special kind of event in two respects. In the standard formulation of the theory the picture is that an experimenter "puts a question to nature" with the help of a "measuring instrument" which determines a disjoint set of possible answers. The concept of event should, however, not depend on whether some artfully constructed instrument is installed in some place but refers to anything which can—with a high confidence level—be regarded as the appearance of a fact. Our knowledge about it may be inferred indirectly from traces in a rock or the theory of stellar evolution. The point is that in such cases the menu of alternative possibilities is not decided by an experimenter but must be inherent already in the quantum state and that we cannot use the Bohr–Heisenberg cut between an experimental arrangement and a "physical system" we want to study. Related to this is the point that an observation result such as the click of a detector is a very coarse event; otherwise it could not be a common experience of many spectators. Under what circumstances can we speak of finer events, considered as individual facts? How does the "total quantum state" determine the menu of alternative possible events if "the observer" does not focus on a particular question? Since this concerns indirect information the answer needs some extrapolation along the imaginary axis.

As a prototype of an individual finer event let us consider the interaction of a single electron with a single atom. If we can isolate this process (whatever that means precisely) we expect that after the process we shall encounter one of several alternative situations: an ion plus two electrons or an atom plus one electron and (perhaps) some photons with characteristic energies. Standard collision theory tells us that in an experiment where we determine the final reaction products by an array of detectors we shall find one of these alternatives realized in each individual case. The formal structure of the theory suggests, however, that if, instead of installing detectors, we consider other kinds of later measurements we might find interference effects between the mentioned alternatives, telling us that no decision between these alternatives has occurred. This argument is misleading. It results from an over-idealization of the term "observable". In a self-consistent treatment we should not speak of an abstract observable represented by some operator in a Hilbert space but have to specify the observation procedure. This includes the positioning of hardware in space-time which may possible interact at some later time with the reaction products and this specification is part of the relevant total state. For an interference effect the obtainable contrast is essential. Given a state and a putative division of the subsequent development into alternative individual histories there is a quantitative measure for the contrast in an interference between the assumed alternatives. It is the deviation from Griffiths' "consistency condition for the histories". The qualitative demand that the process be adequately isolated means precisely that this deviation is negligible under the prevailing circumstances. There are two conclusions to be drawn. First, that there is a holistic feature in Quantum Physics. In other words, the division of "reality" into separate elements called events depends on circumstances which may involve a possibly rather wide environment. Secondly, an appropriate division is determined by the state. It leads to the enumeration of alternative histories, each composed of a sequel of events. In the example considered (given adequate isolation) the ionization or excitation with photon emission can be regarded as alternatively possible real events which cannot be revoked by a "subsequent measurement". They have become facts and a fact is irreversible. Past facts are subsumed in the "state". According to existing experience and theory they only determine possibilities for the subsequent development. The universe does not appear to be a clock work. This brings an asymmetry between past and future. "Reality" consists of past facts. The future is open. Thus we come to an evolutionary picture of reality, similar to that described by A.N. Whitehead many years ago.

I have to close now though there are many issues not addressed. Let me just briefly mention some. There are the attempts to eliminate probabilities and return to a deterministic theory with hidden variables. Personally I regard this as a wrong track. Existing proposals like Bohm's particle trajectories are no adequate tool for the description of particle transmutations (or even photon emission). Concerning the "many worlds picture" of Everett we must bear in mind that the branching of alternative histories reaches the macroscopic level (for instance by the click of one specific detector among many) and thus only one particular history becomes the common experience we call reality. Each emerging fact annihilates the other possibilities and changes the state.

Finally there is the question of localization of events. If we do not assume space-time to be an a priori given 4-dimensional continuum then events (rather than objects) play the essential role in the development of a theory of space-time. Some information concerning the sharpness of localization in space-time of simple events can be obtained from existing theory.

Chapter 5
Physics and Our Intuitive Outlook on Time

Christoph von der Malsburg

Abstract This discussion is an attempt to reconcile our ideas of physical time with those of psychological time. Based on accepted arguments from relativity and on a much less accepted interpretation of quantum phenomena I am adopting a picture of physical time which accords equal and full reality status to all moments in time. This seems to be in sharp conflict with our intuitive outlook, according to which the future has no reality yet and is open to the decisions of our free will. I will show that this conflict is due to a flawed concept of free will and its relationship to determinism.

1 Our Intuitive Outlook on Time

The past, for us, is the accumulation of erstwhile present moments. It exists only in our memory in the form of immutable historical facts. The future does not exist yet. It is open to the decisions of our free will, we have a choice in sculpting it according to our intentions, making, for instance, present sacrifices for the sake of future benefits. The present moment divides time into past and future. It is a thin slice of reality between what is no longer and what is not yet. If $\Psi(t)$, $t \in (0, t_e)$, describes the history of the world, seen from a vantage point outside time, with t the time parameter, t_e the end of times and $\Psi(t)$ a full description of the world at t, the present moment at time T would correspond to something like $\Psi(t)\delta(t - T)$, where Dirac's δ-function lends reality only to an infinitely thin slice of time centered on T.

C. von der Malsburg
Institut für Neuroinformatik and Department of Physics and Astronomy, Ruhr-Universität Bochum, Bochum, Germany

C. von der Malsburg
Dept. of Computer Science, and Program in Neurosciences, University of Southern California, Los Angeles, USA

Present address:
C. von der Malsburg (✉)
Frankfurt Institute for Advanced Studies, Frankfurt am Main, Germany
e-mail: malsburg@fias-frankfurt.de

S. Albeverio, P. Blanchard (eds.), *Direction of Time*,
DOI 10.1007/978-3-319-02798-2_5,
© Springer International Publishing Switzerland 2014

Only this present moment is real, the past being no more, the future not yet. It is as if the machine at the foundation of the world was just sufficient to represent one moment in time and had to go through the succession $\Psi(t)$ of states, one moment T at a time, continually changing its constitution to represent those moments. The present state of the universe has been generated by a chain of states. Each of these states interacts only with itself, and neither the past, being but a memory, nor the future, not existing yet, take part in the dynamic game. Reality—present reality—is a dynamic, ever changing entity which takes us along in its flow.

There is no doubt that this outlook on the nature of time is dominating our thinking and speaking about the reality of this world. And yet this perspective on things engenders enormous difficulties on all ends. I will summarize some arguments that have been raised by physicists over the course of almost a century, adding up to a radically different view, according to which the full entity $\Psi(t)$, $t \in (0, t_e)$, has "simultaneous" reality, if the word "simultaneous" is not taken to refer to "same t" but meaning that physical processes at any moment $\Psi(t)$ of the universe can be influenced directly by physical processes at any other moment, present, past or future: simultaneity *sub specie aeternitatis*, to borrow a phrase from another realm. This view of a universe with eternal reality makes it look like a fixed recording and seems to clash sharply with our outlook on time and life. Although there are important physical arguments in favor of the eternal universe, physicists are essentially ignoring it. Even the original advocates have found their proposal to be starkly counter-intuitive. This paper examines this perceived clash in the light of our concepts of mind processes.

1.1 Physics

A direct reflection of intuitive time in physical theory is field physics. In this version, the description $\Psi(t)$ of the state of the universe at time t includes fields that summarize all retarded signals from past events. The full entity $\Psi(t)$, with time parameter t running from 0 to the end of times, is only a figment of historical nature. To speak of reality one has to pick a moment T and form a time slice $\Psi(t)\delta(t - T)$. This time slice contains all the information necessary to produce the next moment, $T + \varepsilon$, ε an arbitrarily small and positive time increment. In this way, the full history $\Psi(t)$, $t \in (0, t_e)$, of the universe is fabricated incrementally as a wave of real moments.

This is a version of physics that corresponds closely to our intuitive notion of time. But as everybody knows, there are problems with it. One is relativity, according to which the definition of the historical moment T is different for different observers. As we believe that the reality of all observers is to be given equal status, this forces on us the conclusion that more than just the one global moment defined by my own here-now must have "simultaneous" reality.

This problem of the definition of a time slice that can be called this moment is heightened to extreme acuteness in the realm of quantum phenomena, which by their very nature need non-local communication. An excited atom emits a photon that is eventually absorbed by another atom. Interference effects show that the emitted wave is extended in space and reaches many potential absorbers, which may be widely dispersed in the universe. But the photon is absorbed in only one place. This necessitates communication between the potential absorbers. It is as with airline seats, which can be assigned to only one passenger, necessitating communication between all customers potentially interested in the seat and a central computer which eventually assigns it.

As I am aware that this view of the matter is not shared by the majority of physicists, in fact is shared by hardly anyone, a paragraph or two of justification may be in order before progressing with the main argument. The standard battleground of the argument is the EPR experiment [4]. The formalism correctly describing the experiments makes use of a projection operator, which is applied to a description immediately before the absorption event, $|\psi\rangle = \sum_i a_i |\psi_i\rangle$, and which projects out from it one of the possible outcomes $|\psi_i\rangle$, realized with probability $|a_i|^2$. (In my simple one-photon exchange example, $|\psi\rangle$ would correspond to the electromagnetic wave before the absorption event, whereas the $|\psi_i\rangle$ would describe photon states localized at potential absorbers, coefficients a_i corresponding to their amplitudes.)

The problem with this picture is that a projection operator is not of this world. It does not correspond to any physical force or entity we know of, and, worse, it is applied non-locally at a moment in time, which, as remarked above, is a problematic notion in a relativistic universe. In addition, it is not clear when the projection operator is to be applied. A suggestion would be to apply it at the first potential absorption event (the first absorber atom encountered by the wave in our canonical example) and see it realized as a collapse of the wave function, but in view of relativism there is no unique first absorber. This difficulty precludes an interpretation in which quantum theory's projection operator is seen physically realized by a system of instant (that is, non-retarded) communication. Attempts to interpret the projection operator as acting only on our state of knowledge are untenable, as is the assumption that the actual decision in the EPR experiment has taken place already before the quanta parted as this would lead to signal correlations [1] which contradict experiment (for a review see, e.g., [14]).

John Cramer [2], see also [11], has proposed a system of communication between potential absorbers that is overcoming these difficulties. According to it, the transmission of a photon from an emitter to an absorber is organized by a handshake involving in addition to the usual retarded wave an advanced wave that is following the space-time trajectory of the retarded wave backwards in time. All potential absorbers send back these "confirmation waves" to the one emitter, participating there in a tug-of-war. As result, only one absorber receives all the energy of the emitter, all other potential absorbers having to give up any share of it they already had received, the whole process happening in the typical 10^{-15} sec of a photon emission event. There is no conflict with relativity, as all communication takes place along the light cone (or, in the case of transmission of massive particles, along time-like trajectories) and no concept of simultaneity is ever invoked. As Cramer's extensive discussion makes clear, all the philosophical fog surrounding quantum experiments, real and in Gedanken, clears away if this picture is adopted.

Cramer or Lewis are not the first to invoke advanced signals. These are perfectly consistent solutions to the underlying wave equations (if they are of second order in time, as, for instance, Maxwell's equations—for first-order equations like Dirac's a complementary equation has to be invoked to lead to advanced solutions), and there is a long history of discussions taking them seriously [3, 5, 11, 16, 19, 20]. Dirac [3], for instance, invoked them to derive the formula of radiative damping, for which there is no other explanation.

It is true that Cramer's picture (which he himself designates as a mere interpretation, but which is to be taken seriously as a physical theory to make sense at all) raises a number of issues, some of physical, some of psychological nature (the latter being the point of this communication), none of which, however, seems impossible to solve. The first is the necessity to create a full dynamical formulation, complete with the non-linearities involved in the decision process and the influence of all the virtual interactions and failed absorption events that shape the electromagnetic wave (turning, in the example of the double-slit experiment, the original spherical wave into the two focused beams emanating from the slits). As a very encouraging start, Carver Mead in a deceptively modest booklet [13] gives a deterministic dynamic description of the exchange of a photon between emitter and absorber atoms.

Another problem is the necessary explanation of the fact that, in spite of the seemingly time-symmetrical nature of the transaction, quanta always end up in the future, never in the past. Although energy is shipped forwards and backwards in time during the organization of the transaction, upon completion of it no trace of these signals is left, the only effect being the transmission of a photon to a later point in time. There have been several attempts to explain this asymmetry (e.g., [7, 8, 19]), all trying to link this electromagnetic time arrow to the cosmological one, but none of them convincing yet.

A related issue, very central to the argument here, is the question whether information can be sent over space-like distances or backwards in time. There is, of course, no evidence for this, and there is a proof [6] that, given the time-tested quantum mechanical formalism, no information can be exchanged between the local measurements involved in EPR experiments in spite of entanglement and the effectively space-like communication implicit in the Cramer picture. Consequently, there is no challenge here to the physical picture that all the information we get through direct signals comes from the past, or more precisely, the interior of the past light cone. (My wording is cautious here, because on the basis of reasonable assumptions about the stability of the world we can deduce much of what is outside this light cone, a thought not followed up here.) It seems not totally excluded, however, that some experiments can be devised to directly prove the existence of advanced signals, and it may even be possible to obtain some information about the future, as the recoil of photon emission indicates the direction in which to find the absorber, and an uneven distribution of absorbers in the distant future (analogous to the uneven distribution of emitters of the cosmological background radiation) might be revealed by an analogous telescope into the future.

1.2 The Eternal Universe

The upshot of all of these thoughts is a breathtakingly different view of the reality of this universe. Tetrode [16] and Fokker [5] formulated the exchange of electromagnetic interactions between charged particles in terms of a variational principle according to which the actual world corresponds to the stationary points of an action integral (Eq. 5 in Tetrode [16]; p. 389 of Fokker [5]; or Eq. 1 in Wheeler and Feynman [20]) that contains interaction terms for all pairs of charges and all pairs of space-time coordinates lying on the same light cone, that is, being connected by signals traveling with the speed of light. A similar picture is to be painted for gravity [21]. Fields containing energy-momentum and summing up all information of the past as far as needed to make the future, are non-existent in this picture. One speaks of "action at a distance" (distance in space and time).

According to this picture, the present moment loses all of its special reality status, and the total history $\Psi(t)$, $t \in (0, t_e)$, is simultaneously real, simultaneous from a vantage point external to the perspective of our time. From this vantage point, the universe is a totally static, eternal entity, rigid as a crystal, metaphorically speaking.

The universe as an eternal entity forces on us another, curious conclusion. The earliest moments of our universe, to the extent that radiation sent out then eventually is absorbed at the end of time, are already in direct communication with those last moments of the universe. This makes it impossible to see the creation of the universe in terms of a wave progressing from a moment of original creation to the present and on to later moments. At the birth of the universe, its end was already present. This raises the question how our universe "was" created. In a second time, different from ours, along which a baby universe progresses from imperfect consistency to full consistency in terms of all interactions, realizing the variational principle of Tetrode and Fokker in terms of actual variation?

Once fully formulated, the action-at-a-distance picture of physics, complete with unidirectional energy propagation, will have to be equivalent to familiar field physics in essential aspects. In particular, it will have the same asymmetry in time. Physics is very successful in describing events with the help of field and wave equations that express propagation and causation only forward in time. The field equations make, however, only probabilistic predictions. This residue of indeterminacy is eliminated by Cramer's advanced confirmation signals. In this sense, our familiar picture of forward causation seems to be totally untouched by the eternal universe perspective (if indeed the view holds up that no information whatsoever can be transmitted backwards in time). This coexistence of forward causation with a globally entangled description of the universe produces the eerie feeling of living on a theatrical stage. Just as all the scenes of a cinematic film have simultaneous and equal reality while the film is in the box, when viewed it gives the impression of a logical, causal flow. But as the film has been produced off-line in random sequence, this causal flow is a deliberate illusion, created by cunning direction and editing.

Are there any loopholes that could save us from the eternal universe perspective? In Wheeler and Feynman's [19, 20] classical version of absorber theory (this term referring to the idea that emission of a photon necessitates the presence of an absorber as much as absorption of a photon necessitates the presence of an emitter, a thought already clearly expressed by Lewis [11]), only rather general properties of the future absorber needed to be postulated, leaving ample space for different world histories if only they keep these general properties intact. If, however, individual quanta are involved, and if nature is accurately making sure that no single quantum was either lost or absorbed twice under violation of energy conservation, there seems little chance of avoiding the picture that the future is specified "now" down to the last quantum jump of every atom. (In the Hanbury–Brown–Twiss effect, see [10], several emitters of compatible frequency collaborate to transfer photons to several absorbers, so that a picture seems possible in which intermediate bundles of photons shield the emitters from the absorbers, but it is not clear whether detailed communication between all participants can indeed be obviated in such multiple-photon transfer events.) As no such loopholes are in sight, let us proceed with the view that the universe is "simultaneously" real in all its detail; that it has eternal reality.

2 The Eternal Universe and Our Intuitive Notion of Time

Although it is recommended to us by strong arguments, the perspective of the eternal universe is not in the textbooks. The reason for this very likely is the apparent clash with intuition, which has been remarked by almost all authors advocating and discussing this view. Just as the intuitive reality of a solid and immobile Earth at our feet undoubtedly was a factor in delaying the Copernican revolution for more than a century, strong intuitive counterarguments will first have to be put out of the way before the reality of the eternal universe has a chance of being accepted.

2.1 The Present Moment

Einstein regretted that our intuitive notion of a moment in time is not reflected in the theories of physics (see the citation in the chapter by H. Lyre). In the eternal universe the moment is bereaved of special significance altogether. Let us examine whether this creates a serious conflict with our perception of time.

We of course have no direct access to the reality of the world but see it through signals and, little appreciated outside of neuroscience, through reconstructions. It needs complex processes for the brain to make sense of the signals that reach it. What we take to be a directly perceived reality in front of us is a construction whose substance is mostly conjured up from memory. Although we subscribe to it in our everyday life, it is a naive illusion that the reality of our immediate environment should swim directly into our mind through our senses. Already from physics' point of view there is the problem that signals arrive with delays and were it not for benign continuity on a time scale attuned to the pace of our own reactions it would be all too evident that our perception can at best be a perception of the past, not the present.

But the problem goes much deeper. Our brain process is a succession of activity states. These all have rather little content but leave behind traces that help to direct and shape further activity states. The process is structured such as to converge on globally ordered brain states in which a rich array of subsystems each reflect the same reality in their own language in mutual consistency [17, 18]. We perceive this consistency as consciousness: awareness of the reality at the focus of our attention. This coordination of sensory signals with memory items, representations and interpretations comprises predictions of possible future events, complete with potential actions to take.

The attainment of this coherent state, the reconstruction of reality, is a time-consuming process which would not make any sense were it not for structural continuity of our environment on a time scale slower than the brain process. (Indeed, the mind's reconstruction of reality has its very natural extension in the process of science, which may take centuries, and again would not make sense in the absence of structural stability in the world.) We can deal with rapid processes (our auditory system, for instance, does so routinely), but the analysis and proper representation of rapid temporal relations has, of course, to come after the event, our representation of rapid temporal sequences has to be symbolic (like the persistent oscilloscope trace of a nanosecond signal) and the laws and mechanisms involved in the translation from the actual rapid process to the off-line representation must be stable. When our arm is tapped simultaneously at two different points along its length we perceive the two taps as simultaneous although the signals arrive at our brain at different times. This would be impossible if temporal processes in our environment had to be represented in the brain literally, as exact temporal replicas. If we want to be precise about a brief moment—a set of simultaneous events—it has to be a past moment, the present time being employed to reconstruct and contemplate it.

Let it be remarked that also when dealing with very slow sequences of events we have to represent them symbolically to project them onto the time scale of our thought processes. And let it also be remarked that we can perceive and generate historical records of our own thought process, but this perception fails on a time scale faster than a fifth or a tenths of a second.

From these considerations it can be concluded that we do not perceive time in any direct sense, our conscious representations "flowing with the time," but that we deal with time in a symbolic, indirect way, hovering back and forth around the clock's time as we represent it. Not even the timing of our mind's process is perceived and represented in any direct sense. In consequence, the physical time parameter t and the infinitely short δ-moment of the introduction are mere constructions and are not accessible directly. In the psychological literature there is something called the "psychological moment," something evolving on a time scale of a fifth or a tenth of a second, something we perceive as indivisible and elementary and which comprises the coherent perception of a chunk of reality. But there is no reason to ask for a reflection of this in relativistic physics, no reason to ask for a concept of simultaneity, at least not beyond the temporal resolution of our senses and the signal horizon of our immediate perception. The intuitive notion of a moment in time mentioned in the introduction is an idealization that does not stand up to scrutiny. A more realistic notion of time perception creates no clash whatsoever with the idea of an eternal universe.

To the contrary, in its microcosm our brain treats time in a way that has many similarities with the Tetrode–Fokker picture of the universe. Our reconstruction of a sequence of events (or indeed our perception of one) is achieved in an iterative optimization process, distantly analogous to the variation of Tetrode–Fokker's action, involving signals going forward and backward in (imagined) time, all along trying to do justice to forward causality, achieving it dynamically without being bound to it kinematically.

2.2 Asymmetry of Time and Free Will

All the signals that reach our senses come from the past light cone or its interior. To the extent that we can reconstruct certain knowledge about the external world it relates a brain state to the past. As the signals reaching us never convey any information about the future, all we know about it are predictions and imaginations. Recognizing our predictions as unreliable we see the future as uncertain. To the extent, however, that we believe to have certain knowledge and predictive power in a given situation, as is sometimes the case, we consider the future as inevitable and certain. The perceived reality status of the future is therefore merely a function of what we know, and there is no basis for the conclusion that the future is uncertain in any deeper sense. Unexpected movements, e.g., of other animate players, may necessitate quick updates of our predictions so that our imagined future is often subject to sudden changes, but this change takes place only in our head, just as an unexpected move in a film that we see changes our expectations for the rest, although, of course, that rest of the film had been set in concrete before we went to the cinema.

We would be inclined to accept this conclusion lightly, the same way we accept the reality of a far country or planet in spite of uncertain knowledge about it, if it was not for another issue of great impact on our outlook on life, an issue which I presume to be the reason the eternal universe first strikes us as a horrible vision worse than a prison life sentence: We live with the idea that by our own acts of free will we can change the future in a real sense[1] and an eternal universe would take away that freedom from us. Free will is an idea of fundamental importance to us, it establishes the sovereignty of our self and consciousness. If we could not change the future, we feel, we might as well subjugate to fatalism—do nothing and wait for what is coming anyway.

---

[1]This idea of changing the future by a local act of will, by a free decision, would re-introduce the concept of a distinguished moment: the branching point were the decision happened. But this version would not create a conflict with relativity and Einstein causality as my decisions are localized not only in time but also in space and all consequences are confined to the interior of the future light cone.

There is an old and very deeply ingrained sentiment that free will and determinism are not compatible with each other. For this matter, the eternal universe is just an acute version of determinism: complete and unconditional determinism that cannot even occasionally be punctuated, e.g., by superior intervention. Spinoza [15] had this view of a completely deterministic world (and of God, the two being identical with each other), and he was and is deeply hated for it.

Before we can deal with this apparent contradiction we have to briefly discuss the concept of determinism and perceptions thereof. Mechanistic determinism speaks of a system whose inner workings never leave the minutest choice in its progression through time. This was the determinism that Spinoza spoke about, and this was what Einstein meant when saying that God would not play dice. Now, in the conventional view of physics its system, described by deterministic field equations, is randomly changed by quantum chance, although under many circumstances these quantum decisions have only imperceptible effects. In Cramer's transaction version of quantum mechanics even this quantum uncertainty is eliminated with the help of advanced signals, re-establishing complete determinism (although the above wording has to be changed, as the universe is not determined in its forward progression in time but as a totally rigid array of retarded and advances signals criss-crossing the universe from one end to the other, both in space and in time). It is, metaphorically speaking, as if the universe had already run its whole course and we were dealing with a recording.

There is little disagreement about what (mechanistic) determinism is, but what about its meaning for our life? Laplacian predictability has it that if the inner workings of a system as well as the initial state were known, all future states could be predicted with certainty and precision. Laplacian predictability is, however, a mere figment, for a number of reasons of a principled nature. There is, on the other hand, practical predictability, the concrete possibility to know enough about initial states and inner workings and have enough reasoning power to be able to make useful predictions with some certainty and accuracy. This is what our brain does all day and is the basis for our survival. Of course, (practical) predictability is only possible on the basis of at least some degree of determinism.

Modern digital computers have become a powerful metaphor in reasoning about determinism (and free will). They are specifically built to be deterministic by their inner workings, that is, to be totally insensitive to uncontrolled random influences, so that in their isolated domain Laplacian predictability can be realized. They can be put repeatedly into a defined initial state to always run through the same sequence of states, and a second machine can be set up to predict this sequence in full detail. Computers display also two other properties that are often associated with determinism. Clockwork regularity (displayed, for instance, by running the same program with the same initial state repeatedly) and hetero-determination. The latter is the phenomenon, possible only within very specific organized arrangements in this world, that one system tightly controls another system—in the computer, the programmer hetero-determining the process in the machine with the help of a program that the machine follows step by step. Another example of hetero-determination is the imposition of political will.

What about my free will in the eternal universe? Should I fatalistically forsake all effort and let the inevitable future happen? Not at all. My present actions participate in the course of events: the universe is forward causal and my actions do influence the future. Without my presence and without the specific decisions that I am taking, my efforts, my caring, my sacrifices, the future would be different. The future is in complete agreement with forward causality, and so sequences of events are not observed in which I do not react to visual signals and yet my body avoids obstacles. So, can I change the future? Sure I can, but what I change is not the real future but the potential future of my imaginations and predictions. Thus, there is no contradiction between free will and determinism in the external world. To the contrary, the acting individual needs at least some degree of determinism in order to be able to predict and act accordingly.

We all recognize that our freedom to act is constrained by physical law: we cannot will to lift ourselves into the air, for instance, and we do not see this as a contradiction. Also the loose determinism of social law is obeyed by us most of the time, although we cherish the idea of being able to disobey in principle. In the political realm it is important for us to be free from hetero-determinism as much as possible. This latter feeling may be one of the psychological sources of the perceived contradiction between free will and determinism, although there is no deep conceptual link to mechanistic determinism here. An actual limitation of the freedom of my will is given in cases where my actions have no influence on some course of events. In such cases we may as well be fatalistic, but it is rather the absence of deterministic links that is to be blamed here.

As long as we take an external view—treating the acting individual as a unity without analyzing its inner workings—there is no contradiction whatsoever between free will and determinism. As soon, however, as we start to enquire about the inner working of our mind, an irresolvable problem seems to arise, a problem that has been commented upon endlessly. I argue that this problem goes back to a logical self-contradiction that arises if we apply the concept of an act of free will to the thought act itself.

Here are the incompatible statements. On the one hand, the act of free will is to be illuminated by insight into possible courses of future events and an evaluation of them in the light of my preferences and values. On the other hand, as expressed, for instance, by Kant in his Critique of Practical Reason, free will should not be caused itself but should be an original cause and mover.

If the act of free will preceded the willed thought altogether, that thought could not be pre-meditated in the light of alternatives and goals because this pre-meditation would need thought itself. Our feeling of committing a defined act of will cannot precede the thought that formulates its substance, can only arise rather late along with that formulation (this is a logical conclusion, but see also the experiments by Libet et al. [12], that show that the subjectively perceived moment of free decision comes significantly after brain signals on the basis of which the eventual action can be predicted reliably). A tenable account of the situation is that the judgment that a particular thought corresponds to an act of free will arises along with that thought as integral part of the same creative process. This judgment is a deduction, not a cause, and is based on such signs as the absence of external stimuli and the existence of habitual patterns or related preceding thoughts. If someone wants to commit, for the sake of argument, a totally deliberate thought act, this act of will may be first and the substance of the resulting thought second, but this substance will not be the consequence of the original deliberate act but will have its origin in processes going on accidentally at the time in the brain.

The fixation on an ultimate-cause aspect of free will is a cultural tradition without any fundamental necessity. Julian Jaynes [9] argues that in early historic times humans did not see the origin of their decisions in their own minds but rather in voices, experienced literally or in a metaphorical sense. In our cultural circle and time we feel accountable for our decisions and are ready to explain the reasoning behind them (although, as evident in certain neurological conditions, these explanations may sometimes be pure confabulations, having little to do with real mental causes). We see our mind as an indivisible unity and not as a complex process of collaborating subsystems (which it, of course, really is). It is the high efficiency of the brain processes in constructing coherent mind states that creates this illusion of unity [17]. However, to the extent that we insist in the unity of the mind to be a primitive concept we cannot simultaneously reason about its inner working in terms of cause and effect, about the question whether the process of free will is prime mover or secondary attribute. Arguing about the mind process means arguing about a very complex mechanistic system composed of billions of neural elements, and any degree of order, any basis for simple statements about the whole system, must come at the end of a process of organization, not at the start.

Accepting this view of my mind as an incredibly complex array of minute elementary mechanisms I am grateful for every bit of determinism in it. I do not want my decisions to be random, but to be instructed as far as possible by judgment about desirability of outcome. The stricter the logic, that is, the inner working, of my mind the better. But, in view of this determinism, can my decisions still be called free in some sense? What about punishing a murderer if he did not have a choice anyway? There is a widespread belief that here is a deep conflict that needs to be resolved. Determinism of my brain is taken as an infringement on the sovereignty of my self, some kind of hetero-determination by my synapses and neurons. Starting with Pasqual Jordan, thinkers have grabbed an opportunity seen in quantum physics: quantum chance as a loophole out of my mind's determinism. But I do not want to throw dice to make my choices—I want them to be reasoned![2]

It is my suspicion that the generally perceived contradiction between free will and determinism has little to do with determinism as such but rather with the attributes occasionally associated but not necessarily connected with determinism, as discussed above. It would be totally unacceptable, a slap in the face of our ego's glory, if due to the inner workings of our brain we were hetero-determined by primitive instincts, were reduced to clockwork regularity, or subject to the ridicule of

---

[2]The neurophysiologist John Eccles saw in quantum chance the instrument through which an immaterial mind could purposefully influence the mechanistic brain, seen as "the mind's computer." However, this would not solve the dilemma but simply shift it to another domain, the mind as distinct from the brain, whatever that could be.

practical predictability, so that our every move could be foreseen! The spectre of narrowly schematic and therefore predictable behavior could come in several forms—genetically determined behavior, addiction, ingrained habits, or a view that sees our brain in close analogy to our present types of computer programs. Free will, then, means deviation from an otherwise deterministic course, deterministic in this restricted sense.

In consequence we have to subdivide our mind into two subsystems, a lower tier that is narrowly constrained in its behavior, plus an upper one that brings additional mechanisms into play that are free of the constraints of the lower subsystem and that can modify and overrule whatever that level would have done on its own.[3] Indeed this subdivision of our brain and mind into tiers has been formulated in the literature in various ways. The comparative neuroanatomist Edinger spoke, in the 19th century, of the paleoencephalon, the "fish brain", buried in ours, complete with all primitive instincts necessary for simple survival, but unable of differentiated behavior in complex situations, an ability that we owe to the neoencephalon, parts of our brain that are evolutionarily younger, especially the cerebral cortex. Freud has dissected our mind into three parts, the *id*, the *ego* and the *super-ego*. The *id* comprises primitive drives and instincts. These are dominated by the *ego* to give us consistent behavior in line with a well-reasoned set of goals. And the *ego* is modified by a *super-ego* that incorporates the societal influences of norms and ethics. Judges try to come to an assessment whether the criminal is endowed with an upper tier of moral values and of considerations of guilt and punishment. If not, the perpetrator is "deterministic" (that is, victim of the inevitability of a bare lower tier, dominated by lack of intelligence or overwhelming drives), in which case the verdict may ordain treatment and confinement rather than punishment.

The essential point here is that also the upper tier is deterministic, for the reasons given earlier, although due to its complexity it will have no clockwork regularity and easy predictability. This point is more clouded than enlightened by the analogy to the algorithmically controlled computer. There, the upper tier is not in the machine at all but resides in the mind of the programmer, who alone takes into consideration goals and judgments to adapt the machine to the intended application. At the present time we humans are very meticulous about holding the reins in our own hands instead of giving the computer the freedom to develop its own set of goals and decide accordingly. Maybe we should keep it that way, but then we should not complain about a computer incapable of flexible response to the exigencies of situations as they arise. If, on the other hand, we wanted a computer that came close to us in terms of intelligence and situation-awareness we would have to give it an upper tier in terms of motives and a repertoire of reaction patterns that freed it from the "genetic determinism" implicit in algorithmic off-line control by human programmers. To the extent that our brain is indeed deterministic it can be simulated on a computer, but the organization of its program would have to be very different from the machine-like entities we are used to now.

---

[3]This is just like in the external perspective, where we see a situation unfold in a predictable way and intervene with our own decisions to alter the course of action, the external situation corresponding to the lower tier, our presence and influence to the upper.

In summary, a deterministic and an eternal universe would be an unsupportable prison only if we and our mind's mechanisms were not an integral part of it. But as the universe obeys forward causality, we are very concretely participating in forming the future.

3 Conclusion

I am adopting here the view that instead of field physics a more convincing description of the dynamics of our world can be formulated by an action principle along the lines sketched by Tetrode and Fokker, admitting that much further work needs to be done to fill in important conceptual lacunas. The most convincing argument in favor of this view is that on this basis quantum phenomena can be understood in a simple and straightforward way, as worked out by Cramer [2]. According to this action principle, events at all space-time points are stitched together by a tangle of advanced and retarded signals from the beginning to the end of time. Thus, the earliest times of this universe as much as the present moment could not have a definite shape without also all later events of the universe's history being equally definite. However, for reasons that are not clear yet, the relation between past and future is not symmetric, energy and information always being transferred into the future, the effects of advanced signals being subtle and difficult to detect, making for a world that has forward causality.

Although many physical arguments speak for this perspective, it has not attracted widespread attention, let alone acceptance. The main reason for this may lie in the apparent incompatibility with our traditional intuitive outlook on time. Most outrageous seems the proposition that the future, being in instant interaction with events now, has a definite, immutable reality and form, down to the minutest atomic detail. This seems to bereave us of all freedom to act and shape the future. It is the point of this paper to show in a logical analysis of our concept of free will that there is no contradiction and that, to the contrary, free will is unthinkable without determinism, that is, a definite future, the more definite the better. Our existence, structure and behavior are factors that contribute to shaping the future (and to the extent that advanced signals have effects, also the past).

To give the status of reality only to the present moment is just another expression of the extreme egocentric perspective that our civilization has developed. There is not really any fundamental difficulty in attributing to my own youth or my own old age the same reality status as to the moment in which I am writing this, even if they are not accessible to my mind now. In fact, I experience this as a rather relaxing thought (and find it surprising that Barbour, this volume, should come to the opposite conclusion of putting even more emphasis on the here-now).

Acknowledgements I would like to thank Carver Mead for sharing his ideas and deep insights with me in a number of intensely delightful and memorable discussions, discussions that were instrumental in precipitating the views on the nature of quantum phenomena that I am expressing here.

References

1. Bell, J.S.: On the problem of hidden variables in quantum mechanics. Rev. Mod. Phys. **38**, 447–452 (1966)
2. Cramer, J.G.: The transactional interpretation of quantum mechanics. Rev. Mod. Phys. **58**, 647–687 (1986)
3. Dirac, P.A.M.: Classical theory of radiating electrons. Proc. R. Soc. Lond. A **167**, 148–169 (1938)
4. Einstein, A., Podolsky, B., Rosen, N.: Can quantum-mechanical description of physical reality be considered complete? Phys. Rev. **47**, 777–780 (1935)
5. Fokker, A.D.: Ein invarianter Variationssatz für die Bewegung mehrerer elektrischer Massenteilchen. Z. Phys. **58**, 386–393 (1929)
6. Ghirardi, G.C., Rimini, A., Weber, T.: A general argument against superluminal transmission through the quantum mechanical measurement process. Lett. Nuovo Cimento **27**, 293–298 (1980)
7. Hogarth, J.E.: Cosmological considerations of the absorber theory of radiation. Proc. R. Soc. A **267**, 365–383 (1962)
8. Hoyle, F., Narlikar, J.V.: Time symmetric electrodynamics and the arrow of time in cosmology. Proc. R. Soc. **277**, 1 (1963)
9. Jaynes, J.: The Origin of Consciousness in the Breakdown of the Bicameral Mind. Houghton Mifflin, Boston (1976)
10. Klauder, J.R., Sudarshan, E.C.G.: Fundamentals of Quantum Optics. Benjamin, Elmsford (1968)
11. Lewis, G.N.: The nature of light. Proc. Natl. Acad. Sci. USA **12**, 22–29 (1926)
12. Libet, B., Gleason, C.A., Wright, E.W., Pearl, D.K.: Time of conscious intention to act in relation to onset of cerebral activity (Readiness-potential): the unconscious initiation of a freely voluntary act. Brain **106**, 623–642 (1983)
13. Mead, C.A.: Collective Electrodynamics – Quantum Foundations of Electromagentism. MIT Press, Cambridge (2000)
14. Redhead, M.: Incompleteness, Nonlocality and Realism. Clarendon, Oxford (1987)
15. Spinoza, B.: The Ethics. Dover, New York (1955)
16. Tetrode, H.: Über den Wirkungszusammenhang der Welt. Eine Erweiterung der klassichen Dynamik. Z. Phys. **10**, 317–328 (1922)
17. von der Malsburg, C.: The coherence definition of consciousness. In: Ito, M., Miyashita, Y., Rolls, E.T. (eds.) Cognition, Computation and Consciousness, pp. 193–204. Oxford University Press, London (1997)
18. von der Malsburg, C.: How are neural signals related to each other and to the world? J. Conscious. Stud. **9**, 47–60 (2002)
19. Wheeler, J.A., Feynman, R.P.: Interaction with the absorber as the mechanism of radiation. Rev. Mod. Phys. **17**, 157–181 (1945)
20. Wheeler, J.A., Feynman, R.P.: Classical electrodynamics in terms of direct interparticle action. Rev. Mod. Phys. **21**, 425–433 (1949)
21. Woodward, J.F., Mahood, T.: What is the cause of inertia? Found. Phys. **29**, 899–930 (1999)

Chapter 6
The Direction of Time in Quantum Mechanics

Roland Omnès

Abstract The quantum decoherence effect, which is known to destroy quantum interferences at a macroscopic level, is also the most efficient cause of irreversibility. Most of this paper is devoted to a non-technical description of the present theoretical state of the art concerning this effect. It implies a privileged direction of time, which coincides with the one usually associated with thermodynamics. A few considerations concerning consistent quantum histories are also added, because they introduce a third direction of time, the logical one, which coincides with the first two and can be shown moreover to be universal, i.e. the same for two arbitrary regions in the universe.

Keywords Quantum decoherence · Irreversibility · Quantum histories · Logical direction of time

Quantum physics is time-reversal invariant, except for a super-weak interaction with no practical consequence outside of cosmology. Classical dynamics is also invariant, as long as friction is not taken into account. It has been now established that classical physics is a direct consequence of the quantum principles, so that the exact place where irreversibility enters in physics becomes an interesting question. The present communication will be devoted to it.

The results I am going to discuss are a by-product of the renewal of interest for the interpretation of quantum mechanics, which opened some new vistas on various fundamental questions. Different people in different places have built up the kind of interpretation I am thinking of (the so-called "new dogma" as it was called in a somewhat derogative way), so that slightly different versions of it exist. All of them agree in any case about two basic ingredients: the existence of a decoherence effect and an explicit derivation of classical physics from the principles of quantum mechanics, which are enough for our present purpose.

There is less agreement concerning the interest and the importance of another ingredient of interpretation, namely "consistent histories", which were invented by

R. Omnès (✉)
Laboratoire de Physique Théorique, Université de Paris-Sud, 91405 Orsay, France
e-mail: roomnes@wanadoo.fr

S. Albeverio, P. Blanchard (eds.), *Direction of Time*,
DOI 10.1007/978-3-319-02798-2_6,

Griffiths [4] for clarifying the logical background of quantum mechanics and which connected with decoherence in the idea of "decohering histories" by Gell-Mann and Hartle [3]. I am personally a "consistent historian" because of an inclination for logic, but I am aware that some people claim that they can understand quantum mechanics without the help of histories. Griffiths' careful and systematic investigations have shown in any case how his approach can get rid of every so-called "paradox" of quantum mechanics and I highly recommend his recent book on this topic [5]. I wish also to notice that nobody has yet provided a fully consistent derivation of the Copenhagen measurement rules and of all the features of classical physics—including the logical ones—, except using the logic of histories [11, 13].

Anyway, interpretation *per se* is not really necessary for understanding the origin of a direction of time, although its basic ingredients are. Decoherence is by far the most important one and I will devote most of this communication to it, relying for this purpose on some recent work of mine in which the relation with irreversibility is particularly explicit. Consistent histories will be invoked only for a side issue which has only a philosophical interest, namely to show that, according to the principles of quantum mechanics, the direction of time must everywhere be the same in the universe, including the parts of it we cannot see.

1 Decoherence and Irreversibility

I shall presume some acquaintance with the idea of decoherence, which goes back in its dynamical form 30 years ago with the work of Hans Dieter Zeh [16]. The effect relies on a property of most macroscopic systems, which is that their gross features can be described by some "collective" coordinates, or let us say collective observables.

This notion goes back to Lagrange and most physicists consider it more or less as obvious. The collective variables are more or less the relevant ones for macroscopic dynamics. Their choice depends on how precisely one wants to apply dynamics, ultimately at the level of individual atoms (although not their electrons) according to the Born–Oppenheimer representation. This arbitrariness has a smell of non-objectivity, reminding us that no direct construction of the collective observables from the basic principles of quantum mechanics has yet been made. This is still in my opinion the weakest point in our understanding of quantum physics, but we will have to do with it.

Let us therefore assume that a macroscopic system can be described by a complete system of commuting observables, including some relevant collective ones, the non-collective ones being defined as the rest of them. It is more convenient to think of the macroscopic system as consisting of two abstract coupled subsystems, namely:

– A collective subsystem, associated with its own Hilbert space to which the collective observables belong.

– An "environment", associated with the multitude of degrees of freedom inside the macroscopic system and in the outside part of the world with which it interacts.

More explicitly, the total Hamiltonian of the system is often written as

$$H = H_c + H_e + H_{\text{int}}, \tag{1}$$

although this is a bit restrictive. Here H_c is a collective Hamiltonian, acting only in the collective Hilbert space, H_e is similarly an environment observable and H_{int} their coupling. The ideas of thermodynamics invite us to see H_e as having the internal energy as a mean value and H_{int} as responsible for the energy exchanges between large-scale motion and internal thermal motions.

Decoherence, as nicely said by Zurek, is the result of the perturbation of the environment by the collective subsystem. Its main features originate from the high sensitivity of the environment to any perturbation, resulting itself from the closeness of the energy levels of H_e. Its theory always relies on the so-called "reduced density matrix"

$$\rho_r = \text{Tr}_e\, \rho. \tag{2}$$

This is the trace of the full matrix density ρ with respect to the environment degrees of freedom so that ρ_r still contains complete information about the collective degrees of freedom: it tells us everything we can see.

It is often convenient to introduce a set X of Lagrange collective coordinates (observables) and to write down ρ_r in the corresponding basis $|x\rangle$ as

$$\rho_r\left(x', x''\right) = \langle x'|\rho_r|x''\rangle. \tag{3}$$

Decoherence is then often characterized as a very rapid convergence of $\rho_r(x', x'')$ toward an almost diagonal form: it vanishes rapidly when x' and x'' differ macroscopically, exponentially in time and in the distance $(x' - x'')^2$. It must be mentioned that this assimilation of decoherence with approximate diagonalization is not universal in a straightforward way, but this is a tricky question and I must refer to a more complete discussion of it to be found elsewhere.

1.1 The Theory of Irreversible Processes

The existence of a close relation between decoherence and irreversibility is universally accepted. There was, however, no strict proof of it before a theory I proposed some time ago and developed in detail recently. It relies on a standard "projection method" to deal with quantum and classical irreversible processes, which goes essentially as follows [1, 6, 10, 18].

Suppose we are interested in the average values $a_j(t) = \text{Tr}\{A_j \rho(t)\}$ of some set of "relevant" observables A_j. This is a typical observational situation where the relevant quantities are the ones observation can assert whereas, in a macroscopic system, most other physical quantities remain inaccessible. The time evolution of the average quantities $a(t)$ is a typical problem where one may expect irreversibility.

If one were able to construct a test density operator ρ_0 giving exactly the average values, i.e. such that

$$a_j(t) = \text{Tr}\{A_j \rho_0(t)\}, \tag{4}$$

the problem would have been solved. Then, one looks for the less cumbersome test density, i.e. the one giving all the required information about the interesting quantities $a(t)$ and nothing else. Information theory tells us that it has the form

$$\rho_0 = \exp\left\{ -\sum_j \lambda_j A_j \right\}, \tag{5}$$

where the numerical coefficients λ are Lagrange multipliers.

From there on, the method for obtaining the test density operator with minimum information becomes technical and I will not enter in it. It relies of course upon the Schrödinger equation governing the evolution of the complete density operator ρ. It is a very nice exercise in algebraic virtuosity ending with a "master equation" for the time derivative of ρ_0. This kind of result would seem to solve completely the problem if there were not some caveats. The master equation is very formal; it involves the detailed interactions of all the particles in the system and, in a nutshell, the possibility of writing down explicitly this equation is in general tantamount to solving the basic Schrödinger equation itself. This is not surprising after all, since only algebra has been used and algebra has nothing to do with irreversibility. Physical restrictions must enter somewhere and they do in practical applications. Some approximations or restrictions to models always enter therefore into the pattern at one point or another: The restriction to classical physics, to two-particle collisions and the neglect of velocity correlations after scattering yield for instance the Boltzmann equation as a special case of a master equation.

1.2 The Case of Decoherence

This projection method works out very nicely in the case of decoherence, but we will have to be more specific for understanding that [14]. One considers a macroscopic system. To choose the relevant observables will be the essential point, but the splitting of the system into a collective subsystem and an environment makes that clear. From a classical standpoint, one would say that all the "Lagrange" collective coordinates and momenta, as well as any function of them, may be considered as relevant. They should be envisioned of course from a quantum standpoint, i.e. as so many quantum observables. They are in some sense too many, but one may resort to a subset from which all of them can be derived and the most convenient choice is the set of all the operators $|x'\rangle\langle x''|$. Since the environment is defined as not accessible to observation, no environment observable is relevant, except one. This is the internal energy operator H_e, because its value tells us something of the global state of matter.

Knowing the relevant observables in the situation at hand, one can write down the test density operator ρ_0. It has a very simple form, namely a product $\rho_r \rho_e$ where

ρ_r is the reduced density operator we are interested in and ρ_e is a state of the environment at thermal equilibrium. Notice that the occurrence of a thermal distribution does not mean that thermal equilibrium really holds in this irreversible situation; it stands only for a bookkeeping of the internal energy and its evolution.

A last technical point must be mentioned. The coupling term H_{int} in (1) may involve sizable effects, which are of consequence at a macroscopic level. One may think as an example of the pressure exerted by an external gas on a wall. The position of the wall is relevant for direct observation; the molecules in the gas are not but their pressure is. Its effect belongs conceptually to the collective world and one can make that explicit through a simple change in the interaction Hamiltonian. One replaces it by

$$H_{int} \rightarrow \mathrm{Tr}_e(H_{int}\rho_e) \otimes I_e + H'_{int}, \tag{6}$$

where, without entering into the mathematical details, the first term is the part of the interaction with collective effects and the second term consists only of statistical fluctuations. These fluctuations are individually very small and they can be dealt with most often by means of perturbation calculus. This simple trick is very convenient, because it implies that the projection method is not purely formal in that case. The master equation can be written down explicitly and its consequences analyzed. Since we are interested only in these consequences, we may skip the equation itself.

1.3 The General Behavior of Decoherence

The master equation predicts both decoherence and the standard dissipation effects ad expected from thermodynamics. The results are quantitative and decoherence is by far much quicker than dissipation. An interesting special case occurs when there exists a set of collective coordinates X commuting with the fluctuating coupling H'_{int}. I will not discuss here the physical conditions under which this situation occurs, except for mentioning that they were always assumed or automatically satisfied in the standard models of decoherence [2, 7–9, 15]. The effect turns then into an exponential tendency toward approximate diagonalization in the $|x\rangle$ basis, as previously described. But a matrix can only be diagonal, or approximately so, in a unique basis and a question concerning the behavior of decoherence in the general case therefore occurred: is there always a "pointer basis" in which the reduced density matrix tends to become diagonal [17]?

The present method answers this question positively. The states becoming separated through decoherence are "classical states", for instance with definite position *and* momentum in the case of a ball or a speck of dust. The existence of such states (with uncertainties much above the limits of the uncertainty relations) was shown earlier but, once again, I will not discuss that in detail. It involves a whole theory of "classicality" relying on non-trivial mathematics [12], and the final result is so clean that it would be a pity to spoil it by not too obvious equations.

One can then conclude generally that there exists an irreversible decoherence effect separating very rapidly the classical collective states of a macroscopic system, destroying any mutual interference among them, so that they are left with definite probabilities of a classical type. This result is of course valid under the conditions that were already known from the study of models, i.e. non-zero dissipation (ordinary light and some superconducting devices being the main exceptions).

1.4 The Direction of Time

These results are obtained by integrating the master equation between an initial time 0 and a later time t. It should be no surprise that if one starts formally from the same initial state at time 0 and integrate from there to an *earlier* time $-t$, one obtains identical results. Disorder, either in internal motion or in the phase of wave functions, is always found to increase. The only new points are the existence of decoherence itself and the fact that it acts in the same time direction than dissipation.

The interpretation of the arrow of time is accordingly the same as Boltzmann's, except that decoherence refines it. A state occurring naturally or which is prepared in the laboratory can never be controlled down to the details of internal motion or down to quantum phases (which are much less accessible to preparation). The probability of returning to an initial state with the same thermodynamical characteristics is still given by $\exp(-\Delta S)$, where ΔS is the increase of entropy. The probability P of returning to the initial *quantum* state is, however, much smaller.

One may consider the example of a piston with mass M in a cylinder, a gas at temperature T inside the cylinder being taken as the environment. The piston is held by a spring so that its natural frequency of oscillation is ω whereas the damping time of these oscillations is τ. If the piston is initially in a coherent quantum state, the probability P of returning to that state before any sizable dissipation has taken place is of the order of

$$P \propto \sqrt{\hbar \omega \tau / T t},$$

after a time t, assuming high temperature ($T \gg \hbar \omega$) and that decoherence has begun to act (a dew collisions of the piston with gas molecules are enough for that). If on the other hand the initial state of the piston is a superposition of two coherent states with the same average momentum and average positions separated by a distance a, one gets for the probability of return to the initial quantum state

$$P \approx \exp(-M T a^2 t / \hbar \tau).$$

One may then conclude that decoherence *and* dissipation always proceed from an easily prepared state to a more disordered one, which cannot be prepared except through the process it underwent. The extremely small probability of a chance return to the initial thermodynamical state is controlled as usual by the change in entropy, but a return to an initial (collective) quantum state is forbidden much more rapidly with much smaller probabilities. The direction of time is therefore established through quantum processes (decoherence), much more efficiently than through the much slower thermodynamical processes.

1.5 The Logical Direction of Time

There are two other interesting arrows of time. One of them is cosmological, meaning essentially that the universe is expanding (or able to contract in some mathematical models). It would be easy to recast the previous analysis in the framework of a Robertson–Walker universe and it is obvious that there is no connection between the cosmological direction and the microscopic one. Of course, the real initial conditions were such that expansion went along with decoherence and dissipation, but this is a matter of circumstance and not a matter of principle.

The other direction is associated with the logic of quantum mechanics, as it occurs in consistent histories. Roughly speaking, a history is a sequence of properties describing the behavior of a system [5]. They may be classical properties and/or quantum properties, each of them being associated formally with a projection operator in Hilbert space. A set of equations involving these operators (the consistency conditions) ensures that every history belonging to a family of mutually exclusive ones has a well-defined probability. It can then be shown that classical logic holds among the possible propositions resulting from the description. When the histories refer only to classical properties of macroscopic objects, the corresponding logical setup generates the standard interpretation of classical physics. This interpretation is by the way so intuitive and straightforward that most people do not realize it exists as an interpretation and they simply call it common sense.

There is again a direction of time in the logic of histories: a time-reversed history looks very much like a motion picture running backward. But the validity of the consistency conditions relies in most practical cases on decoherence [3], so that in view of what we just saw, only one direction in a series of events or in the corresponding history can satisfy these conditions. Another way of saying that is that there is a logical direction of time and it coincides with the two previous ones, in decoherence and thermodynamics.

The existence of this third interpretation of the time direction implies some specific consequences, which do not follow directly from the first two. One of them is the universality of the arrow of time. It may be checked to begin with that we have the following. (i) Any part of a history (involving at least two instants of time) and the whole history must have the same logical direction of time if consistency and logic can hold. (ii) Any consistent set of histories can be extended arbitrarily far in the past and the future. Then let us consider two•regions of space-time A and B with a space-like separation. If there is a common region 1 in the past of A and B, the consistency of histories linking 1, respectively, with A and B implies a common logical direction of time for two different families of histories describing, respectively, some events in the two regions. Notice that this result also implies a common direction of time for thermodynamics. When there is no such common region 1 in the past of A and B, one can always envision a series of intermediate regions such as: 1 in the past of A, 2 in the future of 1, 3 in the past of 2 and so on, until B is reached.

The directions of time of "common sense", thermodynamics, and decoherence must then be the same in two different regions of the universe regions, whatever these regions, and they are therefore universal. This is as far as I know the only example of a physical result that can be established globally on a basis of consistency for the entire universe, without any restriction of causality. It is of course perfectly useless but, since we had several interesting communications of a philosophical character during this meeting, I thought it was worth mentioning.

References

1. Balian, R., Alhassid, Y., Reinhardt, H.: Phys. Rep. **131**, 1 (1986)
2. Caldeira, A.O., Leggett, A.J.: Physica A **121**, 587 (1983)
3. Gell-Mann, M., Hartle, J.B.: In: Zurek, W.H. (ed.) Complexity, Entropy and the Physics of Information, pp. 425–458. Addison-Wesley, Reading (1990)
4. Griffiths, R.B.: J. Stat. Phys. **36**, 219 (1984)
5. Griffiths, R.B.: Consistent Quantum Theory. Cambridge University Press, Cambridge (2002)
6. Haake, F.: In: Quantum Statistics in Optics and Solid-State Physics. Springer Tracts in Modern Physics, vol. 66, p. 98 (1973)
7. Hepp, K., Lieb, E.H.: Helv. Phys. Acta **46**, 573 (1994)
8. Hu, B.L., Paz, J.P., Zhang, Y.: Phys. Rev. D **45**, 2843 (1992)
9. Joos, E., Zeh, H.D.: Z. Phys. B **59**, 229 (1985)
10. Nakajima, S.: Prog. Theor. Phys. **20**, 948 (1958)
11. Omnès, R.: Rev. Mod. Phys. **64**, 339–382 (1992)
12. Omnès, R.: J. Math. Phys. **38**, 687 (1997)
13. Omnès, R.: Understanding Quantum Mechanics. Princeton University Press, Princeton (1999)
14. Omnès, R.: Phys. Rev. A **65**, 052119 (2002)
15. Unruh, W.G., Zurek, W.H.: Phys. Rev. D **40**, 1071 (1989)
16. Zeh, H.D.: Found. Phys. **1**, 69 (1970)
17. Zurek, W.H.: Phys. Rev. D **26**, 1862 (1982)
18. Zwanzig, R.: Physica **30**, 1109 (1964)

Chapter 7
Two Arrows of Time in Nonlocal Particle Dynamics

Roderich Tumulka

Abstract Considering what the world would be like if backwards causation were possible is usually mind-bending. Here I discuss something that is easier to study: a toy model that incorporates a very restricted sort of backwards causation. It defines particle world lines by means of a kind of differential delay equation with negative delay. The model presumably prohibits signaling to the past and superluminal signaling, but allows nonlocality while being fully covariant. And that is what constitutes the model's value: it is an explicit example of the possibility of Lorentz-invariant nonlocality. That is surprising in so far as many authors thought that nonlocality, in particular nonlocal laws for particle world lines, must conflict with relativity. The development of this model was inspired by the search for a fully covariant version of Bohmian mechanics.

Keywords Bohmian mechanics · Relativity · Quantum nonlocality · Backwards causation · Differential delay equations

In this paper I will introduce to you a dynamical system—a law of motion for point particles—that has been invented [5] as a toy model based on Bohmian mechanics. Bohmian mechanics is a version of quantum mechanics with particle trajectories; see [4] for an introduction and overview. What makes this toy model remarkable is that it has two arrows of time, and that precisely its having two arrows of time is what allows it to perform what it was designed for: to have effects travel faster than light from their causes (in short, *nonlocality*) without breaking Lorentz invariance. Why should anyone desire such a behavior of a dynamical system? Because Bell's nonlocality theorem [1] teaches us that any dynamical system violating Bell's inequality must be nonlocal in this sense. And Bell's inequality is, after all, violated in nature.

It is easy to come up with a nonlocal theory if one assumes that one of the Lorentz frames is preferred to the others: simply assume a mechanism of cause and effect

---

R. Tumulka (✉)
Department of Mathematics, Rutgers University, 110 Frelinghuysen Road, Piscataway, NJ 08854-8019, USA
e-mail: tumulka@math.rutgers.edu

S. Albeverio, P. Blanchard (eds.), *Direction of Time*,
DOI 10.1007/978-3-319-02798-2_7,

(an interaction in the widest sense) that operates *instantaneously* in the preferred frame. That is what nonrelativistic theories usually do. In other frames, these non-local effects will either travel at a superluminal ($>c$) but finite velocity or precede their causes by a short time span. Of course, causal loops cannot arise since in the preferred frame effects never precede causes; yet the entire notion of a preferred frame is against the spirit of relativity. Without a preferred frame, to find a nonlocal law of motion is tricky, and much agonizing has been spent on this. About one way to achieve this you will learn below.

Let us come back first to the two arrows of time. They are opposite arrows, in fact. But unlike the arrows considered in Lawrence Schulman's contribution to this volume, they are not both thermodynamic arrows. One of the two is the thermody-namic arrow. Let us call it Θ. It arises, as emphasized first by Ludwig Boltzmann and in this conference by Schulman, not from whichever asymmetry in the micro-scopic laws of motion, but from boundary conditions. That is, from the condition that the initial state of the universe be taken from a particular subset of phase space (corresponding to, say, a certain low entropy macrostate), while the final state is not subjected to any such conditions—except in some scenarios studied by Schulman. The dynamical laws considered in discussions of the thermodynamic arrow of time are usually time reversal invariant. But not so ours! It explicitly breaks time symme-try, and that is how another arrow of time comes in: an arrow of microscopic time asymmetry, let us call it C. Such an arrow must be assumed before writing down the equation of motion, which will be (6) below. In addition, the equation of motion is easier to solve in the direction C than in the other direction. Does not it seem ugly and unnatural to introduce a time asymmetry? Sure, but we will see it buys us something: Lorentz-invariant nonlocality.

Recall that such an arrow is simply absent in Newtonian mechanics and other time symmetric theories. So it is not surprising that the microscopic arrow C is not the source of the macroscopic time arrow Θ, even more, the direction of Θ is completely independent of the direction of C. Θ depends on boundary conditions, and not on the details of the microscopic law of motion. In our case, Θ will in-deed be opposite to C. Since inhabitants of a hypothetical universe will regard the thermodynamic arrow as their natural time arrow, related to macroscopic causation, to memory, and to apparent free will, you should always think of Θ as pointing towards the future, whereas C is pointing to what we call the past.

It is time to say what the equation of motion is. The equation is intended to be as close to Bohmian mechanics as possible, to be an immediate generalization, and to have Bohmian mechanics as its nonrelativistic limit. To remind you of how Bohmian mechanics works, you take the wave function (which is supposed to evolve according to Schrödinger's equation—without ever having to collapse), plug in the positions of all the particles (here is where a notion of simultaneity comes in), and from that you compute the velocity of any particle by applying a certain formula, Bohm's law of motion, which amounts to dividing the probability current by the probability density. Now, for a Lorentz-invariant version, we first have to worry about the wave function.

There are three respects in which the wave function of nonrelativistic quan-tum mechanics (or Bohmian mechanics, for that matter) conflicts with relativity:

(a) the dispersion relation $E = p^2/2m$ at the basis of the Schrödinger equation is nonrelativistic, (b) the wave function is a function of $3N$ position coordinates but only one time coordinate, (c) the collapse of the wave function is supposed instantaneous. While (a) has long been solved by means of the Klein–Gordon or Dirac equation, it is too early for enthusiasm since we still face (b) and (c). We will worry about (c) later, and focus on (b) now. The obvious answer is to introduce a wave function ψ of $4N$ coordinates, that is, one time coordinate for each particle, in other words ψ is a function on (space-time)^N. You get back the nonrelativistic function of $3N + 1$ coordinates after picking a frame and setting all time coordinates equal. Such multi-time wave functions were first considered by Dirac et al. in 1932 [2], but what they did not mention was that the N time evolution equations

$$i\hbar \frac{\partial \psi}{\partial t_i} = H_i \psi \quad \text{for } i \in \{1, \ldots, N\} \tag{1}$$

needed for determining ψ from initial data at $t = 0$ do not always possess solutions. They are usually inconsistent. They are only consistent if the following condition is satisfied:

$$[H_i, H_j] = 0 \quad \text{for } i \neq j. \tag{2}$$

This is easy to achieve for non-interacting particles and tricky in the presence of interaction. Indeed, to my knowledge it has never been attempted to write down consistent multi-time equations for many interacting particles, although this would seem an obvious and highly relevant problem if one desires a manifestly covariant formulation of relativistic quantum mechanics. We will here, however, stay on the easy side and simply consider a system of non-interacting particles. We take the multi-time equations to be Dirac equations in an external field A_μ,

$$1 \otimes \cdots \otimes \underbrace{\gamma^\mu}_{i\text{th place}} \otimes \cdots \otimes 1 \left(i\frac{\partial}{\partial x_i^\mu} - eA_\mu(x_i) \right) \psi = m\psi \tag{3}$$

where $\psi : \text{(space-time)}^N \to (\mathbb{C}^4)^{\otimes N}$, and e and m are charge and mass, respectively. The corresponding Hamiltonians commute trivially since the derivatives act on different coordinates and the matrices on different indices.

Such a multi-time Dirac wave function naturally defines a tensor field

$$J^{\mu_1 \cdots \mu_N} := \bar{\psi} \gamma^{\mu_1} \otimes \cdots \otimes \gamma^{\mu_N} \psi, \tag{4}$$

and according to the original Bohmian law of motion (for Dirac wave functions), the 4-velocity of particle i is, in the preferred frame,

$$\frac{dQ_i^\mu}{ds} \propto J^{0 \ldots \overset{i}{\mu} \ldots 0}(Q_1, \ldots, Q_N) \tag{5}$$

where only the ith index of J is nonzero, and $Q_i^\mu(s)$ is the world line parameterized by proper time, or indeed by any other parameter since a law of motion need only

Fig. 1 How to choose the N
space-time points where to
evaluate the wave function, as
described in the text

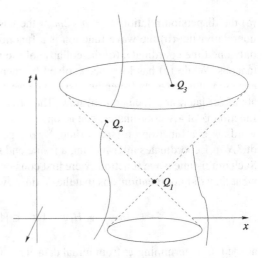

(and (5) does only) specify the *direction* in space-time of the tangent to the world line. The coordinates taken for the other particles are their positions *at the same time*, $Q_j^0 = Q_i^0$. Instead of a Lorentz frame, one can take any foliation of space-time into spacelike hypersurfaces for the purpose of defining simultaneity-at-a-distance [3]. The theory I am about to describe, in contrast, uses the hypersurfaces naturally given by the Lorentzian structure on space-time: the light cones. More precisely: the future light cones—and that is how the time asymmetry comes in.

So here are the steps: first solve (3), so you know ψ on (space-time)^N. Then, compute the tensor field J on (space-time)^N according to (4). For determining the velocity of particle i at space-time point Q_i, find the points Q_j, $j \neq i$, where the other particles cross the future light cone of Q_i, as depicted in Fig. 1. Plug these N space-time points into the field J and get a single tensor. Find out what the 4-velocities u_j^μ of the other particles at Q_j, $j \neq i$, are. Use these to contract all but one index of J. We postulate that the resulting vector is, up to an irrelevant proportionality factor, the 4-velocity we have been looking for:

$$\frac{dQ_i^{\mu_i}}{ds} \propto J^{\mu_1 \ldots \mu_N}(Q_1, \ldots, Q_N) \prod_{j \neq i} u_{j\mu_j}(Q_j). \tag{6}$$

One can show [5] that this 4-velocity is always timelike or null.

This law of motion is what can be called an ordinary differential equation with advanced arguments, or a differential delay equation with negative delay, because the velocity depends on the positions (and velocities) of other particles at future times, indeed with a *variable* delay span $Q_j^0 - Q_i^0$. It may seem to complicate things considerably that what happens here depends on the *future* rather than past behavior of the other particles, but that is an artifact of perspective: look at the equation of motion (6) in the other time direction, that is, in the direction C, and notice it now has only *retarded* arguments. That is a more familiar sort of differential delay equation that gives rise to no logical or causal problems. So this theory, although involving a mechanism of backwards causation, is provably paradox free, since no causal loops can arise: first solve the wave equation for ψ in the usual direction Θ, then solve the equation of motion in the opposite direction C.

Unfortunately, there is no obvious probability measure on the set of solutions to (6). This is different from the situation in Bohmian mechanics, where the $|\psi|^2$ distribution is conserved, a fact crucial for the probability predictions of that theory. The lack of such a measure for the model considered here makes it impossible to say whether or not this theory violates Bell's inequality, which is a relation between probabilities. But this law of motion takes what is perhaps the biggest hurdle on the way towards a fully covariant law of motion conserving the $|\psi|^2$ distribution, by fulfilling what Bell's theorem says is a necessary condition: nonlocality. I should add that in the nonrelativistic limit, the future light cone approaches the hyperplane $t = $ const. and the law of motion approaches the original Bohmian law of motion (5), conserving $|\psi|^2$.

How does nonlocality come about in this model? That has to do with the two arrows of time, pointing in opposite directions. Had we chosen them to point in the same direction, the theory would have been local, because what happens at Q_i would only depend on (what we call) the past light cone. But in this model, we evaluate ψ on the future light cone of Q_i, which means ψ has, in its multi-time evolution, gone through all the external fields at spacelike separation from Q_i. And that is how the velocity at Q_i may be influenced by the field imposed by an experimenter at spacelike separation from Q_i.

And what is the story then about problem (c) above, the instantaneous collapse? The first thing to say is that collapse is not among the basic rules of this model, or any Bohmian theory. That simply disposes of problem (c). But something more should be said, since the collapse rule can be derived in Bohmian mechanics: even if the wave function of Schrödinger's cat remains forever a superposition, *the cat itself* (formed by the particles) is either dead or alive, with probabilities determined by $|\psi|^2$. Moreover, since the wave packet of the dead cat (i.e., the corresponding term in the superposition) and that of the live cat have disjoint supports in configuration space, the wave packet of the dead cat does not influence the motion of the live cat (nor vice versa). In the model we are concerned with here, everything just said still applies, except that the model does not define any probabilities.

The model thus shows that a relativistic theory of particle world lines can indeed be nonlocal. Let me also point to another consequence: It has often been claimed that Bell's nonlocality proof excludes relativistic Bohm-type theories. This claim has always been inappropriate because Bell's proof actually shows that *any* serious version of quantum mechanics, Bohm-like or not, must be nonlocal; now we see that the claim is also inappropriate in another way, as nonlocality actually does not imply a conflict with relativity. Finally, let me add that a fully covariant version has been developed for a different quantum theory without observers, the GRW theory [6]. Also this model uses time-asymmetric laws, but not backwards causation.

To this day, thinking about time, time's arrows, and relativity remains a source of the unexpected.

Acknowledgements I wish to thank Sheldon Goldstein for his comments on a draft of this paper.

References

1. Bell, J.S.: Speakable and Unspeakable in Quantum Mechanics. Cambridge University Press, Cambridge (1987)
2. Dirac, P.A.M., Fock, V.A., Podolsky, B.: On quantum electrodynamics. Phys. Z. Sowjetunion **2**, 468 (1932). Reprinted in Schwinger, J. (ed.): Quantum Electrodynamics. Dover Publishing, New York (1958)
3. Dürr, D., Goldstein, S., Münch-Berndl, K., Zanghì, N.: Hypersurface Bohm–Dirac models. Phys. Rev. A **60**, 2729 (1999). quant-ph/9801070
4. Goldstein, S.: Bohmian mechanics. In: Zalta, E.N. (ed.) Stanford Encyclopedia of Philosophy (2001). Published online by Stanford University. http://plato.stanford.edu/entries/qm-bohm/
5. Goldstein, S., Tumulka, R.: Opposite arrows of time can reconcile relativity and nonlocality. Class. Quantum Gravity **20**, 557–564 (2003). quant-ph/0105040
6. Tumulka, R.: A relativistic version of the Ghirardi–Rimini–Weber model. J. Stat. Phys. **125**, 821–840 (2006). quant-ph/0406094

Chapter 8
Boundary Conditions, Time Reversal and Measurements

J.C. Zambrini

Abstract This contribution is divided into two parts. In the first one, we argue that the idea of time reversal in Quantum Mechanics is considerably more subtle than generally thought. For example, it is not even possible to make sense of Feynman's reinterpretation of the Heisenberg uncertainty principle without a good grasp of it. In the second part, more speculative, we discuss the importance of "randomizing" some times, in Quantum Mechanics, as a preliminary step before the expected conciliation with General Relativity.

1 The (Deterministic) Time We Know

There are basically two levels of analysis, in theoretical physics, of the issue of time-reversal (TR) symmetry:

(A) It is a trivial issue.
(B) It is one of the most vexing issues of Theoretical Physics.

There is no need to allude to the devastating problems associated with the Wave Function of the Universe to see how limited is the first opinion. As a matter of fact, it is sufficient to pick the most offensively trivial system of classical mechanics: the one dimensional free particle (of mass 1), whose second order (Newton's) dynamical law is

$$\frac{d^2}{dt^2}q = 0 \tag{1}$$

According to (A) there is no more in the statement that this law is invariant (or symmetric) under time reversal than the trivial observation:

"If $q(t)$ solves (1) so does $\hat{q}(t) = q(-t)$, $\forall t \in \mathbb{R}$".

One can as well define a time-reversal operator T, acting on the state of the system, here $\xi = (q, p) \in S = \mathbb{R}^2$ by $T(q, p) = (q, -p)$. Then, since the Hamiltonian flow

J.C. Zambrini (✉)
GFMUL, Av. Prof. Gama Pinto 2, 1649-003 Lisbon, Portugal
e-mail: zambrini@cii.fc.ul.pt

S. Albeverio, P. Blanchard (eds.), *Direction of Time*,
DOI 10.1007/978-3-319-02798-2_8,
© Springer International Publishing Switzerland 2014

(with Hamiltonian $h(q, p) = \frac{1}{2}p^2$) is given by $U_t : S \to S, (q, p) \mapsto (q + pt, p)$, the fact that $T\hat{q}(t) = q(t)$ and $T\hat{p}(t) = -p(t)$ can be rewritten as

$$U_{-t} = T^{-1}U_t T \qquad (2)$$

Of course, the time reversed $(\hat{q}(\cdot), \hat{p}(\cdot))$ are not really used physically. Instead, the previous formula allows us to extend the dynamical information available about the future, i.e. $t \in [0, \infty[$, into the past $t \in]-\infty, 0]$, given the initial condition, i.e the state, at $t = 0$. (The initial time is, of course, arbitrary.)

This way to think about time symmetry of physical laws of nature is, in fact, universal since it is thought that (almost) all fundamental laws are invariant under time reversal for the appropriate operation T, which depends on the considered domain of physics (for instance, in classical electrodynamics, if $(\vec{E}, \vec{B})$ denotes, respectively, the electric and magnetic fields then $T(\vec{E}, \vec{B}) = (\vec{E}, -\vec{B})$.

A substantial part of the discussions on physical interpretations of the time-reversal symmetry amounts to ponder over the operational meaning, if any, of the mathematical procedure given before. Is it physically realistic to transfer our dynamical information from the future to the past. (Or the other way around!) What is the meaning of such a transfer in the lab?

In any time reversal, initial conditions become final ones and this may easily conflict with our naïve (intuitive) concept of causality. It is a trivial observation that initial boundary conditions are, practically speaking, more easy to deal with than final ones. But one tends to use excessively this argument to eliminate (or ignore) some solutions of the laws of motion which are precisely needed to show the invariance of the theory under TR! An example is the propagation of classical waves where we tend to ignore the advanced solution and retain only the retarded one, more in accordance with "causality".

We can, of course, give at once boundary conditions at two different times but the associated boundary value problem is, in general, considerably more subtle than the traditional (Cauchy) problem. Consistency conditions are needed between those data, and we may easily loose the existence and uniqueness of the solution.

Let us come back to our trivial mechanical example, but regarded now as a boundary value problem. Since nothing in it depends on the choice of initial instant we shall consider any time interval $I = [s, u]$ and pick a reference time t in between. According to the classical Hamilton–Jacobi theory, we have now a dual description of the dynamics on I, when the boundary data of (1) become

$$q(s) = x \quad \text{and} \quad q(u) = z \qquad (3)$$

According to the first description, say the "causal" one, we have to consider a family of solutions of the (free) Hamiltonian equations with (past) boundary conditions:

$$q(s) = x, \qquad p(s) = \nabla S_s^*(x) \qquad (4)$$

where S_s^* is regular enough to define an initial Lagrangian manifold in phase space (we shall need, in fact, singular manifold for our example). This family of solutions is described by the action with initial condition, regarded as function of the final point (q, t):

$$S_L^*(q, t) = S_s^*(x) + \int_{x,s}^{q,t} L \, d\tau \tag{5}$$

for any t in I, where L is the Lagrangian of the system (for our Hamiltonian $h(q, p) = \frac{p^2}{2}$, $L(q, \dot{q})$ reduces to $\frac{1}{2}|\dot{q}|^2$ and the integral is computed along the characteristics connecting x and $q(t) = q$, q being regarded as variable). As a function, S_L^* solves the Hamilton–Jacobi (HJ) equation

$$\begin{cases} \frac{\partial S_L^*}{\partial t} + h(q, \nabla S_L^*) = 0 & t \in I \\ S_L^*(q, s) = S_s^*(q) \end{cases} \tag{6}$$

Clearly, this first order equation chooses definitely an arrow of time. But how come, since the resulting free dynamics does not? Even stranger, the time-symmetric Newton's equation results from the gradient of the "irreversible" HJ equation! There is no paradox here, however, but the explanation may be more interesting and general than expected. In the Hamilton–Jacobi framework, we had to ignore half of the boundary conditions (3), the future one. But we could have done a symmetric selection and keep the future information of (3). Then the relevant family of solutions would be described by an action with this final condition and regarded as a function of the initial point (q, t):

$$S_L(q, t) = S_u(z) + \int_{q,t}^{z,u} L \, d\tau \tag{7}$$

that is, the solution of

$$\begin{cases} -\frac{\partial S_L}{\partial t} + h(q, -\nabla S_L) = 0 & t \in I \\ S_L(q, u) = S_u(q) \end{cases} \tag{8}$$

This HJ equation can be regarded as the time reversed of (6) on I, because $dS_L^* = L \, dt$ and $dS_L = -L \, dt$. But, since our trivial boundary value dynamical system (1) and (3) has clearly an unique solution $t \mapsto q(t)$, $\forall t \in I$, some consistency condition is needed between (6) and (8). It is the following.

For any $t \in]s, u[$ along this solution

$$p_*(q(t), t) = \frac{\partial S_L}{\partial q}(x, s, q, t)\Big|_{q=q(t)} = -\frac{\partial S_L}{\partial q}(q, t, z, u)\Big|_{q=q(t)} = p(q(t), t) \tag{9}$$

expressing the smoothness of the trajectory, $\forall t \in I$. Notice that because our conditions (3) at the boundary ∂I are trivial, here, we can just use Hamilton's principal function and, then, drop the $*$ on the l.h.s. action without ambiguity.

So our trivial (time homogeneous) boundary value problem (1) and (3) involves, in the Hamilton–Jacobi perspective, two distinct momenta needed to take the arbitrary given data at ∂I into consideration. And our second order homogeneous problem can be solved via two time dependent first order problems. Since the Hamilton

Fig. 1 The two distinct momenta (or velocities) at a given time t

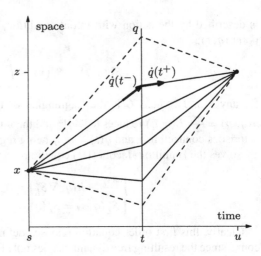

Fig. 2 The track of a quantum particle according to Heisenberg [1]

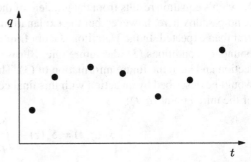

principal function reduces here to $S(q_1, t_1, q_2, t_2) = \frac{1}{2} \frac{|q_2 - q_1|^2}{t_2 - t_1}$, they can be written as a left hand differential $d_* q(\tau) = \dot{q}(\tau^-) d\tau$ where (cf. Fig. 1)

$$\begin{cases} d_* q = \frac{q - x}{\tau - s} d\tau = p_*^x(q, \tau) d\tau & s \leq \tau < t \\ q(t) = q \end{cases} \tag{10}$$

and a right hand differential $dq(\tau) = \dot{q}(\tau^+) d\tau$:

$$\begin{cases} dq = \frac{z - q}{u - \tau} d\tau = p^z(q, \tau) d\tau & t < \tau \leq u \\ q(t) = q \end{cases} \tag{11}$$

The consistency relation (9) determines uniquely the solution of (1) and (3).

According to Heisenberg we are not allowed to preserve any such space-time view for the quantum version of our trivial dynamical system, i.e for the one dimensional quantum free particle. We are even told why this is impossible; because the track of our quantum particle looks like (see [1]) Fig. 2.

But, 20 years after Heisenberg, Feynman has shown that this radicalism was not necessary [2]. One should just relax the classical hypothesis of smoothness of the trajectories. The building block of the Feynman reinterpretation is the concept of transition element (or amplitude) on I:

$$\langle \varphi | \mathbb{I} \psi \rangle_{S_L} = \int \int \psi_s(x) K(x, u - s, z) \bar{\varphi}_u(z) \, dx \, dz$$

$$= \int \int_{\Omega_x^z} \int \psi_s(x) e^{\frac{i}{\hbar} S_L[\omega(\cdot); u-s]} \bar{\varphi}_u(z) \, \mathcal{D}\omega \, dx \, dz \qquad (12)$$

where S_L is the classical action, regarded now as a functional along Feynman's quantum paths $\omega \in \Omega_x^z = \{\omega \in C([s, u]; \mathbb{R}) \text{ such that } \omega(s) = x, \ \omega(u) = z\}$, $\mathbb{I}$ denotes the identity operator, $\hbar$ is Planck's quantum of action and $\mathcal{D}\omega$ denotes the symbolic product $\prod_{s \leq \tau \leq u} d\omega(\tau)$.

The definition (12) involves boundary conditions $\{\psi_s, \varphi_u\}$, two states in $L^2(\mathbb{R})$ at two different times. When those states are arbitrary, the transition element has no probabilistic interpretation; it is just a (complex) scalar product of vectors. But we can, in particular, propagate a single state ψ_s to its future value by $\varphi_u = \exp(-\frac{i}{\hbar}(u - s)H)\psi_s = \int \psi_s(x) K(x, u - s, z) \, dx$, where H is the quantization of the Hamiltonian h. Then the integrand of (12) reduces to Born's probability density of the initial (or final) wave function. The integral kernel propagating forward (causally!) the initial probability $|\varphi_s(x)|^2$ in I is

$$P_F(s, x, t, dz) = (\bar{\varphi}_s(x))^{-1} K(x, u - s, z) \bar{\varphi}_u(z) \, dz \qquad (13)$$

for all x s.t. $\varphi_s(x) \neq 0$. But we could as well propagate backward in time Born's final probability density $|\psi_u(z)|^2$, via the kernel

$$P_B(s, dx, u, z) = \psi_s(x) K(x, u - s, z)(\psi_u(z))^{-1} \, dx \qquad (14)$$

Notice that if we were allowed to regard $t \mapsto \omega(t)$ as a well defined Markovian (stochastic) process, then, using (13) and (14), the integrand of Feynman's transition element (12) would satisfy a "detailed balance condition", one of the statistical expressions of equilibrium:

$$dx \, |\varphi_s(x)|^2 P_F(s, x, u, dz) = P_B(s, dx, u, z) |\psi_u(z)|^2 \, dz \qquad (15)$$

Of course, now, the classical consistency condition (9) in I cannot be true anymore since it means that the realized (extremum) trajectory is smooth everywhere in I. Moreover it uses a (dual) concept of momentum apparently obsolete in the quantum context.

Still a quantum deformation of (9) is available. It has been discovered by Feynman, in a time discretized way, as the following kinematical property (see [3]):

$$\left\langle \omega(t) \left(\frac{\omega(t) - \omega(t - \Delta t)}{\Delta t} \right) \right\rangle_{S_L} - \left\langle \left(\frac{\omega(t + \Delta t) - \omega(t)}{\Delta t} \right) \omega(t) \right\rangle_{S_L} = i\hbar \qquad (16)$$

where $\langle\cdot\rangle_{S_L}$ denotes the "expectation" with respect to the above-mentioned "process". If we were allowed to take the limit $\Delta t \to 0$ in (16) then the first time derivative in (16) should be a left hand one, like $p_*(q, t)$ before, and the second one a right hand derivative like $p(q, t)$. It is, therefore intuitively clear that the only way the difference on the l.h.s. of (16) could be non-zero, for our free quantum dynamics, is when $t \mapsto \omega(t)$ becomes very irregular. This is, indeed Feynman's way to show that the quantum trajectories are Brownian like. The beauty of (16) is that it is the space-time version of $QP - PQ = i\hbar$, i.e. of Heisenberg's uncertainty principle motivated by Fig. 2!

What is definitely missing for a probabilistic understanding of Feynman's ideas is the stochastic process itself and, therefore, the expectation $\langle\cdot\rangle_{S_L}$. But using (13) and (14) it is a simple exercise to find its profile: $\omega(t)$ should be a diffusion process (for Hamiltonians like the one considered here) with drift, or mean velocity, $i\hbar\frac{\nabla\bar{\psi}_t}{\bar{\psi}_t}$ (or $-i\hbar\frac{\nabla\psi_t}{\psi_t}$) and diffusion constant $i\hbar$, like the r.h.s. of (16). Following St. Anselm, however, we regard the existence as an important part of the perfection and so we feel compelled to look for what can really makes sense in Feynman's point of view.

Besides the existence problem there is another one showing us the way: to give boundary conditions at ∂I is not usual in the classical theory of stochastic processes. The future data excludes, for instance, the basic class of processes with independent increments (like Brownian or Poisson processes). On the other hand, the separation between past and future is sharp, here; this suggests that the process should still be Markovian. Coming back to our trivial example, we shall keep the classical drifts of (10) and (11) and just add a mathematically decent noise to Feynman's picture, namely, for $t \in I$

$$\begin{cases} d_*X(t) = \sqrt{\hbar}\,d_*W_*(t) + p_*^x\big(X(t), t\big)\,dt \\ X(u) = z \end{cases} \tag{17}$$

and

$$\begin{cases} dX(t) = \sqrt{\hbar}\,dW(t) + p^z\big(X(t), t\big)\,dt \\ X(s) = x \end{cases} \tag{18}$$

where W_* and W denote, respectively, Brownian motions adapted to our dual description. The diffusion coefficient $\sqrt{\hbar}$ is imposed by the above mentioned profile. Anyone of these (Itô's stochastic) differential equations can be solved explicitly. Their common solution is a Gaussian process, whose mean solves our classical boundary value problem (1) and (3). Its covariance is the one computed by Feynman using $\omega(t)$, after the substitution $t \mapsto it$ (the "Euclidean" or "Wick" rotation). The ("Bernstein") process $X(t)$ is Markovian, not of independent increment, but invariant under time reversal in the same sense as (1). The probabilistic counterpart of Feynman's kinematical property (16) in terms of the well defined expectation $E[\cdot]$ of $X(t)$ is

$$E\big[X(t) \cdot p_*^x(X(t), t) - p^z(X(t), t) \cdot X(t)\big] = \hbar \tag{16'}$$

If we relax the boundary conditions δ_x, δ_z at ∂I and give, instead, a pair of (strictly positive) probability densities at time s and u, the construction survives and provides all the well defined processes realizing Feynman's idea (12) of transition element on I. This is also the case if our starting classical particle is not free anymore but subjected to a force $F(q) = -\nabla V(q)$, for most of the potentials V of physical interest (see [4]).

It is interesting to reconsider Feynman's approach to the one-slit experiment in this new perspective. The introduction of a slit in the picture corresponds to a measurement of position of the (free) particle. To say that a particle, starting originally form the origin, has to be localized in the slit at a given time T in the future is a conditioning, in the traditional probabilistic sense [5, 6]. Then one verifies that this conditioning introduces indeed an irreversibility in an otherwise perfectly time-symmetric framework [7].

In general, any such process $X(t)$, $t \in I$, associated with an Hamiltonian H as before can be found in an interval A with the probability $P(X(t) \in A) = \int_A \eta^* \eta(q, t) \, dq$ where η and η^* are positive solutions of

$$\begin{cases} -\hbar \frac{\partial \eta^*}{\partial t} = H \eta^* \\ \eta^*(q, s) = \eta_s^*(q) \end{cases} \quad \text{and} \quad \begin{cases} +\hbar \frac{\partial \eta}{\partial t} = H \eta \\ \eta(q, u) = \eta_u(q) \end{cases} \qquad (19)$$

One checks easily that the drifts of $X(t)$ are the Euclidean translation of Feynman's ones. This is not a surprise since its above probability constitutes manifestly the Euclidean counterpart of (Born's) probabilistic interpretation of the state ψ_t. In this sense, our boundary value problem (19) mimics the way probability arises in quantum theory. Is it accidental?

A crucial theoretical test is to look for symmetries. Here, this means that knowing the pair (η, η^*) determining $X(t)$ we look for another one $(\eta_\alpha, \eta_{\alpha^*})$ determining $X^\alpha(t)$, for any α in $\mathbb{R}$. But then, clearly, we should have, $\forall t \in I$,

$$1 = \int_{\mathbb{R}} \eta \eta^* \, dq = \int_{\mathbb{R}} \eta \eta^* \frac{\eta_\alpha}{\eta} \frac{\eta_\alpha^*}{\eta^*} \, dq \equiv E\left[h^\alpha h_*^\alpha(X(t), t)\right]$$

The probabilists are familiar with such transformations $X(t) \rightarrow X^\alpha(t)$. They are called Doob's h-transforms (our notations in the last expectation are not arbitrary) and allow us to produce a large collection of Euclidean counterpart of quantum unitary transformations. The first integrals associated with those symmetries are martingales of $X(t)$. The concept of martingale is the closest analogue of constant of motion for a stochastic process. It is also, interestingly enough, the cornerstone of the mathematical theory of stochastic processes [5, 6].

The good surprise of this way to interpret Feynman is that it enables us to guess new quantum symmetries. Let us consider again our free particle. A particular one-parameter family of solutions of the second equation (19), for instance, is

$$\eta_\alpha(q, t) = e^{\frac{1}{\hbar}(\alpha q - \frac{\alpha^2}{2} t)} \eta(q - \alpha t, t), \quad \forall \alpha \in \mathbb{R} \qquad (20)$$

The simplest free solution is $\eta = 1$. Then the drift of Feynman's associated "diffusion" is zero, so we know that he is talking really about the Brownian motion. The relation (20) can be understood as $\eta_\alpha = e^{-\alpha N} 1$, for $N = t\frac{\partial}{\partial q} - \frac{q}{\hbar}$. $\eta_\alpha(q, t)$ is what the probabilists call the "exponential martingale". So

$$h_\alpha(q, t) = \eta_\alpha(q, t) = e^{\frac{1}{\hbar}(\alpha q - \frac{\alpha^2}{2}t)} = 1 + \frac{\alpha}{\hbar}q + \frac{\alpha^2}{2\hbar^2}(q^2 - \hbar t) + \frac{\alpha^3}{3!\hbar^3}(q^3 - 3\hbar t q) + \cdots$$

By successive differentiations with respect to α, at $\alpha = 0$, we find the collection of martingales of the Brownian motion. The quantum translation of this observation is that

$$Q(t), \quad Q^2(t) + i\hbar t, \quad Q^3(t) + 3i\hbar t\, Q(t), \quad \text{etc.} \tag{21}$$

for $Q(t)$ the position observable, in the sense of Heisenberg representation, are constants of the free quantum motion. Trivial as it is, this remark if far from being common knowledge.

The perspective sketched here (cf. [7] for more about this "Euclidean Quantum Mechanics" founded on Schrödinger's suggestion in [8], forgotten until the mid-1980s but periodically rediscovered since then: cf. L. Schulman's contribution in this volume, for example) suggests that it is indeed possible to think about quantum physics in probabilistic terms but that this is a rather subtle exercise. In part because, after A.N. Kolmogorov, the theory of stochastic processes itself has developed with an arrow of time in it, which is not natural in a quantum perspective. But the subtle exercise in question can be illuminating, for this reason, in probability theory and in quantum physics, since it leads us to question some generally accepted ideas.

One of the rewards of such a line of thought is precisely the fact that, on the Euclidean side, the problem belongs to regular statistical mechanics. It has been shown long ago (cf. [4] and references therein) that the unique difference with the "usual" construction of Markovian processes like $X(t)$ lies in our boundary conditions. As said before, to determine $X(t)$, $t \in I$ we need, in general, to give a probability density ρ_s at $t = s$ and another one, ρ_u at $t = u$. From this follows, indeed, a quantum-like structure suggesting, as we said, new results on the physical side. Is it a modest expression of the "Eternal Universe" mentioned by C. von der Malsburg? Or is it that, somehow, to understand better the structure of the probabilistic interpretation of quantum mechanics, one needs to think about a classical experiment already done, in our past? After all, it is not true that, for such a finished experiment, the nonlocality is much less shocking?

2 The (Random) Time We Would Like to Know

This section will be more speculative but will try to touch upon the heart of our subject matter: not only the direction of time, but its own nature.

It is not necessary, here, to elaborate on the fact that the two pillars of Modern Physics, i.e. General Relativity and Quantum Physics are irreconcilable. In fact, short after the heroic period of creation of the second theory, it was frequent to read very critical comments about the status of time in elementary Quantum Mechanics. For example, E. Schrödinger:

> Cette notion (beaucoup trop classique) de temps est un grave manque de conséquence dans la mécanique quantique ... abstraction faite des postulats de relativité. [8, p. 293]

or J. von Neumann emphasizing the:

> Chief weakness of Quantum Mechanics: its non-relativistic character. While the space coordinate is represented by an operator, the time is an ordinary number parameter. [9, p. 354]

It may seem strange that, 70 years after, this issue is manifestly not regarded anymore as worrying by most scientists (but cf. [10–12], for example). Is it, as suggested by T. Kuhn, that Theoretical Physics did not leave, yet, one of these long periods of "normal science" where the community tends to ignore difficulties seriously challenging accepted theories?

As well known, the difficulty in question is already obvious if one tries to understand the possible interpretations of Heisenberg's uncertainty principle when the canonically conjugate observables of position Q and momentum P are replaced by time T and energy H.

It was shown by W. Pauli, in his famous 1958 Encyclopedia of Physics article, that since the first version of uncertainty relation requires the spectra of both Q and P to be unlimited and the one of H should be, realistically, bounded below, T cannot be an observable in von Neumann sense.

Although the names of some famous scientists are associated with various attempts to puzzle out Pauli's observation, it is fair to say that no indisputable progress has been made on this basic issue.

But what about Feynman's formulation of Heisenberg's uncertainty principle? It is revealing that the father of path integral does not have anything like (16) to suggest as counterpart of the informal (Hilbert space) time-energy commutation relation. And, indeed, he complains that his framework "does not exhibit the important relationship between the Hamiltonian and time ([3, Sect. 7.7]).

Taking for granted that (16') is the mathematically consistent version of (16) it is clear that, to make sense of such a time-energy relation, we should have some random times to start with.

There is little hope to ever construct those directly in the Hilbert space framework of Quantum Theory, for two kinds of reasons. The first one is that we do not know at all where to look for observables which are not (denselly defined) self-adjoint operators in Hilbert space, i.e. von Neumann's observables.

The second one is related with the very shaky status of probability theory in Quantum Physics. This framework is supposed to describe quantitatively the ultimate kind of unpredictable phenomena, only accessible to a statistical analysis. And, indeed, the theory does this quite well, with a remarkable level of a precision in its statistical predictions. But, as far as probability theory is concerned, Quantum Mechanics in an embarrassing mystery: all the ingredients needed to construct a decent mathematical model of random experiments are missing. The above-mentioned existence problem with Feynman's "stochastic process" is, unfortunately, typical. The situation gets only worse when more complicated quantum systems are considered.

On the other hand, when the stochastic processes make sense, the concept of random time is a tool immediately available. According to Kai Lai Chung, in point of fact, "this is the single tool that separates probabilistic methods from others, without which the theory of Markov processes would lose much of its strength and depth" ([7, p. 80]).

Feynman is by far the theoretical physicist who tried hardest to turn Quantum Mechanics and Quantum Field Theory into theories involving fundamentally the tools of Stochastic Analysis instead of the ones of elementary (Newton–Leibniz) calculus [3, 13]. The failure of his "probabilistic" approach (the "Path Integral" approach) is very relative. Relative, in particular, to the scientific community in charge of its assessment. Many physicists do not understand why an approach allowing systematically to guess new results is not taken more seriously by some mathematicians. Those, however, would invariably answer that none of Feynman's path integrals (or processes) do exist.

Our hunch is that, using the well defined counterpart of Feynman's approach sketched in the first section, it will be possible to construct specific random times, corresponding to realistic experimental conditions.

Now, of course, such times would not be the quantum times we would like to know. Our Euclidean counterpart is only an analogue of Quantum Mechanics. But it seems to be a pretty good analogue; for example the "new" quantum constants of the free motion listed in (21) have been discovered directly via our probabilistic analogy. As a matter of fact, they are a very special case of a quantum Theorem of Noether providing systematically richer informations on quantum symmetries than the textbooks results on that matter (cf. [7, 14]). The same should happen with random times. Although such times are, indeed, immediately available on the theoretical (Euclidean) side, the algorithms involved in their computations are sophisticated, plunging into the heart of the theory of Markov processes and properties of their trajectories. Nothing, certainly, that Hilbert spaces should help us to discover.

If, as expected, a natural randomization of some specific times is possible, this new breach into determinism could open the way to the more radical ones needed to think simultaneously about Quantum Physics and General Relativity.

Acknowledgements It is special pleasure to thank the organizing committee of this really interdisciplinary meeting. Their initiative was exceptional and I hope that it will not remain so.

References

1. Heisenberg, W.: Über den anschaulichen Inhalt der quantentheoretischen Kinematik und Mechanik. Z. Phys. **43**, 172 (1927). Reprinted in Wheeler, J.A., Zurek, W.H. (eds.) Quantum Theory and Measurement. Princeton Series in Physics. Princeton University Press, Princeton (1983)

2. Feynman, R.P.: The space-time approach to non-relativistis quantum mechanics. Rev. Mod. Phys. **20**, 367 (1948)
3. Feynman, R.P., Hibbs, A.R.: Quantum Mechanics and Path Integrals. McGraw-Hill, New York (1965)
4. Albeverio, S., Yasue, K., Zambrini, J.C.: Euclidean quantum mechanics: analytical approach. Ann. Inst. Henri Poincaré **49**(3), 259 (1989)
5. Karlin, S., Taylor, H.M.: A First Course in Stochastic Processes. Academic Press, San Diego (1975)
6. Karlin, S., Taylor, H.M.: A Second Course in Stochastic Processes. Academic Press, San Diego (1981)
7. Chung, K.L., Zambrini, J.C.: Introduction to Random Time and Quantum Randomness, 2 expanded edn. World Scientific, Singapore (2003)
8. Schrödinger, E.: Sur la théorie relativiste de l'électron et l'interprétation de la mécanique quantique. Ann. Inst. Henri Poincaré **2**, 269 (1932)
9. von Neumann, J.: Mathematical Foundations of Quantum Mechanics. Princeton University Press, Princeton (1955)
10. Blanchard, Ph., Jadczyk, A.: Time and events. Int. J. Theor. Phys. **37**(1), 227–233 (1998)
11. Zeh, H.D.: The Physical Basis of the Direction of Time, 4th edn. Springer, Berlin (2001)
12. Muga, J.G., Sala Mayato, R., Egusquiza, I.L. (eds.): Time in Quantum Mechanics. LNP, vol. 72. Springer, Berlin (2002)
13. Malliavin, P.: Stochastic Analysis. Grund. der Math. Wiss., vol. 313. Springer, Berlin (1997)
14. Albeverio, S., Rezende, J., Zambrini, J.C.: Probability and quantum symmetries II. The theorem of Noether in quantum mechanics. J. Math. Phys. **47**, 062107 (2006)

Chapter 9
On Abuse of Time-Metaphors

Anindita Niyog Balslev

Abstract Given that 'time' is a pertinent theme for initiating a fresh conversation in a multi-cultural and multi-disciplinary context, this paper warns against the abuse of such time-metaphors as 'cyclic' and 'linear'. It points out how exactly in cultural discourses the misleading usages actually create havoc. Focusing on diverse models of time in major traditions that can be found in the history of Indian and Western philosophy, it shows that uninformed and simplistic usages of these time-metaphors give rise to misunderstandings in the domains of meeting of cultures and encounter of world-religions. It also draws attention to the scientific discourse, where these metaphors are equally in vogue but not always precise, encouraging the participants of scientific, philosophical, and religious traditions to exercise caution.

Keywords Time-metaphors · Cyclic · Linear

1 Introduction

Let me begin by thanking the organizers for inviting me to this conference where almost all the speakers are from the field of natural sciences. Given that over the years I have been involved with the deliberations regarding the theme of time that are documented in the history of Indian and Western philosophy, I indeed welcome this opportunity in order to draw your attention to the use and abuse of time-metaphors that are rampant in the literature, both in humanities and natural sciences. Indeed, there is a need for a greater conceptual clarification of time-metaphors that are inextricably intertwined with the question of 'direction of time'—the theme of this conference.

A perusal of the global history of ideas makes it apparent that reflections on time has a longer record than most other themes that have been topics of continued intellectual scrutiny across the boundaries of cultures. In fact, a conceptual preoccupation with this theme of time can be traced back within the frame of both Indian and Western thought to a period when a clear-cut disciplinary boundary between

A.N. Balslev (✉)
Elsdynvets 52, 8270 Høyberg, Denmark
e-mail: aninditabalslev@hotmail.com

S. Albeverio, P. Blanchard (eds.), *Direction of Time*,
DOI 10.1007/978-3-319-02798-2_9,

scientific, philosophical, and religious thinking was yet to be drawn. Those who are acquainted with this record are aware of the fact that concern for time has played an important role not only in the study of nature entailing physical processes but also as well in the attempts to explore a range of issues including the large theme of consciousness that play a decisive role in the study of cultures. In his impressive anthology on time, Charles Sherover rightly observes that "whether we are thinking of the nature of Nature or the nature of the self, we cannot escape thinking of the nature of time" [1].

Evidently I cannot here go into the details of the network of ideas that could be pertinently addressed demonstrating how in the philosophical literature the category of time is treated as ontologically real or as merely a conceptual construction or as inseparably intertwined with space and matter etc. and then analyze the large question concerning 'direction of time'. I only intend within the short compass of this paper to focus on such metaphorical designations as 'cyclic' and 'linear' time, precisely in order to caution against ambiguous and misleading usages of these metaphors. These expressions which are used profusely in the literature cutting across the boundaries of disciplines have had tremendous negative impact in the context of meeting of cultures and encounter of world-religions—a fact which may come to you as a surprise. We will see, in what follows, how this practice, which I call 'abuse of time-metaphors', has actually given rise to stereotypes and clichés that block an authentic understanding of the cognitive record and conceptual experience of time in cross-cultural contexts.

2 Time in Different Cultures

As a general background it may be recapitulated that at a rather early date in history the philosophical and religious traditions stemming from diverse cultures have been concerned with the enigmatic problem of time and have also anticipated many of the crucial issues which, since then, have remained topics for debates and discussions. A search to find answers to such questions as: whether time is a category independent of physical processes or not, whether it is observer-dependent or not, or how it is to be envisaged with regard to the notion of causality, or even how to formulate notions of timelessness and eternity—have given rise to a large number of views about time. Consequently the richness and enormity of the material that is available to us on this topic is simply overwhelming.

Given that in the global context one comes across a wide variety of views on time documented in the texts of highly articulate traditions of thought, references here are exclusively to views and conceptualizations from the Indo-European literature on the subject. As mentioned before, I am specifically seeking to draw attention to certain metaphorical designations of conceptual interpretations of time that are ambiguous or even directly misleading in a cross-cultural context. It is precisely because the differences in time-experience of diverse traditions are described by the use of such metaphorical expressions as 'cyclic and linear time', which one needs to examine how these metaphors have been used as well as abused. Since, as has been observed earlier, the theme of time is vital not only for the study of nature but also that of cultures, it needs to be specially noticed that the metaphors of cyclicity and linearity are utilized not only in natural and human sciences but also in the discourses of religious traditions. However, prior to focusing on certain misleading usages of time-metaphors and discussing why caution must be exercised in this respect, a few general observations may be made regarding the contending models of time in major traditions of thought.

A survey of the history of Indian and Western philosophy lays bare before a reader a great variety of views on time. Time being a fundamental concern, any effort to adequately comprehend these views show that this theme needs to be studied not only in isolation but in relation to other major concepts such as space, matter, motion etc. A reading of philosophical records unfailingly makes it evident that in any given conceptual system, specific views on time have important bearing on such concepts as those of being, non-being, causality, change and becoming and even that of consciousness. It can be seen that an analysis of time-experience entails conceptualizations not only of measurement of time but also of the three time-phases—past, present, and future.

However, the primary philosophical controversies that need to be looked at center around the basic positions that are taken regarding the idea of time itself. It is in this process of theory-making that thinkers have wondered about whether time is real or appearance, static or dynamic, discrete or a continuum. They further asked: Is time a distinct entity apart from processes, changes, and events? If it is, what makes time-divisions possible? Does time 'flow' (as in Newton's rendering) or is that merely a metaphor for describing conventional usage? Is time a mental construal? Can it be that the subjective experience of change is simply identical with the process of concrete becoming (à la Sankhya) which does not require the postulation of a category called time? More radically put, is there an objective category of time or is it a subjective construction or even illusory? Is time a relational concept (Leibniz)? There are plenty of examples of such views in Western and Indian philosophy either in support of or in opposition to any of these alternatives that are posed here. It is in connection with these various efforts at theory-making, one comes to employ notions of recurrence, unrepeatability, reversibility, and irreversibility. The implications of these notions are to be evaluated in conceptual settings that operate with different sets of philosophical interpretations of time.

In order to perceive the diversity in interpretations, let us take note of a few views that have emerged in different cultural contexts. The ideas of absolute time and relative time, for example, are present in philosophical as well as in scientific discourses. Recall Newton's view that dominated Western thought until the relativistic notion appeared on the conceptual scene. Newton, influenced by his predecessor Galileo's finding on motion, claimed that, to put it in his own words:

> Absolute, true and mathematical time, of itself, and from its own nature, flows equably without regard to anything external, and by another name is called duration; relative, apparent, and common time is some sensible and external measure of duration by means of motion, which is commonly used instead of true time; such as an hour, a day, a month, or a year. [2]

The old Vaisesika school of India advocated the view of absolute time (mahakala) which is said to be unitary (eka), all-pervasive (vibhu) and indivisible (akhanda). They claimed that the so-called time divisions, thereby the three time-phases, are made with reference to any standard motion, such as the solar motion, which was regarded as merely a conventional practice. Time per se (mahakala), they said, knows of no beginning or end or even of any movement. They argued that all that is contingent is in time, time does not rest in anything else (anasrita). Likewise all that changes is in time, time itself does not change or flow (niskriya). With regard to the latter observation, the position is at variance with that of Newton's reading that time "flows equably". The Vaisesika philosophers insisted that a philosophical analysis of the experiences of priority and posteriority, simultaneity and succession, quickness and slowness point to the ontological reality of time, without which these events are neither possible nor conceivable. Their ontology operated with the idea that without time we will be confronted with a static universe where no event can at all take place.

However, this notion of absolute time had to face the challenge of not only the Buddhist and the Jaina philosophers (who otherwise propounded views of 'discrete time' that come in radically different versions), it was also under attack from the quarters of other schools belonging to the Upanisadic tradition itself. I mention some examples below in order to show the complexities in the conceptual scenario and the eventual futility of describing all the views from the Indian sources under the single caption of 'cyclic time'.

Sankhya, held to be the oldest of all schools of Indian philosophy, had found it futile to postulate a notion of absolute time in its cosmological speculations that entailed notions of repeated creation and dissolution. However, it championed metaphysical dualism and saw all these processes as an interplay of two ultimate principles—Purusa and Prakrti, the former as constant and conscious and the latter as ever-changing and insentient Nature. The school accepted the ontological reality of change, entailing the notions of past, present, and future, upholding that such temporal usages can be explained with reference to the different stages of the unfoldment or evolution of Nature. They claimed that there is no need for the notion of an empty time as a separate category. According to Sankhya, nature is dynamic to its core, space-time-matter are combined in the same principle. The later advocates of Sankhya polemised against the Vaisesika philosophers by noting that the view of absolute, unitary time was not of much help as the Vaisesika philosophers themselves could not account for the experience of past, present, and future without a qualifying adjunct such as the solar motion. In other words, if the view of absolute time cannot account for such conventional temporal usages, then why not accept solar motion (that is, any standard motion) to be sufficient for that purpose and accept that time is an aspect of the causal process, as Sankhya advocated. The Yoga school, which joins Sankhya in its rejection of any notion of an absolute, unitary time, also accounts for the three time-phases as the potential/not-yet (future), the manifest/actual (present) and the sub-latent/no-longer (past) stages of the process that causally unfolds in nature. However, the Yoga school puts forward a discrete view. Since no two moments can be said to exist simultaneously, they maintained that such notions as collection of moments or an objective series is a mental structuring, a subjective construction, devoid of any reality. A still more radical stand that was held within the Upanisadic tradition itself is that by the school of Advaita Vedanta. This school advocated that the problem of change qua time was a problem of appearance, while projecting the notion of a timeless reality as ontological—an idea not unknown in soteriologies across cultures [3].

In all these systems, it is worth noting, causality is emphasized as a fundamental concept. There are several theories of causality, interpreting ideas of sequence, antecedence, and consequence entailing various speculations on the theme of time which has relevance for a comprehensive discussion on the issue of 'direction of time'.

3 Time Metaphors

Let us now turn to the question of time-metaphors. Note that the two dominant poles of human experience concerning recurrence as well as unrepeatability of events have often been expressed through these imageries of cycles and lines/arrows across cultures. These are present in the everyday discourse such as in the case of awareness of the recurrence of cycles of seasons or the irreversibility of the process that leads us from birth to death. Leaving aside all the complexities and technicalities that the questions of reversibility and irreversibility of cosmological processes with all their implications pose to enquirers of certain special sciences, let me draw attention at this point to the sort of readings that have been made while depicting diverse cultural experiences of time with the help of such simple metaphorical designations as cyclic and linear time. It is commonplace to say, in a confrontation of the pre-Christian Greek, the Judaeo-Christian and the Indian traditions regarding time, that the Greek/Hellenic and the Indian traditions have cherished a cyclic conception of time whereas the Judaeo-Christian traditions have maintained a linear conception. Precisely because of such widely used designations, the implications of time-metaphors become a pressing issue for anyone involved in cross-cultural studies or concerned with the endeavor toward an authentic encounter of world-religions. A further investigation into this question also shows—which I have discussed elsewhere in greater detail [4]—why there is an urgent need for a proper propagation of the wide range of theoretical moves concerning time in various traditions as well as an exploration of the important bearings of metaphors in connection with time-experience.

Before examining the reach of the appellations of cyclic versus linear time, let us ask, what does the metaphor 'cyclic' or 'linear' entail about a notion of time? These descriptions, obviously drawn on the analogy of geometrical figures, are intended symbolically to represent poles of time-experience, which entail the notions of recurrence and reversibility as well as unrepeatability and irreversibility. However, it is indeed surprising that granted that in a multi-cultural and inter-religious framework, these designations of 'cyclic and linear time' are repeated endlessly, to the extent that these have virtually become clichés, it is not quite so easy to lay one's hand on a proper formulation of the conceptual contents of these metaphorical usages. Although some loose observations are obtainable regarding their significance, these are not sufficient for a proper evaluation. It has been said, for example, that the 'linear' notion of time in the Biblical context implies that time has a beginning and an end, whereas the cyclic does not grant that. Some consider 'cyclic' time as advocating reversibility, others point out that the order of events as such is irreversible, but any future event is also to be seen as a past event since the beginning and end coincide in a cycle etc. Some even conclude that 'linear' time makes room for progress and history meaningful, freeing it from mechanical recurrences. It is evident that any attempt at a precise formulation would indicate the complexities and difficulties that are involved in such geometrical representations of the understanding of time in different traditions.

These readings do not seem to be of great help. It can indeed be argued that a more precise use of time-metaphors can aid and enhance the process of understanding both nature and cultures. On the other hand, misleading usages of the same can cause grave misunderstandings in the context of meeting of cultures and that of world-religions that have their own distinct conceptual histories, as will be indicated below. This is also why a further investigation into the question is needed in order to explore the important bearings of such metaphors. Time-experience is as such enigmatic, it does not help to further confuse the situation through the use of time-metaphors that leads to setting up traditions as though these are 'diagrammatically opposed'.

A careful scrutiny of these designations led me to the important passage of St. Augustine. In his well-known work, entitled, *The City of God*, Augustine refers to a specific Greek view, which he describes as that of 'circular time'—a view that he challenges. This passage is one of the most interesting of early documents that enables the reader to grasp and eventually analyze the specific implications of what Augustine and those who followed after him understood by that time-metaphor. Referring to a specific model, which of course any informed scholar knows to have been present only as one among various other contending models in ancient Greece, Augustine writes:

> ...Those others think, the same measures of time and the same events in time are repeated in circular fashion. On the basis of this cyclic theory, it is argued, for example, that just as in a certain age the philosopher Plato taught his students in the city of Athens and in the school called the Academy, so during countless past ages, at very prolonged and definite intervals, the same Plato, the same city and the same school with the same students had existed again and again... [5]

The appellation of 'circular or cyclic time', as is clear from this passage, is to be read as entailing the idea of exact mechanical recurrence of not only cosmological processes but also that of individual destinies. Obviously there cannot be room for any genuine progress, let alone a sense of history or that of salvation in such a worldview. Augustine, quite understandably, repudiates the position while highlighting the Christian contribution to the religious interpretation of time.

The record of this discussion regarding what is entailed in the notion of cyclic time in the Greek context is helpful for investigating the pertinence of the label of cyclic time in the Indian conceptual context. This enables one also to judge whether it makes any sense to ascribe such a view to the Indian conceptual world, given that the import of the label of "circular time", as Augustine describes, is not only repetition of cosmological processes but a mechanical recurrence of particular phenomena and of specific events, involving human destinies as well.

It is indeed in this process of encountering such interpretations of these metaphors that one cannot but take note of how cycles and arrows/lines gradually cease to be simple time-metaphors projecting notions of reversibility and irreversibility, but come to get associated with such concepts as those of history, of progress and even of salvation. A possible interpretation of direction of time in the context of study of nature thus comes to show a different face in the context of study of cultures through such metaphorical usages.

Let us look closer into the world of Indian thought for obtaining a clearer picture. Note that the idea of creation ex nihilo is absent in the Indian traditions whereas the notion of world-cycle is a general feature of Indian conceptual world. This idea is also present in Indian mythology and philosophy, as it has been in ancient Greece. However, the Indian conceptual world has its own distinctness, its own complexities. In the literature of the Puranas for example, one encounters a grandiose conception of the cosmological process where the universe is conceived as undergoing repeated creation and dissolution. The time-span of a world-cycle is calculated in terms of billions of human years, divided, and sub-divided into periods denoted as 'manvantaras', 'mahayugas', 'yugas', etc.

However, it is important to notice that the world-cycles in the Indian context can be compared to one another only in terms of generic similarity just as one day resembles another, but the idea of exact repetition involving the return of the particulars does not occur in the texts. The Greek model of 'circular time', referred to by Augustine, is wholly deterministic. It is supportive of the notion of pre-destination which nullifies the power of efficacy of human actions and consequently renders a fatalistic picture of the human situation. This is not the case in the Indian renditions. It is important to note that the idea of mukti or salvation, which comes in many versions, as well as the idea of karma, emphasizing efficacy of human actions, are pan-Indian concepts which leave no room for an interpretation as projected by that specific Greek model of 'circular time'. Granted that there are some common ideas present in the ancient conceptual worlds of the Greeks and the Indians such as those of world-cycles, transmigration, developed in strict adherence with the principle of ex nihilo nihil fit, there are some distinct features of each that must not be lost sight of. However, a lack of awareness regarding the spectrum of views on time that developed in Indian thought combined with the presence of certain similar ideas that are also found in ancient Greece, misled even some of the significant western culture historians (as Toynbee), theologians (as Tillich) in the West in their rendition of Indian conception of time. In this connection, it is also particularly important to make a clear conceptual distinction between the idea of 'cyclic time' and that of 'cosmological cycles'.

Recall that a review of the history of Indian philosophy shows that while sharing the idea of repeated creation and dissolution in their cosmological speculations, a number of schools actually held sharply divergent views with regard to time, some of which have been discussed above. Indeed, a global history of ideas discloses that no major philosophical tradition has an unanimous view of time. A cursory glance at the history of Western philosophy discloses to a curious reader a wide range of views, such as the notion of absolute time, time as a relational concept, time as process etc. Similarly, there are a number of contending views about time documented in the history of Indian thought. The Upanisadic, the Buddhist, and the Jaina traditions all know of internal divergences in that respect. The contrast of ideas is indeed awe-inspiring—at one end of the scale there is a unitary view of time whereas at the other end a pluralistic view. Some have maintained the objective, independent reality of time, while others have counteracted this stand urging that even to assert the reality of change does not necessarily require the postulation of time as an independent ontological category; still others have maintained time to be phenomenal, having no ontological status. One even encounters, as in the Buddhist tradition, the startling assertion that being and time (the moment and the momentary) coalesce ontologically and their separation is nothing more than an arbitrary linguistic convention. History of philosophy in India and in the West has equally witnessed schools of thought that supported a view of ultimate reality as free of all change and becoming at the ontological level. This is exemplified in the Eleatic school of Parmenides as well as in Advaita Vedanta. However, there are others who strongly objected to that idea, such as the Buddhists as well Heraclitus and his followers, insisting that there is nothing whatsoever which is exempt from change.

However, such well-known culture historians as Arnold Toynbee, for example, seem to be totally unaware of the diversity of views regarding time and the implications of the notion of 'cosmological cycles' in the Indian context [6].

Toynbee characterizes the cyclic image of Hellenic and Indic civilization as 'a counsel of despair for humanity' and remarks in his A Study of History, 'This philosophy of sheer recurrence, which intrigued, without ever quite captivating the Hellenic genius, came to dominate contemporary Indic minds'.

How a built-in theoretical bias as one's starting point, which often is due to a lack of information, blocks cross-cultural and inter-religious communication is demonstrated when Toynbee further asks:

Are these 'vain repetitions' of the Gentiles really the law of the universe and, therefore, incidentally the law of the histories of civilizations? If we find that the answer is in the affirmative, we can hardly escape the conclusion that we are the perpetual victims of an everlasting cosmic practical joke, which condemns us to endure our sufferings and to overcome our difficulties and to purify ourselves of our sins—only to know in advance that the automatic and inevitable lapse of a certain meaningless measure of time cannot fail to stultify all our human exertions by reproducing the same situation again and again ad infinitum just as if we have never exerted ourselves at all.

All these clearly show that instead of continuing with stereotypes and clichés that any given tradition upholds concerning the 'otherness' of other traditions, it is now crucial that we create opportunities which allow us, as this conference has done, to obtain deeper insights into the historic consciousness of major thought-traditions. Time is not only a multi-dimensional issue, it is of great significance for the self-understanding of cultures in which religions play a major role.

Record show that effort is made to defend and to maintain a cohesive understanding of religious interpretation of time in diverse traditions in different cultural soils. St. Augustine, for example, while being deeply engaged in reflection on time, came up with a triumphant declaration that he has finally found an answer to the "Why not sooner" question through the comprehension that God does not create in time but is Himself its source, its creator.

Again, Leibniz, while forwarding a relational concept of time, claimed that he was providing not only an alternative to the Newtonian concept of absolute time but was also answering the "Why not sooner" question. If one asks such a question, Leibniz wrote,

> We should reply that his inference would be true if time were something apart from temporal things, for it would be impossible that there should be reasons why things should have been applied to certain instants rather than to others, when their succession remained the same. But this itself proves that instants apart from things are nothing, and that they only consist in the successive order of things.

This is also why today when certain physicists and cosmologists speak of such ideas as 'commencement of time' or even of an 'absolute beginning' of the universe in support of a specific religious tradition and try to circulate these ideas in the forums of 'science-religion dialogue', some effort seems necessary to articulate clearly the conceptual implications of such positions in a manner that, philosophically speaking, do not appear as questionable in a multi-religious context.

A conceptual transparency needs to be achieved through an open discussion in a multi-disciplinary framework so that a proper diffusion of ideas can rectify the distortions and help appreciate the insights into this large and abstruse question of time [7]. A fuller exploration of the implications of major metaphors in scientific discourses is also to be welcome for achieving clarity and precision about these issues.

One could perhaps mention in this connection that in the current attempts to construe physical theories, it should be made clear whether irreversibility of time and irreversibility of processes are held to be identical or different. The usages of the metaphor of arrow (recall on the very first day of the conference it was mentioned that in a given scheme one could classify seven ways of using it) and even of the word 'time' in different contexts of discussions leave one wondering about whether the referents of these words have the same conceptual content in each case. It is not always clear whether time is meant as a distinct principle or as inseparable from processes—both views are documented in the history of philosophy. It will be philosophically more interesting if physicists could interpret a remark such as that of Wheeler that the 'concept of time is a human invention' (Wheeler) or the implications of the discussions in physical sciences where the unidirectionality of time has been perceived to be an assumption which underlies classical and relativistic physics as Schrodinger noted. In his work, entitled, *Mind and Matter*, he observed that "The theory of relativity ... however revolutionary leaves untouched the unidirectional flow of time which it presupposes while the statistical theory constructs it from the order of the events".

As long as attempts to account for the irreversibility of time continue to be a topic for an ongoing enquiry by physicists, an analysis of the idea concerning the 'direction of time', as conveyed by the metaphor of 'arrow', is also called for. The differences among the leading physicists such as Wheeler, Prigogine, and Penrose show the great difficulties involved in the search for an understanding of an underlying physical reality which may explain the so-called 'arrow of time'.

Cycles and arrows are major metaphors which form part and parcel of not only everyday discourse in various contexts, they appear and reappear—sometimes assuming technical significance—as in the frame of specific disciplines such as physics, cosmology etc. All these usages are modes of representations of the two fundamental poles of human experience viz. recurrence and irreversibility. It is short-sightedness to think that any cultural tradition makes exclusive use of one metaphor at the expense of the other. Stephen Jay Gould recognizes that. While acknowledging the arrow as the major metaphor of the Western culture, he reclaims the place of time's cycles and quotes from the Book of the Ecclesiastes in order to confirm that time's cycles is an idea that has a religious foundation. Cycles and arrows, he says, are "so central to intellectual (and practical) life that western people who hope to understand history must wrestle intimately with both" [7].

I am persuaded to think that greater academic involvement with the question of time, cutting across the boundaries of disciplines and cultures, will lead to a deeper understanding of this multi-layered theme. Time indeed is one among the most pertinent topics for initiating a fresh conversation among the participants of scientific, philosophical, and religious traditions.

References

1. Sherover, C.M.: The Human Experience of Time. Northwestern University Press, New York (1975)
2. Newton, I.: Philosophiae Naturalis Principia Mathematica. J. Societatis Regiae ac Typis J. Streater, London (1687)
3. Balslev, A.N.: A Study of Time in Indian Philosophy, 3rd edn. Munshiram Manoharlal Publishers, New Delhi (2009)
4. Balslev, A.N. (ed.): Religion and Time. Brill, Leiden (1993)
5. Augustine, St.: De Civitate Dei, trans. by P. Levine, Cambridge, Mass (1966)
6. Toynbee, A.: A Study of History (abridged). Oxford University Press, New York (1972)
7. Gould, S.J.: Time's Arrows and Time's Cycles. Harvard University Press, Cambridge (1987)

Chapter 10
The Direction of Time Ensured by Cosmology

Hervé Barreau

Abstract Cosmology gives us two ways for considering the direction of time, which means an overall development of physical reality from past to the future. Firstly there is an overall development if, coming from a beginning, the evolution of the universe is a function of the cosmic time which gives us the age of the universe. Secondly there is an overall development if cosmic time cannot only date the eras of the universe but also explain the birth of the main complex structures we find inside it. In this second manner, which presupposes the first, the expansion of the universe during cosmic time is a cooling factor which permitted the breaking of the symmetries discovered by theoreticians while studying the different interactions. These breakings were responsible not only of the four distinct interactions but also of the various degrees of physical reality. Some metaphysical reflections are unavoidable in the view of such a history.

Keywords Life-story · Irreversibility · Thermodynamics · Branch systems · Cosmic time · Unification and desintegration of primitive forces · Anthropic principle

We speak of irreversibility for physical processes and of the direction for time. This means that characterizing the irreversibility of these processes and assigning a direction in time are different things. The difference lies in the fact that processes are irreversible in time considered as a natural or space-time dimension, whereas time itself is conceived as a global development, extending only from the past to the future, from a beginning to an end or to infinity. The physicist's problem is to decide whether it is legitimate to believe in such a global development as it was postulated by Newtonian physics in the form of real time, mathematical and absolute.

This problem is particularly evident in the study of nature. In our own individual lifetime the difference between the irreversible processes which we experience and our overall development scarcely presents a problem. Our age makes the dif-

H. Barreau (✉)
CNRS, 23 rue Goethe, 67000, Strasbourg, France
e-mail: hbarreau@noos.fr

S. Albeverio, P. Blanchard (eds.), *Direction of Time*,
DOI 10.1007/978-3-319-02798-2_10,
© Springer International Publishing Switzerland 2014

ference. Of course it is easy to isolate ongoing and irreversible processes such as the progress of a career or a scientific achievement, but it is scarcely more difficult to assign a tangible and immutable place to any event in a lifetime. Historical biographers are guided in their work by the fact that every human life follows a unique course from birth to death and the impact and effect of events, predictable or otherwise, on the life of the individual depends first and foremost on the exact time location upon which they take their place. If a historian cannot date an event in a life precisely he cannot claim to have reconstituted the unique life history of a particular individual, but this task is possible of achievement in principle and we can always give reasons why it has not been achieved in a particular case (generally by lack of documents).

Could this be the same for the study of Nature? We see the difference immediately. In the biography of an individual it is his legitimate postulated identity which enables us to assign him a life history, tangible to the extent that it has been lived; nothing allows us, a priori, to assign such a life history to Nature, based on a unique support, oriented in a single and irreversible direction. We can even say that, despite Newton's reassuring stance in making time an absolute development, modern physics has conspired to cast doubts on the unity of a global process by linking time to concrete processes subject to specific laws. With Einstein's relativity theories, the doubt even seems to have given way to an opposing belief. In the context of special relativity, each reference system refracts the development of the others in its proper perspective, so that the temporal order of two space-like separated events may differ from one reference point to another. In the context of general relativity the space-time curve imposed by the presence of mass-energy obliges us even to measure time differently at each point in space-time, so much so that individual futures can no longer be compared. For Einstein himself, these individual futures were pure illusion, at least in the final expression of his thinking.

Nevertheless it is relativity itself which has enabled the reintegration of a universal past and future and, thereby, of a single direction for time. This is a paradox worth considering. On one hand, in linking the measurements of time ever closer to phyisical processes, relativity has obliged us to abandon the belief in Newtonian metaphysical time; on the other, in rethinking the approach to cosmological problems, it allows us to look at the universe as a developing whole, a global process with a unique history and oriented in a single direction, like a living being.

We can see, therefore, that there are two problems regarding the direction of physical time as revealed by cosmology and we must take account of the fact that the solution of the first authorizes the onset of the second. The first problem is this: how can we be sure that relativist cosmology offers us a unique universe with a beginning and a future, thus confirming the irreversibility of the processes which unfold in it? The second problem is the following: on this basis, can we imagine solutions to the various problems in theoretical physics, such as the diversification of forces and the formation of chemical elements, for which we have been unable to offer a plausible justification up to now? We will see how relativist cosmology answers these two questions.

1 The Cosmic Arrow of Time

The answer to the first question invites us to review the historical stages of contemporary cosmology and the successes it has obtained from our point of view.

First of all, to be sure, we must consider the universe as a whole. The cosmological principle that the universe is, on the whole, homogeneous and isotropic, is sufficient. It is a reasonable assumption and Einstein himself had recourse to it in his *Cosmological Considerations* (1917). However, in predicating a static and homogeneous density of matter in a finished though endless universe, Einstein came up against the problem that Newtonian cosmology had failed to resolve even in his time: the intensity of the gravitational field must increase to infinity. Einstein met this difficulty, firstly, by distinguishing time from the three spatial dimensions, and secondly, by imagining a cosmological constant Λ which must prevent the universe from collapsing in on itself—hence the image of a "cylindrical" universe with neither beginning nor end. We should remember that time, distinguished from the three spatial dimensions, becomes "cosmic time", to use the terme applied to it by H. Weyl in 1923.

However, in 1922, Friedmann noted that there was no static solution to the ten equations of the new relativist theory of gravity, as applied to the universe as a whole. This, furthermore, was one of the reasons for which Einstein had been obliged to introduce a cosmological constant. Friedmann suggested various models of relativist cosmology (taken up later by Robertson and Walker), by supposing that the average density of the matter which filled the universe varied with time. Three-dimensional space corresponding to the model would be spherical and closed, hyperbolic and open or else Euclidian, depending on the value of a cosmological constant k.

Up to then cosmology had been purely speculative, as in Gödel's later cyclical model. It was Lemaître who, in 1927, imagined putting the expanding universe model which he had adopted in correspondence with the distancing of the galaxies, demonstrated by the red shift of their atomic spectra, which he interpreted as a Doppler effect. This distancing of the galaxies, which Hubble had demonstrated in 1924, received outstanding confirmation in 1929. An expanding universe has been the relativist cosmological model ever since.

It was very fortunate that this expanding universe model was confirmed in 1965 by the discovery of fossil cosmological radiation of 2.7 K. It was also confirmed from other sources, such as the probable age of stars and galaxies and the proportion of chemical elements in the cosmos, not to speak of the Olbers paradox regarding the scattered brilliance of stars in the cosmos, which is explained by this expansion. All these facts enabled us to calculate the age of the universe, estimated between 13 and 15 billion years.

What concerns us more closely in our investigation into the direction of time is that the expansion theory (the "Big Bang") itself offers the direction we lost when we were obliged to abandon Newton's absolute time, which science rejected as metaphysical. Expansion involves the dispersal of the initial energy and an increase in the global entropy of the universe as it proceeds to cool. In our opinion it is this cosmic need for increasing entropy which must be regarded as the ultimate source of any detectable increase in entropy in any experiment, regardless of the level.

This question is worth discussing, as physics, a science based on experiment (even more than on principles, no matter how necessary and reasonable these must be), is in a delicate situation regarding the direction of time. For all human experience, whether physical or moral, is irreversible. We cannot imagine that the repetition of the same act can constitute an objection to the general application of such a law, as this repetition creates a habit. In any event we must acknowledge that everything we observe in the universe bears the stamp of irreversibility from the very fact we observe it, as we never observe exactly the same event twice. It is significant, furthermore, that Bertrand Russell thought that our belief in the direction of time was based on the generalization of this subjective experience of irreversibility. However, this means that the belief is based on the future-oriented nature of this experience, as we noted earlier, and this is obviously not the experience of the entire universe. This basis has always seemed insufficient to physicists who aim at an "objective" science dealing with physical reality, and therefore independent of human observation, and it is understandable that they were not content with classic thermodynamics, which is phenomenological, even in the expression of its second law on the increase of entropy in a closed system. They wanted to base thermodynamics on statistical mechanics, which also brought new results. But with regard to the second law, in particular, the recourse to statistical mechanics is both despairing and desperate and Boltzmann, among others, lost his robust faith in it. When we have effectively reduced entropy to a probability, we have not taken a step towards objectivity—we have aggravated the problem. The question is not that the notion of probability introduces an approximate and subjective element of knowledge, as this subjectivity can almost be reduced to objectivity if we examine it closely. It is that probability, regarded in its most objective aspect as a carefully defined mathematical notion, is indifferent to the flow of time. Probability makes entropy intemporal. We have not gained the direction of time—in fact we have lost it. The Ehrenfests [1] and van der Waals [2] were aware of this, as was the philosopher Reichenbach, who was confident at first of the habitual direction of causality, and finally became sceptical regarding the theoretical advantage to be gained from the physical experience of time. All these authors believe that entropic growth is due to the initial conditions of a system which finds itself relatively isolated thereafter. We must therefore introduce the theory of *branch systems* developed by Reichenbach [3] and Grünbaum [4] to complete the mechanistic theory of entropy. It must be admitted that the branch systems in which entropy increases find themselves in a state of relatively low entropy when separated from their environment. For the greater majority of them, therefore, there is no doubt at all about the direction of time.

It is the expansion of the universe, therefore, which provides the most general framework and ensures the effectiveness of the speculative branch systems theory. Entropy was low in the universe in the past—it is the general dispersal of energy which will entail a general rise in entropy and a fall in the intensity of cosmic radiation in the future. Even if the Big Bang is followed by a Big Crunch for obvious gravitational reasons, the second universe born of the first (if rebirth is possible), will be generally in a higher entropic state than the first from the outset, as Tolman [5, 6] had demonstrated. As all forms of physical irreversibility appear to be closely linked to thermodynamic irreversibility, we see that the direction of cosmic time involves all forms of irreversibility and ensures the prevalence of the cosmic time arrow, the only thing which gives it direction. This is the "master arrow time" as Professor Zeh wrote [7].

2 The Effects of the Cosmic Time Arrow

Relativity had consequences for research in two directions: cosmology, as we have just seen, and the unification of physical interactions, as we will also see, as this question brings us back to cosmology and the cosmic time arrow. In this direction, however, the first attemps were disappointing. Einstein devoted the last 30 years of his life to them without success. H. Weyl failed to unite relativist gravitational interactions with classic electromagnetic interaction, though he discovered the promissing role of gauge symmetries during his attempt. Things took on a new perspective when quantum field theory was accepted as the appropriate theoretical framework for union. Glashow, Salam, and Weinberg were therefore able to unify quantum electrodynamics and weak nuclear interaction in 1967. Then came the idea of combining them with strong nuclear interaction in a "great unification" theory. Then it was hoped to combine this unification with gravity by using string theory. However, combining fermions and bosons in "great unification" called for a particular kind of symmetry, the supersymmetry—hence the superstring theories arose around 1985, later revived in their combination in the M theory and their application to gravity in 1995. Theoretically speaking, therefore, we seem to be close to unifying the four fundamental forces and we refer to a "theory of everything", but it is impossible to obtain experimental data. This is because we are dealing with the conditions which must present at the very beginning of the universe, even before the "Big Bang". Thus cosmology obtains a new role, which we can describe, if not explain exactly, and this description runs like the following scheme.

The Big Bang we place after the "Planck's Time", when the age of the universe was 10^{-42} s in the classic model. It is clear that general relativity and quantum theory must have been intermingled during this "Planck's Time", as the theoreticians imagine. They believe that this theory implies that the primordial strings were twisted through 10 or 11 spatial dimensions, which were rolled in on themselves. The "Big Bang" can then be explained by the fact that strings which twisted through only three spatial dimensions could collide and be annihilated (between strings and anti-strings) and could free these three dimensions, which then dilated to produce an universe. It was the first break in symmetry and it inaugurated cosmic time. Planck's "nut" was broken and the universe as we know it was born with cosmic arrow pointing towards expansion. It is to be observed that with the theory of quantum gravity we obtain the same phenomenon, as Professor Kieffer has here demonstrated.

What we must remember in such a hypothesis is that the direction of time was not only the prime director for all the time arrows that experimental or theoretical physics can detect, as we have tried to show earlier, but that it also governed all the remarkable events which shaped our universe; the universe cooled as it expanded and the progressive drop in temperature was responsible for all the structures we know today.

The first question relates to the disintegration of forces by successive breaks in symmetry. It seems that gravity was separated from the other forces when temperature fell to 10^{32} K and the strong nuclear force separated from the electro-weak at 10^{28} K. These two disruptions gave rise to a phenomenon of very sudden expansion, which we term "inflationary" and which Guth and Englert postulated in the 1980s to explain the similarity between very distant regions of the universe which we cannot connect by causal influence because of the finitude of c. This brings us to 10^{-32} s in the age of the universe. The X bosons and anti-X bosons, which were exchanged when the strong nuclear force combined with the electro-weak force, disintegrated over two short but different periods of time, and this difference was sufficient to annihilate the antiquarks by fusion with quarks, leaving a persistent residue of surplus quarks. This can explain why anti-matter only exists in cosmic radiation and under the artificial conditions of some giant accelerators. The weak nuclear force separated from the electromagnetic force at a temperature of 10^{16} K and quarks fused to form nucleons (i.e. protons and neutrons) at 10^{12} K. This brings us to around 10^{-7} s in the age of the universe. This era has been called the "particular" or "first second" era [8]. As far as we are concerned it is the era when the four fundamental forces were separated one after the other.

Thereafter we must wait until the temperature reaches 10^{10} K for the beginning of a new era, known as the "nuclear era", and the formation of deuterium nuclei, and subsequently of tritium, helium 3 and lithium 6, all of which were formed from deuterium. We call this the primordial nucleosynthesis; it lasted for the first three minutes of the life of the universe. We know that the nucleosynthesis of the other elements occurred later inside the stars.

During the third or "radiative" period the temperature fell by several million degrees to 10,000 K. Photons were emitted constantly and absorbed by electrons, which formed a cloud which was independent of that of the protons and the few nuclei formed during the primordial nucleosynthesis. At the end of this era hydrogen ceased to be ionized and became "atomic"; electrons began to orbit protons and lost their power to interact with photons, which they left to circulate in space thereafter. Such was the beginning of diffuse cosmic radiation, which appeared when the universe was 300,000 years old and which we discovered in 1965.

We could continue this genesis of the universe as we know it by describing the fourth "material" ou "stellar" era and the formation of the galaxies, the stars and the planets which might support life but as far as the direction of time is concerned it has nothing further to teach us about its evident power to generate the forces and shape the principal structures of our environment. These were conditioned by the fall in temperature during the expansion and we have seen that this was decisive. This seems to be the "true story" of the birth of our universe, and we see that the expansion and cooling time was the regulator.

3 Conclusion

In conclusion I would like to return to our initial comparison between physical/cosmological time and personal time. The unity of time for a person lies in the fact that it continues from his/her birth to his/her death, constantly enriching him/her, constantly changed by his/her memories, his/her anticipations ans his/her experience of present events. Apparently there is nothing like this in cosmic time. Nevertheless the universe also has an age, and we have seen that this age is very important because the progressive cooling factor.

Also some reflexions seem appropriate and in some manner inescapable. When we consider the marvels which were created as the universe cooled we cannot help but share Dyson's opinion: "When we look at the universe and identify the multiple accidents of physics and astronomy which have worked together for our benefit, it all seems to have happened as if the universe must somehow have known that we had to appear" [9]. Steven Weinberg echoes this comment, even if he does not agree with it, when he writes: "It is almost impossible for human beings not to believe that they have a special relationship with the universe, that life is not just the grotesque result of a series of accidents extending back into the past to the first three minutes, that somehow we were intended from the beginning" [10].

Obviously such reflexions are metaphysical and do not belong to physical science. Nevertheless they have their use in assessing our must speculative physical theories. Because they are mathematical these theories have a tendency to neglect the course of time and to see it only as an accident, but Brian Greene, who not only contributed to the string theory, but also to our thinking about its scope, does not believe that such theories can replace the considerations which stem from the *anthropic principle* and from the imagination of *multiuniverses*, which he places (quite rightly as it seems to me) in the same thought register [11]. Certainly it is doubtful that a future theory can encompass all the initial conditions necessary for the development of the universe. On the other hand it is obvious that this evolution presents us with cosmic time, which produced formidable results from limited resources. Far from tending to discourage us, the vision of an *elegant universe* should enable us to give the appreciation it deserves to this time direction. For the future of earth and mankind, this time direction is largely in our hands and we seem to be called upon to use it in a manner worthy of the great epoch of which we are the heirs. In any event, we can no longer profess ignorance of this heritage.

References

1. Ehrenfest, P., Ehrenfest, T.: Begriffliche Grundlagen der statistischen Auffassung in der Mechanik. Encyklopädie der mathematischen Wissenschaften, vol. IV, 2, II, pp. 41–51 (1909)
2. van der Waals, J.D.: Über die Erklärung der Naturgesetze auf statistischen Grunlage. Z. Phys. **12**, 547–549 (1911)
3. Reichenbach, H.: The Direction of Time, pp. 135–143. University of California Press, Berkeley (1971)

4. Grünbaum, A.: Philosophical Problems of Space and Time, pp. 254–264. Reidel, Dordrecht (1973)
5. Tolman, R.C.: Relativity, Thermodynamics and Cosmology. Clarendon, Oxford (1934)
6. Davies, P.C.: The Physics of Time Asymmetry, pp. 80–109, 185–202. University of California Press, Berkeley (1977)
7. Zeh, H.D.: The Physical Basis of the Direction of Time, p. 124. Springer, Berlin (1989)
8. Audouze, J.: L'Univers. Que sais-je?, vol. 687, pp. 105–111. PUF, Paris (1997)
9. Dyson, F.J.: Sci. Am. **225**, 51 (1971)
10. Weinberg, S.: Les trois premières minutes de l'univers (french edition), p. 179. Seuil, Paris (1978)
11. Greene, B.: L'Univers élégant (french edition), pp. 394–401. Laffont, Paris (2000)

Chapter 11
Asymmetries, Irreversibility, and Dynamics of Time

Luciano Boi

> *Non in tempore sed* cum *tempore Deus creavit cœlum et terram.*
> *Saint Augustine*

> *A moment not out of time, but in time, in what we call history:*
> *transecting, bisecting the world of time, a moment in time but*
> *not like a moment of time.*
> *T.S. Eliot*

Abstract The paper wants to address some conceptual issues concerned with the finding of the fundamental role played by the phenomenon of breaking symmetry in different natural processes. We also shall discuss a certain theoretical problem that poses the asymmetrical nature of time in a manifold of scientific domains. In the second part of the paper, we describe some fundamental features of the action of time in the framework of dynamical systems and irreversibility. This article is aimed at showing some features of the dynamics of time into diverse subjects of physics, as well as of our perception of psychological time.

Keywords Invariance violation · Arrows of time · Symmetry and symmetry breaking · Dynamical systems · Bifurcations · Irreversibility · Entropy · Geometrical modelling · Psychological time

1 From *CPT* Invariance Violation and Cosmic Asymmetry to the Fundamental Concept of Entropy

Let us start with a brief review of some fundamental concepts of modern physics into which the role of time enters as a fundamental part of the study of nature. We shall

This is a revised version of the lecture presented in the International Symposium on *The Direction of Time: The Role of Reversibility/Irreversibility in the Study of Nature*, held at the *Zentrum für interdisziplinäre Forschung* (ZiF), Universität Bielefeld, January 14–18, 2002.

L. Boi (✉)
École des Hautes Études en Sciences Sociales, Centre de Mathématiques, 54, boulevard Raspail, 75006 Paris, France
e-mail: luciano.boi@ehess.fr

S. Albeverio, P. Blanchard (eds.), *Direction of Time*,
DOI 10.1007/978-3-319-02798-2_11,
© Springer International Publishing Switzerland 2014

start with a few remarks on the development of thermodynamics, before we expose the second law and the related concept of entropy where for the first time the possibility that time could be irreversible appears. The other context in which enters the notion of "*arrow of time*", i.e. the fact that the physical laws governing the universe should not be invariant with respect to time reversal, is cosmology and particularly the quantum theory of space-time singularities, which leads one to consider the existence of a cosmic asymmetry between matter and antimatter as the realistic scenario followed by the universe since its origins. Finally, there has been a very surprising result in quantum field theory: the 1956 discovery of parity (P) non-conservation in weak interaction phenomena. Even more surprising was the discovery of CP violation in 1964, which shattered the illusion concerning the fundamental nature of CPT theorem, that is, the belief that the invariance of time reversal transformation, of charge conjugation and of space inversion or mirror symmetry are the general principles to be satisfied by the equations of motion—hence a firm root in the foundations of physics, and opened up questions concerning its origin and its profound implication for our conception of physics and nature. These questions have not yet been answered satisfactorily despite an enormous effort in theoretical and experimental physics. Nevertheless, the developments of physics and of the other natural sciences in the last two decades lead to the belief that the violation of CPT invariance is needed to deal with interactions that are not invariant under one or more of these transformations.

One should distinguish two aspects of the violation of the three fundamental symmetries of nature, namely the time invariance (T), the electrical charge invariance (C) and the space inversion invariance (P). The first concern the consequences of T invariance for those properties of matter that depend on electromagnetic and strong interactions, and even on the grosser features of the weak interactions; the other concern the violation of CP invariance and T invariance in some special aspects of the weak interactions. The ability to separate these two aspects rests on the fact that the observed violation is an extremely small effect, not influencing in a (so far) measurable way even high-precision weak interactions measurements other than those specific, particularly sensitive one by means of which the CP violation was discovered. Nevertheless, most physicists believe firmly in the notion of a theory that unifies the electromagnetic, weak, and strong interaction phenomena at some level. At that level the separation of the physical phenomena into two classes should likely become meaningless. And the very fact that the observed violation occurs in such a limited though very meaningful way suggests that the level of unification at which T violation originates in a fundamental way must be very deep indeed. Therefore its elucidation may have profound implications for our understanding of the nature of physical theories. It may also have important implications for cosmology and notably for our actual conception of the structure of space and time.

1.1 Arrows of Time and Their Relations

So in our universe, as we find it, there are at least five *arrows* of time. Physicists do not yet know how they are interrelated. The preferred time direction on the sub-atomic microlevel, in certain weak interactions involving K-mesons, is still a mystery. It may have no connection with the macroscopic arrows, just as the handedness of particles seems to have no connection with the handedness of molecules, and the handedness of molecules in turn has no bearing on the bilateral symmetry of human body. On the macrolevel are four arrows. First, there is the entropy arrow, which has a precise technical definition in both thermodynamic theory and information theory. The notion of entropy was first introduced by the 19th century Austrian physicist Ludwig Boltzmann who founded statistical thermodynamics, whose starting point was the study of a system of gas molecules moving about randomly in a closed container. According to his vision, entropy is the principal foundation for the arrow of time. We can think of it in a rough way as a measure of disorder—the absence of pattern. The "information" content of a system, roughly speaking, is a measure of order (see below for a mathematical definition). The two measures vary inversely. If the entropy of a system goes up, its information content goes down, and vice versa.

We suggest distinguishing between two classes of phenomena and events in which time acts in a fundamental way. One in one case uses the term *geometrical arrow* for those processes in which order is increasing. They are very grounded in historical as well as in biological evolution. The formation of matter, moving in an orderly fashion outward from the site of the big bang, was the first gigantic instance of an event stamped with the geometrical arrow. The evolutions of stars and planets are later examples. The formation of strongly ordered crystals is another example. Finally, the energy radiating from a highly ordered sun allowed the rise and proliferation of life, the most highly patterned thing we know. The *entropy arrow* points opposite ways with respect to order, hence apply to those natural phenomena which evolve towards disorder. Let us now mention the other arrows of time.[1] There is the arrow defined by events radiating from a centre like expanding circular ripples on a pond or energy radiating from a star. This kind of arrows (for example, those concerned with dissipative chaotic systems) seems to derive from the probability of initial or boundary conditions. Third, there is the expansion of the universe, or the cosmic arrow. Fourth, there is the psychological arrow of consciousness. (For some remarks about the last two arrows of time, see below.)

1.2 The Fundamental Principle of Entropy in Thermodynamics Theory

The second law of thermodynamics has "various formulations", but they all lead to the existence of an entropy function whose reason for existence is to tell us which processes can occur and which cannot. We shall reformulate it by referring to the existence of entropy as the second law. The entropy we are talking about is that defined by thermodynamics, and not some analytic quantity that appears in information theory, probability theory and statistical mechanical models. The statement

---

[1] For further interesting reflections on this subject, see [4, 5].

of the first law of thermodynamics is essentially the statement of the principle of the conservation of energy for thermodynamical systems. As such, it may be expressed by stating that the variation in energy of a system during any transformation is equal to the amount of energy that the system receives from its environment. Briefly, it is a concept that provides the connection between mechanics (and things like falling weights) and thermodynamics. The first law arose as the result of the impossibility of constructing a machine that could create energy. However, it places no limitations on the possibility of transforming energy from one form into another. Thus, for instance, on the basis of the first law alone, the possibility of transforming heat into work or work into heat always exists provided the total amount of heat is equivalent to the total amount of work.

The three popular formulation of the second law are: (i) No process is possible the sole result of which is that heat is transformed from a body to a hotter one (postulate of Clasius). (ii) No process is possible the sole result of which is that a body is cooled and work is done (postulate of Kelvin and Planck). (iii) In any neighbourhood of any state there are states that cannot be reached by it by an adiabatic process. All three formulations are supposed to lead to the entropy principle (defined below).

Definition A state Y is adiabatically accessible from a state X, in symbols $X < Y$, if it is possible to change the state from X to Y by means of an interaction with some device consisting of some auxiliary system and a weight in such a way that the auxiliary system returns to its initial state at the end of the process, whereas the weight may have risen or fallen.

We could have (in principle, at least) both $X < Y$ and $Y < X$, and we could call such a process a *reversible adiabatic process*. Let us write $X \ll Y$ if $X < Y$ but not $Y < X$ (written $Y < X$). In this case we say that we can go from X to Y by an *irreversible adiabatic process*. If $X < Y$ and $Y < X$ (i.e., X and Y are connected by a reversible adiabatic process), we say that X and Y are *adiabatically equivalent* and write $X \sim Y$.

Entropy Principle There is a real-valued function on all states of all systems (including compound systems) called entropy, denoted by S, such that:

(a) *Monotonicity*: When X and Y are comparable states, then $X < Y$ if and only if $S(X) \leq S(Y)$.
(b) *Additivity and extensivity*: If X and Y are states of some (possibly different) systems and if (X, Y) denotes the corresponding state in the compound system, then the entropy is additive for these states; i.e., $S(X, Y) = S(X) + S(Y)$. S is also extensive; i.e., for each $\lambda > 0$ and each state X and its scaled copy $\lambda X \in \Gamma^{(\lambda)}$ (where Γ is the space of states of the system) $S(\lambda X) = \lambda S(X)$.

A formulation logically equivalent to (a) is the following pair of statements: $X \sim Y \Rightarrow S(X) = S(Y)$ and $X \ll Y \Rightarrow S(X) < S(Y)$. The last line is especially noteworthy. It says that entropy must increase in an irreversible adiabatic process.

Then, *irreversibility* means that for each $X \in \Gamma$ there is a point $Y \in \Gamma$ such that $X \ll Y$.

The reversibility of time in physical elementary process (both in classical and in quantum mechanics, as well as in the relativistic theories) is commonly accepted and very well established; that means that the fundamental laws of physics are invariant under time reversal. However, it is an obvious fact that most phenomena in Nature distinguish a direction of time; time is irreversible in complex systems. Electromagnetic waves are observed in their retarded form only, where the fields causally follow from their sources. The increase of entropy, as expressed in the second law of thermodynamics, also defines a time direction. This is directly connected with the psychological arrow of time—we remember the past but not the future. In quantum mechanics it is the irreversible measurement process and in cosmology the expansion of the universe, as well as the local growing of inhomogeneities, which determine a direction of time.

1.3 Irreversibility of Complex Systems

In order to make clear the irreversible character of most complex systems, let us consider a simple case of a droplet of ink added to water in a jar. The droplet spreads out rapidly, so that the colour becomes uniform in the entire vessel. Anyone can observe these phenomena. However, no one has ever seen a process developing in the opposite direction: ink particles collecting from the whole volume into a single droplet. Take now an iron rod, heat it and then put it into a vessel with cold water. The rod will cool down, the water will get warmer and their temperatures will become equal. The process always goes this way. Heat is never transferred from cold water to hot iron, raising its temperature still further. This is another example of an irreversible process, similar to the spreading of a droplet. Why does irreversibility always arise in all such processes, even though they are composed of particle motions that are definitely time-reversible? Where and how does reversibility perish?

The answer to that question, as we have seen above, lies in the second law of thermodynamics discovered by the physicists Rudolf Clasius and William Thomson. Their thermodynamic ideas were then developed and extended by Ludwig Boltzmann. He uncovered the meaning of the second law of thermodynamics. Heat is, in fact, the chaotic motion of atoms and molecules of which material bodies consist. Hence the transition of the energy of mechanical motion of individual constituents of the system into heat signifies the transition from the organised motion of large parts of the system to the chaotic motion of the smallest particles; this means that an increase in chaos is inevitable owing to the random motion of particles, unless the system is influenced from outside so as to maintain the level of order. Boltzmann showed that the measure of chaos in a system is a quantity called *entropy*. The greater the chaos, the higher the entropy. The transition of different types of motion of matter into heat means that entropy grows. When all forms of energy have transformed into heat, and this heat has spread uniformly through the system, this state of maximum chaos ceases to change with time and corresponds to maximum entropy.

This is the gist of the matter! In complex systems consisting of many particles or other elements, disorder (chaos) inevitably increases as a result of the random nature of numerous interactions. Entropy is that very measure of the degree of chaos. It is very important that when creating a more ordered state in a system, by influencing it from within a larger system, we inevitably insert additional disorder into this larger system. The laws of thermodynamics state that the "chaos" added to the larger system is inevitably greater than the 'order' introduced into the smaller system. Hence the "chaos", and "entropy", in the whole world must grow, even though order may be established in some parts of the world. One realises then that the second law of thermodynamics is of great importance for the evolution of the universe. Indeed, exchange of energy between the world and "other systems" being impossible, the universe must be treated as an isolated system. Therefore, all types of energy in the universe must ultimately convert to heat spread uniformly through matter, after which all macroscopic motion peters out. Even though the law of conservation of energy is not violated, the energy does not disappear and remains in the form of heat, it 'loses all forces', any possibility of transformation, any possibility of doing the work of motion. This bleak state became known as the 'thermal death' of the universe. The irreversible process in the universe is thus the growth of entropy. The question, however, remain open: can this process entirely dictate the direction of flow of time? I guess that we shall search for some other key feature of time and of space-time if we want to be able to give a satisfactory answer to these questions.

For the moment, we may ask: how can one understand that most phenomena distinguish a direction of time? One of the most interesting answer likely lies in the possibility of very special boundary conditions such as an initial condition of low entropy (see [6]). Such an assumption transcends the Newtonian separation into laws and boundary conditions by also seeking physical explanations for the latter. Where lies the key to the understanding of the irreversibility of time? According to Roger Penrose, it is primarily the high-unoccupied entropy capacity of the gravitational field that allows for the emergence of structure far from thermodynamical equilibrium. As he has stressed, the presence and the apparent structure of space-time singularities contain the key to the solution to one of the long-standing mysteries of physics: the origin of the *arrow of time* [7]. He has emphasised that the statistical notion of entropy is crucial for the discussion of time-symmetry. And if the fundamental local laws are all time-symmetric, then the place to look for the origin of statistical asymmetries is in the boundary conditions. This assumes that the local laws are of the form that, like Newtonian theory, standard Maxwell–Lorentz theory, Hamiltonian theory, Schrödinger's equations, etc., they determine the evolution of the system once we have boundary conditions *either* in the past or in the future. Then the statistical arrow of time can arise via the fact that, for some reason, the *initial* boundary conditions have an overwhelmingly *lower* entropy than do the *final* boundary conditions. Penrose has convincingly showed that the expansion of the universe cannot, in itself, be responsible for the entropy imbalance either. Accordingly, the arrows of entropy and retarded radiation can be explained if a reason is found for the initial state of the universe (big bang singularity) to be of comparatively low entropy and for the final state to be of high entropy. Consequently some low-entropy assumption does need to be imposed on the big bang; that is, the mere fact that the universe expands away from a singularity is in no way sufficient. We need some assumption on initial singularities that rules out those which would lie at the centres of white holes. But what is it in the nature of the big bang that is of 'low entropy'? The answer to this question lies in the unusual nature of gravitational entropy.

Many authors have pointed out that gravity behaves in a somewhat anomalous way with regard to entropy. This is true just as much for Newtonian theory as for general relativity. Thus, in many circumstances in which gravity is involved, a system may behave as though it has a negative specific heat. This is directly true in the case of a black hole emitting Hawking radiation, since the more it emits, the hotter it gets (the energy increase). This is essentially an effect of the universally attractive nature of the gravitational interaction. As a gravitating system "relaxes" more and more, velocities increase and the sources clump together—instead of uniformly spreading throughout space in a more familiar high-entropy arrangement. With other types of forces, their attractive aspects tend to saturate (such as with a system bound electromagnetically), but this is not the case with gravity. Only non-gravitational forces can prevent parts of a gravitationally bound system from collapsing further inwards as the system relaxes. Kinetic energy itself can halt collapse only temporarily. In the absence of significant non-gravitational forces, when dissipative effects come further into play, clumping becomes more and more marked as the entropy increases. Finally, maximum entropy is achieved with collapse to a black hole.

Consider a universe that expands from a "big bang" singularity and then recollapses to an all-embracing final singularity. The entropy in the late stages ought to be much higher than the entropy in the early stages. How does this increase in entropy manifest itself? In what way does the high entropy of the final singularity distinguish it from the big bang, with its comparatively low entropy? We may suppose that, as is apparently the case with the actual universe, the entropy in the initial matter is high. The kinetic energy of the big bang, also, is easily sufficient (at least on average) to overcome the attraction due to gravity, and the universe expands. But then, relentlessly, gravity begins to win out. The precise moment at which it does so, locally, depends upon the degree of irregularity already present, and probably on various other unknown factors. Then clumping occurs, resulting in clusters of galaxies, galaxies themselves, globular clusters, ordinary stars, planets, white dwarfs, neutron stars, black holes, etc. The elaborate and interesting structures that we are familiar with all owe their existence to this clumping, whereby the gravitational potential energy begins to be taken up and the entropy can consequently begin to rise above the apparently very high value that the system had initially. This clumping must be expected to increase; more black holes are formed; smallish black holes swallow material and congeal with each other to form bigger ones. This process accelerates in the final stages of re-collapse when the average density becomes very large again, and one must expect a very irregular and clumpy final state.

As Roger Penrose [3] has emphasised, there is very likely a qualitative relation between gravitational clumping and an entropy increase due to the taking up of gravitational potential energy. In terms of space-time curvature, the absence of clumping

corresponds to the absence of Weyl conformal curvature (since absence of clumping implies spatial isotropy, and hence no gravitational principal null-directions).
When clumping takes place, each clump is surrounded by a region of nonzero Weyl
curvature. As the clumping gets more pronounced owing to gravitational contraction, new regions of empty space appear with Weyl curvature of greatly increased
magnitude. Finally, when gravitational collapse takes place and a black hole forms,
the Weyl curvature in the interior region is larger still and diverges to infinity at the
singularity. In other words, Penrose formulated his Weyl tensor hypothesis that the
Weyl tensor vanishes at singularities in the past but not at those in the future. The
Weyl tensor is that part of the Riemann tensor which is not fixed by the boundary
equations (in which only the Ricci tensor enters) but by the boundary conditions
only. It describes the degrees of freedom of the gravitational field. Since it vanishes
exactly for a homogeneous and isotropic Friedmann universe, it can be taken as a
heuristic measure for inhomogeneity and, therefore, for gravitational entropy.

2 Symmetry and Symmetry Breaking in Nature

2.1 The Meanings of Symmetry

In general terms, what symmetry means is that the (physical) system possesses the
possibility of a change that leaves some aspect of the system unchanged. Symmetry
of the laws of nature concerns conservation. There are a number of conservations,
called "conservation laws", that hold for quasi-isolated systems. The best known of
them are conservation of energy, conservation of linear momentum, conservation of
angular momentum and conservation of electric charge. What is meant is that, if the
initial state of any quasi-isolated physical system is characterised by having definite
values for one or more of those quantities, then any state that evolves naturally
from that initial state will have the same values for those quantities. The conceptual
definition of symmetry can be thus: *Symmetry is immunity to a possible change*. We
can point out the two following essential components of symmetry: 1. Possibility of
change. 2. Immunity. If a change is possible but some aspect of the system is not
immune to it, we have asymmetry. The system can be said to be asymmetric under
the change with respect to that aspect.

The *symmetry principle* is fundamental to the applications of symmetry in science, and especially in physics. It states that *the symmetry group of the cause is a
subgroup of the symmetry group of the effect*. In other words: *the effect is at least
as symmetric as the cause*. However, these principles is in many situations contradicted by the phenomenon of "spontaneous symmetry breaking". There appear to
be cases of physical systems where the effect simply has less symmetry than the
cause, where the symmetry of the cause is possessed by the effect only as a badly
broken symmetry, so that the exact symmetry group of the effect is a subgroup of
the symmetry group of the cause, rather than vice versa. In fact, what is assumed
to be the exact symmetry of the cause is really only an approximate symmetry. Just

how small, symmetry-breaking perturbations of a cause affect the symmetry of the effect? What can be said about the symmetry of an effect relative to the *approximate* symmetry of its cause? That depends on the actual nature of the physical system, on whatever it is that links cause and effect in each case. But we can consider the possibilities.

1. *Stability*. The deviation from the exact symmetry limit of the cause, introduced by the perturbation, is "damped out", so that the approximate symmetry group of the cause is the minimal symmetry group of the effect.
2. *Lability*. The approximate symmetry group of the cause is the minimal approximate symmetry group of the effect, of more or less the same goodness of approximation.
3. *Instability*. The deviation from the exact symmetry limit of the cause, introduced by the perturbation, is "amplified", and the minimal symmetry of the effect is only the exact symmetry of the cause (including perturbation), with the approximate symmetry of the cause appearing in the effect as a badly broken symmetry. That is what is commonly called *spontaneous symmetry breaking*. Thus, although symmetric causes must produce symmetric effects, nearly symmetric causes need not produce nearly symmetric effects: a symmetry problem need have no stable symmetric solutions.

2.2 Examples of Symmetry Breaking

As an example of instability, we can take the solar system, its origin and evolution. Modern theory has the solar system originating as a rotating cloud of approximate axial symmetry and reflection symmetry with respect to a plane perpendicular to its axis. If that state of what is now the solar system is taken as the cause, the present state can be taken as the effect. And any axial symmetry the proto solar system one had has clearly practically disappeared during the course of evolution, leaving the solar system as we now observe it. The random, symmetry-breaking fluctuations in the original cloud grew in importance as the system evolved, until the original axial symmetry became hopelessly broken. Consider now, for another example of spontaneous symmetry breaking, a volume of liquid at rest in a container; such liquid is isotropic, which is to say that its physical properties are independent of direction, hence it is a symmetric system. Now, a small crystal of the frozen liquid thrown into the liquid breaks the symmetry, but is soon melts and isotropy returns. However, when the liquid is cooled to below its freezing point, the situation alters drastically. Let now throw in a crystal, then the supercooled liquid will immediately crystallise and thus become highly anisotropic. If in the subfreezing temperature range the system is unstable for isotropy; any anisotropic perturbation is immediately amplified until the whole volume becomes anisotropic and stays that way. The cooler the liquid (below its freezing point), the greater its instability. The freezing point is the boundary between the temperature range of stability and that of instability.

It must be pointed out that one of the most important upheavals in the scientific vision of nature in the last century has been the discovering that spontaneous breaking symmetries, bifurcations and singularities are three mechanisms which play a fundamental role for the organisation of physical and living matter and the unfolding of natural phenomena. These mechanisms are very deep related, because each time that a physical or living system bifurcs, the immediate consequence is that the symmetry of the system breaks down and instead of that a new broader symmetry will appear. Besides, the fact that a system may bifurc at some moment of his evolution means that its unfolding stops to be (mathematically speaking) continuous or linear and become discontinuous and non-linear. In many situations, this non-linearity (of partial differential equations) lead to the emergence of new order-disorder transition phenomena which exhibits non-equilibrium states mathematically expressible by time-dependent equations, and it is a source of instability, bifurcation and symmetry-breaking phenomena. Many of these macroscopic and local dynamical laws and phenomena manifest time asymmetry or irreversibility, which is a feature of key significance. Let me first mention some examples and fields in which spontaneous symmetry breaking manifests itself as a primary feature of the problem.

Example 1 (Morphogenesis and molecular biology) A striking example of symmetry breaking in a biological system is the breakdown of rotational symmetry in the *Fucus* seaweed egg. At a critical stage in the development of the egg a transition is made from a spherically symmetric membrane potential distribution to a polarised state with an axial symmetry, and a net trans-cellular current leaving one pole and entering the opposite. This phenomenon (or effect) is termed "self-electrophoresis". The net trans-cellular potential gradient is believed to be essential in the development of the asymmetry that leads to dramatically different rhizoid and thallus cells after the first division of the egg. The symmetry breakdown in the *Fucus* egg is of the form rotational invariance to axial invariance. That is, prior to self-electrophoresis the solutions are invariant under the entire rotation group $O(3)$, while the bifurcating solutions are invariant only under a subgroup of rotations about a fixed axis. The solutions thus appear in two-dimensional orbits with one-dimensional isotropy subgroup. This, however, is by no means the only symmetry breakdown that can occur in rotationally invariant systems.

Processes underlying the growth and reproduction of living organisms seem to be governed by a fundamental asymmetrical structure. In particular, sister cells can be born different by an asymmetric cell division. At each stage in its development, a cell in an embryo is presented with a limited set of options according to the state it has attained: the cell travels along a developmental pathway that branches repeatedly. At each branch in the pathway it has to make a choice, and its sequence of choices determines its final destiny. In this way, a complicated array of different cell types is produced. To understand development, we need to know how each choice between options is controlled, and how those options depend on the choices made previously. To reduce the question to his simplest form: how do two cells with the same genome come to be different? When a cell undergoes mitosis, both of the resulting daughter cells receive a precise copy of the mother cell's genome.

Yet those daughters will often have different specialised fates, and, at some point, they or their progeny must acquire different characters. In some cases, the two sister cells are born different as a result of an *asymmetric cell division*, in which some significant set of molecules is divided unequally between the two daughter cells at the time of division. This asymmetrically segregated molecule (or set of molecules) then acts as a *determinant* for one of the cell fates by directly or indirectly altering the pattern of gene expression within the daughter cell that receives it. Asymmetric division are particularly common at the beginning of development, when the fertilised egg divides to give daughter cells with different fates, but they also occur at later stages—in the genesis of nerve cells, for example.

Example 2 (Wave propagation in neural networks) Bifurcation phenomena in simple mathematical models of excitatory inhibitory neural networks have been discussed recently by many peoples (see, for instance [1]). Neural networks are aggregates of nerve cells which interact with other neurones in the network in either an excitatory or inhibitory way, and so it is plausible to expect these networks to exhibit such non-linear collective phenomena as bifurcation, threshold effects, and hysteresis. One can model these networks by a system of equations

$$\mu Y = -Y + S(KY + P) \tag{1}$$

where Y is a two-component vector, S is a non-linear vector-valued function, K is a linear convolution operator, and P is the external stimulus. Equation (1) may be studied in one, two, or three dimensions. Some neurophysiologists seek to model the patterns of activity of the central nervous system by showing how organised space-time neuronal activity patterns can arise through the mechanisms of bifurcation from an initially uniform resting state. They investigate the structure of the bifurcation when two pairs of complex conjugate eigenvalues cross the imaginary axis simultaneously. In that case one gets secondary bifurcation as some of the parameters in the problem are varied. J.D. Cowan and G.B. Ermentrout [2] have treated hallucinatory phenomena from the standpoint of symmetry-breaking bifurcations. Recent experiments on mescaline induced hallucinations have led to the conclusion that most simple hallucinations could be classified into one of four categories: (a) grating, lattice, honeycomb or chessboard; (b) cobweb; (c) funnel, tunnel, cone or vessel; (d) spiral. Cowan and Ermentrout base their analysis on the contention that simple formed hallucinations arise from an instability of the resting state leading to concomitant spatial patterns of activity in the cortex. This instability arises from a combination of enhanced excitatory modulation and decreased inhibition. They demonstrate that such spatial patterns are a property of neural nets with long strong lateral interactions acting to provide a dominant negative feedback. They formalise these postulates into a simple mathematical model and then use bifurcation theory to demonstrate the existence of the relevant spatial patterns.

The relevant spatial patterns are none other than those crystallographic patterns that have already made their appearance in the Bénard problem, with one additional factor. Experimental observations have established that in primates there is a conformal transformation from the retinal field, which is circular, to the cortical field, which has Cartesian (rectangular) symmetry. This implies that the transformation from retinal polar co-ordinates to cortical rectangular co-ordinates must be essentially logarithmic in nature. Such a logarithmic transformation would take a tunnel pattern consisting of concentric circles of activity to a pattern of rolls parallel to the y-axis. Similarly, spirals are transforms of rolls with some other direction. Thus the patterns observed in hallucinatory phenomena are images under the log transformations of the cellular patterns familiar in the analysis of the Bénard problem: hexagons, squares, rectangles, and rolls. One can in fact assume that, as some parameter λ increases, the strength of the excitation increases until, beyond some critical value λ_c, the rest state becomes unstable and gives way to the stationary patterns of spatial activity. Thus, according to this theory, the drug-induced hallucinatory patterns are precisely those which one would see when Euclidean invariance is broken.

Example 3 (Phase transitions in statistical mechanics) The notion of symmetry breaking is fundamental to phase transitions, yet much harder to treat mathematically. Until the renormalisation theories developed in the last two decades, the primary approach to phase transition was, in one way or another, a mean-field approximation coupled with a bifurcation analysis of the mean-field equations. The simplest mean-field theories for critical phenomena were the scalar equations of state, such as the Van der Waals equation for a gas of the Curie–Weiss model for a ferromagnet. In more elaborate theories the state of the ensemble is described, for example, by a single particle density function, and an integral equation is derived for this function by some kind of closure hypothesis for the hierarchy of higher-order (multiple particle) correlation functions. Nevertheless, these approximations are still mean-field theories, and depend, for their validity, on the assumption that fluctuations are negligible; the major difficulty is that in many cases, large fluctuations become important precisely at the critical point. In fact, at a critical point the fluctuations very often diverge to infinity, making the mean-field approximation invalid, and it is this fact which accounts for the deviation of the critical exponents from the "classical exponents" predicted by bifurcation (mean-field models. All this notwithstanding, the bifurcation models do have some areas of validity, and they are generally successful in predicting the symmetry changes actually observed. Landau's theory of second-order phase transitions is a phenomenological description of phase transitions, which is essentially a theory of "symmetry-breaking bifurcations". According to this point of view, the generalised mean-field approximation usually brings us to the formulation of the broken-symmetry problem in terms of the bifurcation on a non-linear integral equation solution for the Bogolyubov quasi-average. Especially the liquid-solid phase transition is considered as a bifurcation of the solution of the equation of Hammerstein type

$$\Phi(r_1) - \mu \int K(r_1, r_2) f\big(\Phi(r_2), r_2\big) dr_2 = 0. \tag{2}$$

The phase transitions of the ensemble are described in terms of bifurcations of this integral equation. In the area of non-equilibrium thermodynamics the operation of the laser can be described by a mean-field theory, which is amenable to a bifurcation analysis. In the Dicke–Haken–Lax model of the laser it is possible to describe the many body photon field by a mean-field theory as N (the number of degrees of freedom) tends to infinity. Thus it is possible in this case to solve a non-linear quantum-mechanical model, far from equilibrium, by reducing the problem to a system of ordinary differential equations for the expectation values of the extensive variables. The onset of laser action in these theories is then described by the bifurcation of time-periodic solutions from the equilibrium solution, that is, so-called Hopf bifurcations.

3 Spontaneous Symmetry Breakdown, Gauge Fields and Particle Physics

Here are some long-standing problems in particle theory: (1) How can we understand the hierarchical structure of the fundamental interactions? Are the strong, medium strong (i.e. $SU(3)$-breaking), electromagnetic, and weak interactions truly independent, or is there some principle that establishes connections between them? (2) How can we construct a renormalisable theory of the weak interactions, one which reproduces the low-energy successes of the Fermi theory but predicts finite higher-order corrections? (3) How can we construct a theory of electromagnetic interactions in which electromagnetic mass differences within isotopic multiplets are finite? (4) How can we reconcile Bjorken scaling in deep inelastic electro-production with quantum field theory? The SLAC-MIT experiments seem to be telling us that the light-cone singularities in the product of two currents are canonical in structure; ordinary perturbation theory, on the other hand, tells us that the canonical structure is spoiled by logarithmic factors, which get worse and worse as we go to higher and higher orders in the perturbation expansion. Are there any theories of the strong interactions for which we can tame the logarithms, sum them up and show they are harmless? Very significant advances have been made on all of these problems in the last 20 years. There now exist a large family of models of the weak and electromagnetic interactions that solve the second and third problem, and there has been discovered a somewhat smaller family of models of the strong interactions that solve the fourth problem. As we shall see, the structure of these models is such that we are beginning to get ideas about the solution of the (very deep) first problem; connections are beginning to appear in unexpected places, and one might optimistically say that we are on the road to the first truly unified theory of the fundamental interactions. All these marvellous developments are based upon the ideas of spontaneous symmetry breakdown and gauge fields.

Let us briefly discuss spontaneous symmetry breakdown, Goldstone bosons, gauge fields, and the Higgs phenomenon in the simplest context, that is, classical field theory. I will have no time to go into the renormalisation problem, nor into the non-Abelian generalisations of the Wald identities, and other aspects of the quantisation of gauge fields. In general, there is no reason why an invariance of the Hamiltonian of a quantum-mechanical system should also be an invariance of the ground state of the system. Thus, for example, the nuclear forces are rotationally invariant, but this does not mean that the ground state of a nucleus is necessarily rotationally invariant (i.e. of spin zero). This is a triviality for nuclei, but it has highly non-trivial consequences if we consider systems which, unlike nuclei, are of infinite spatial extent. The standard example is the Heisenberg ferromagnet, an infinite crystalline array of spin $-1/2$ magnetic dipoles, with spin–spin interactions between nearest neighbours such that neighbouring dipoles tend to align. Even though the Hamiltonian is rotationally invariant, the ground state is not; it is a state in which all the dipoles are aligned in some arbitrary direction, and is infinitely degenerate for an infinite ferromagnet. A little man living inside such a ferromagnet would have a hard time detecting the rotational invariance of the laws of nature; all his experiments would be corrupted by the background magnetic field. If his experimental apparatus interacted only weakly with the background field, he might detect rotational invariance as an approximate symmetry; if it interacted strongly, he might miss it altogether; in any case, he would have no reason to suspect that it was in fact an exact symmetry. Also, the little man would have no hope of detecting directly that the ground state in which he happens to find himself is in fact part of an infinitely degenerate multiplet. Since he is of finite extent (this is the technical meaning of "little"), he can only change the direction of a finite number of dipoles at a time; but to go from one ground state of the ferromagnet to another, he must change the directions of an infinite number of dipoles—an impossible task.

At least at first glance, there appears to be nothing in this picture that cannot be generalised to relativistic quantum mechanics. For the Hamiltonian of a ferromagnet, we can substitute the Hamiltonian of a quantum field theory; for rotational invariance, some internal symmetry; for the ground state of the ferromagnet, the vacuum state; and for the little man, ourselves. That is to say, we conjecture that the laws of nature may possess symmetries that are not manifest to us because the vacuum state is not invariant under them. This situation is usually called "spontaneous breakdown symmetry". Let us investigate spontaneous symmetry breakdown in the case of classical field theory. For simplicity, we will restrict ourselves to theories involving a set of n real scalar fields, which we assemble into a real n-vector, ϕ, with Lagrange density

$$L = 1/2(\partial_\mu\phi)\cdot\left(\partial^\mu\phi\right) - U(\phi), \tag{3}$$

where U is some function of the ϕ_S, but not of their derivatives. We treat these theories purely classically, but use quantum-mechanical language; thus, we call the state of lowest energy "the vacuum", and refer to the quantities which characterise the spectra of small oscillations about the vacuum as "particle masses". For any of these theories, the energy density is

$$H = 1/2(\partial_0\phi)^2 + 1/2(\nabla\phi)^2 + U(\phi). \tag{4}$$

Thus the state of lowest energy is one for which the value of ϕ is a constant, which we denote by $\langle\phi\rangle$. The value of $\langle\phi\rangle$ is determined by the detailed dynamics of the particular theory under investigation, that is to say, by the location of the minimum (or minima) of the potential U. We call $\langle\phi\rangle$ "the vacuum expectation value of ϕ". Within this class of theories, it is easy to find examples for which symmetries are either manifest or spontaneously broken. The simplest one is the theory of a single field for which the potential is

$$U = (\lambda/4!)\phi^4 + (\mu^2/2)\phi^2, \tag{5}$$

where λ is a positive number and μ^2 can be either positive or negative. This theory admits the symmetry

$$\phi \to -\phi. \tag{6}$$

If μ^2 is positive, the potential has one minimum. The vacuum is at $\langle\phi\rangle$ equals zero, the symmetry is manifest, and μ^2 is the mass of the scalar meson. If μ^2 is negative, though, the situation is quite different; the potential has two minima. In this case, it is convenient to introduce the quantity

$$a^2 = -6\mu^2/\lambda, \tag{7}$$

and to rewrite the potential as

$$U = \lambda/4!(\phi^2 - a^2)^2, \tag{8}$$

plus an (irrelevant) constant. It is clear from this formula that the potential now has two minima, at $\phi = \pm a$. Because of the symmetry (6), which one we choose as the vacuum is irrelevant to the resulting physics; however, whichever one we choose, the symmetry is spontaneously broken. Let us choose $\langle\phi\rangle = a$. To investigate physics about the asymmetric vacuum, let us define a new field

$$\phi' = \phi - a. \tag{9}$$

In terms of the new ("shifted") field,

$$U = \lambda/4!(\phi'^2 + 2a\phi')^2 = (\lambda/4!)\phi'^4 + (\lambda a/6)\phi'^3 + (\lambda a^2/6)\phi'^2. \tag{10}$$

We see that the true mass of the meson is $\lambda a^2/3$. Note that a cubic meson self-coupling has appeared as a result of the shift, which would make it hard to detect the hidden symmetry (6) directly.

A new phenomenon appears if we consider the spontaneous breakdown of continuous symmetries. Let us consider the theory of two scalar fields, A and B, with

$$U = \lambda/4![A^2 + B^2 - a^2]^2. \tag{11}$$

This theory admits a continuous group of symmetries isomorphic to the two-dimensional rotation group, $SO(2)$:

$$A \to B \cos \omega + B \sin \omega, \qquad B \to -A \sin \omega + B \cos \omega. \tag{12}$$

The minima of the potential lie on the circle

$$A^2 + B^2 = a^2. \tag{13}$$

Just as before, which of these we choose as the vacuum is irrelevant, but whichever one we choose, the $SO(2)$ internal symmetry is spontaneously broken. Let us choose

$$\langle A \rangle = a, \qquad \langle B \rangle = 0. \tag{14}$$

As before, we shift the fields,

$$\phi' = \phi - \langle \phi \rangle, \tag{15}$$

and find

$$U = 1/4!\left(A'^2 + B'^2 + 2aA'\right)^2. \tag{16}$$

Expanding this, we see that the A-meson has the same mass as before, but the B-meson is massless. Such a massless spin-less meson is called a *Goldstone boson*; for the class of theories under consideration, its appearance does not depend at all on the special form of the potential U, but is a consequence only of the spontaneous breakdown of the continuous $SO(2)$ symmetry group (12). To show this, let us introduce "angular variables",

$$A = \rho \cos \theta, \qquad B = \rho \sin \theta. \tag{17}$$

In terms of these variables, (12) becomes

$$\rho \to \rho\theta \to \theta + \omega, \tag{18}$$

and the Lagrange density becomes

$$L = 1/2(\partial_\mu \rho)^2 + 1/2\rho^2(\partial_\mu \theta)^2 - U(\rho). \tag{19}$$

In terms of these variables, $SO(2)$ invariance is simply the statement that U does not depend on θ. The transformation to angular variables is, of course, ill-defined at the origin, and this is reflected in the singular form of the derivative part of the Lagrange density (19). However, this is of no interest to us, since we wish to do perturbation expansions not about the origin, but about an assumed asymmetric vacuum. With no loss of generality, we can assume this vacuum is at $\langle \rho \rangle = a$, $\langle \theta \rangle = 0$. Introducing shifted fields as before,

$$\rho' = \rho - a, \qquad \theta' = \theta, \tag{20}$$

we find

$$L = 1/2(\partial_\mu\rho')^2 + 1/2(\rho' + a)^2(\partial_\mu\theta')^2 - U(\rho' + a). \tag{21}$$

It is clear from this expression that the θ-meson is massless, just because the θ-field enters the Lagrangian only through its derivatives. This can also be seen purely geometrically, without writing down any formulae. If the vacuum is not invariant under $SO(2)$ rotations, then there is a curve passing through the vacuum along which the potential is constant; this is the curve of points obtained from the vacuum by $SO(2)$ rotations—in terms of our variables, the curve of constant ρ. If we expand the potential around the vacuum, no terms can appear involving the variable that measures displacement along this curve—the θ variable. Hence we always have a massless meson. This argument can easily be generalised to the spontaneous breakdown of a general continuous internal symmetry group.

Summarising, we can make the following remarks relating to the above description:

(i) There is a large family of field theories that display spontaneous breakdown on internal symmetries. If the spontaneously broken symmetry is discrete, this causes no problems; however, if the symmetry is continuous, symmetry breakdown is associated with the appearance of Goldstone bosons. This can be cured by coupling gauge fields to the system and promoting the internal symmetry group to a gauge group; the Goldstone bosons then disappear and the gauge mesons acquire masses. It should be remembered that, at the time of their inventions, both the theory of non-Abelian gauge fields and the theory of spontaneous symmetry breakdown were thought to be theoretically amusing but physically untenable, because both predicted unobserved massless particles, the gauge mesons and the Goldstone bosons. It was only later that it was discovered that each of these diseases was the other's cure.

(ii) What has been done for classical field theory can be extended to some extent into the quantum domain. At least for weak couplings, the phenomenon of spontaneous breakdown of internal symmetries survives substantially unchanged; in particular, all of the equations we have derived can be reinterpreted as the first terms in a systematic quantum expansion.

(iii) Regarding theories with fermions, it is clear that if we couple fermions to the scalar-meson systems we have discussed, either directly (through Yukawa couplings) or indirectly (through gauge field couplings), then the shift in the scalar fields will induce an apparent symmetry-violating term in the fermion part of the Lagrangian. A more interesting question is whether spontaneous symmetry breakdown can occur in a theory without fundamental scalar fields. For example, perhaps bilinear forms in Fermi fields can develop symmetry-breaking vacuum expectation values all by themselves. There is one exactly soluble model without fundamental scalars that displays the full Goldstone–Higgs phenomenon. This is the Schwinger model, quantum electrodynamics of massless fermions in two-dimensional space-time.

(iv) It is important to realise that we can make the effects of spontaneous symmetry breakdown as large or as small as we want, by appropriately fudging the parameters in our models. Thus, in the real world, some of the spontaneously broken symmetries of nature may be totally inaccessible to direct observation. Also, of course, there is no objection to exact or approximate symmetries of the usual kind coexisting with spontaneously broken symmetries. Presumably symmetries such as nucleon number conservation, neither broken nor coupled to a massless gauge meson, are of this sort.

4 Some Mathematical Aspects of Bifurcations, Singularities and Universality

Bifurcation, as a scientific terminology, has been used to describe significant and qualitative changes that occur in the solution curves of a dynamical system, as the key system parameters are varied. Very frequently, it is used to describe the qualitative stability changes of the solution curves of a non-linear dynamical system. In other words, the concept of bifurcation allows studying the branch points in non-linear equations, that is, of singular points of the equations where several solutions come together. It is important in applications because bifurcation phenomena typically accompany the transition to instability when a characteristic parameter passes through a critical value. Most of the dynamical systems naturally depend on parameters. For some special (critical) values of the parameters, say c, the non-generic situations may occur. For example, two stationary points A and B, depending on the parameter c, may collide at $c = 1/4$. If c is decreasing, the unique stationary point existing for $c = 1/4$, is "subdivided" into two points A and B. In such examples, all topological changes of the phase portraits under the change of the parameters are called *bifurcations*. A *phase singularity* is a point at which phase is ambiguous and near which phase takes on all values. In other words, *singularity* means a place where slopes become infinite, where the rate of change of one variable with another exceeds all bounds, and where a big change in an observable is caused by an arbitrarily small change in something else. Various areas of physics (solid state physics, hydrodynamics, fluid mechanics, physical chemistry and statistical physics) are a rich source of instability and bifurcation phenomena. We mention the formation of convection cells in the Bénard problem, which furnishes an excellent example of what is called a "symmetry-breaking instability". Prior to the onset of instability the solution is invariant under the entire group of rigid motions, whereas the bifurcating convective motions are invariant only under a crystallographic subgroup. Symmetry is broken "spontaneously", because the symmetry group of the equations is unchanged, while the bifurcating solutions have a smaller symmetry group.

The *centre manifold theorem* is one of the most useful tools for giving a representation of the solution trajectories of a non-linear dynamical system in a neighbourhood of a non-hyperbolic equilibrium (for further details and a mathematical statement, see [10] and [9]). It permits to understand the transition from stability to instability in many non-linear dynamical systems—the stability may vanish at the criticality appearing with different kinds of bifurcation points and trajectories or other singularities—and the emergence of new periodic or non-periodic (such as in the case of chaotic time evolutions like hydrodynamic turbulence and strange attractors) solutions. The concepts of bifurcation and attractor are very important for understanding the transition from stable dynamical systems to unstable dynamical systems in a large class of natural phenomena.

Closely related to this last question is a recent remarkable discovery—the so-called *Feigenbaum universality*, which is based on the well-known renormalisation-group method in theoretical physics, first used in statistical mechanics and quantum field theory. The problem with which Feigenbaum started consists in studying how dynamical systems depending on a parameter pass from a stable type of motion, which it is natural to call laminar, to an unstable type that involves the appearance of strong statistical properties frequently associated with turbulence. The Feigenbaum universality refers directly to sequences of period-doubling bifurcations. In traditional bifurcation theory it is usual to consider the local behaviour of families of dynamical systems in a neighbourhood of a bifurcation value of the parameter. Here, however, we encounter a completely new problem: the local behaviour of a family of dynamical systems in a neighbourhood of a parameter value where infinitely many parameter bifurcation values accumulate. It should be observed that the form of the trajectories becomes more complicated as the parameter increases for a broad class of one-parameter families of maps of a closed interval into itself, namely, a stable periodic trajectory becomes unstable as the parameter increases, and a stable periodic trajectory with twice the period is created, which attracts all points except for unstable cycles. Feigenbaum observed that the successive parameter values where such bifurcations take place for the family of maps $x \rightarrow \mu x(1-x)$ $(0 \leq \mu \leq 4)$ of [0, 1] into itself converge to a limit at the rate of a geometric progression with the ratio $\delta = 4.6692\ldots$, the famous Feigenbaum constant. He then made analogous calculations with the family $f(x : \mu) = \mu \sin(\pi x)$ and observed here a geometric progression with the same ratio. This led to the natural conjecture that δ does not depend at all on the form of the specific family of maps. Feigenbaum also proposed a theory explaining the universality of δ. It is useful qualitatively to form an intuitive picture of the phenomenon taking place when there is an infinite sequence of period-doubling bifurcations.

5 Brief Remarks on Conservative and Dissipative Systems

Very roughly one can classify the natural phenomena into two great classes: those that do not depend on time, i.e. which are invariant with respect to time changes, and those that depend on time, that is, which transforms in the course of time evolution and, more important, *with* time. The most interesting example of this variation of many natural phenomena and living systems is the spontaneous symmetry breaking, which produces a *qualitative* change in the state of those phenomena and systems.

Consider a (non-linear) oscillator, which is an archetypal system having a behaviour depending on time. Consider further the periodic movement of a physical pendulum; now this movement will stops after a certain time owing to frictions (and other perturbations). In other words, the amplitude of oscillations will decrease inevitably with time. This phenomenon consists in a dissipation of energy, which we found to be very general and to lead to implications for the study of everyday physical experience, and which can be expressed in a precise mathematical formulation. One can first consider the ideal case of a simple pendulum where we have, in addition to the point-like character of the mass, the absence of any kind of friction; yet some physical quantities and properties of the system are conserved.

The damped-oscillator provides a typical example of a dissipative dynamical system; and its most striking dynamical properties may be summarised as follows:

1. For such dissipative systems, there is not in general a time-independent Hamiltonian H, hence, no conservation of the energy of the systems.
2. In some cases, on the other hand, there exists a function of the dynamical variables, called the Lyapounov function, which is positive and monotonically decreasing (with time), which means that the system under consideration undertake an irreversible process.
3. One can also have, in the case of a dissipative system, a domain of evolution much more complicated than a simple decreasing. In any case, every time there is dissipation the equations of movement change by time reversal: therefore, the dynamics of dissipative systems is irreversible.

6 Some Qualitative and Geometrical Properties of Psychological Time

This last reflection is twofold aimed. First, we would like emphasising some properties of time which make up its peculiar structure in the conception of the physical world that governs everyday life. Let us begin by distinguishing *quantitative* from *qualitative* properties of time. In measuring time by the help of clocks we make use of its quantitative, or *metrical*, properties. Such measurements concern the determination of time distances of equal length, represented, for instance, by two consecutive hours; and, in addition, the determination of simultaneity, that is, of equal time values for spatially distant points. The theory of the metrical properties of time has been developed in great detail in modern physics—in particular, in Einstein's theory of relativity. The qualitative, or *topological*, properties of time are fundamental, in that they hold independently of specific procedures of measurement and remain unchanged even if the form of measuring time are varied. They comprise notably those properties that confer upon time its specific nature as different from space and that account for our sensible perception towards time. Following the precise analysis given by Hans Reichenbach [8] (see also [11]), we can formulate in several statements the most evident qualitative properties of time as follows:

(1) *Time goes from the past to the future.* This statement refers to the flow of time; it expresses what we call becoming. Time is not static; it moves. We may regard

the flow of time as the common product of an objective (physical) factor and a subjective one (connected with the structure of human consciousness).

(2) *The present, which divides the past from the future, is now.* The meaning of "now" might express either (or simultaneously) our subjective and intentional approach to time for the "present", or (and) the fact that we see the things around us in a certain spatial perspective. However that may be, this statement appears, from the psychological point of view, rather enigmatic.

(3) *The past never comes back.* This statement appears to be closely connected with the flow of time, that is, with the fact that time flows linearly in one and same direction, in the direction of a straight line, without thus never and nowhere intersects itself; the one-dimensional and linear continuum is the model of this conception of time.

The following three statements are intended to express the differences between the "past" and the "future".

(4) *We cannot change the past, but we can change the future.* The statement means, among other things, that there are some future happenings which we can predict and control though—owing to the random and complex nature of many macroscopic—we cannot predict and control cosmic events, or the weather, or earthquakes; and we are rather poor at controlling human society, which continues to drift from crisis into crisis and from war into war, but there are not events of the past which we can change.

(5) *We can make records of the past, but not of the future.* It is not possible to predict the future from isolated indications. And even if such a prediction from a few isolated causes is possible, it can be made only in approximate terms. Moreover, even the knowledge of the total cause cannot permit sure predictions.

(6) *The past is determined; the future is undetermined.* In some sense, the past consists of established facts, whereas the future does not; and an established fact is something that we cannot change, whereas the future concerns uncertain and questionable facts, and it is open to very different issues.

Then we want to sketch the essential features of a geometric suited model to represent the multidimensional and polycyclical nature of psychological and possibly physical time upon which rests partly our perception of the world. We borrow the fundamental ideas of this model from the mathematical theory of superstrings and from the theory of Calabi–Yau spaces. The central idea is that the space-time structure of the universe may have both extended dimensions and curled-up dimensions. This is an astounding suggestion made in 1919 by the Polish mathematician Theodor Kaluza, and refined some years later by the Swedish physicist Oskar Klein. This means that our spatial universe has dimensions that are large, extended, and easily visible, namely the three spatial dimensions of common experience, but that may also have additional spatial dimensions that are tightly curled up into a very tiny space. For instance, circular loops may exist at every point in the familiar extended dimensions. It is worthy of note that the circular dimension is not merely a circular bump within the familiar extended dimensions; rather, the circular dimension is a new dimension, one that exists at every point in the familiar extended dimensions;

it is a new and independent direction in which some being, if it were small enough, could move.

In the 1980s, it has been showed that one may generalise the Kaluza–Klein theory to higher-dimensional theories with numerous curled-up spatial directions. These extra dimensions are curled up into the surface of a sphere. Of course, beyond proposing a different number of extra dimensions, one can also imagine other shapes for the extra dimensions, for instance, the shape of a torus. And also more complicated possibilities can be imagined in which there are three, four, five, essentially any number of extra spatial dimensions, curled up into a wide spectrum of exotic shapes. In fact, the extra dimensions or the curled-up dimensions, which seem very profoundly to influence basic physical properties of the universe, look like a class of six-dimensional geometrical shapes known as *Calabi–Yau spaces*. Roughly, we have to imagine replacing each of the spheres—which represented two curled-up dimensions—with Calabi–Yau space. That is, at every point in the three familiar extended dimensions, string theory claims that there are six hitherto unexpected dimensions, tightly curled up into one of these rather complicated-looking shapes. These dimensions are an integral and ubiquitous part of the space's structure; they exist everywhere. For instance, if you sweep your hand in a large arc, you are moving not only through the three extended dimensions, but also through these curled-up dimensions. Of course, because the curled-up dimensions are very small, as you move your hand you circumnavigate them an enormous number of times, repeatedly returning to your starting point.

Now, given the requirement of numerous extra dimensions, is it possible that some are additional *time* dimensions, as opposed to additional space dimensions? We all have an understanding of what is means for the universe to have multiple space dimensions, since we live in a world in which we constantly deal with a plurality three. But what would it mean to have multiple times? Would one align with time as we presently experience it psychologically while the other would somehow be "different"? It gets even hard to accept when you think about a curled-up time dimension. Nevertheless, we may think of time not solely as a dimension we can traverse in only one direction with absolute inevitability, never being able to return to an instant after it has passed. At any rate, it might be that curled-up time dimensions have vastly different properties from the familiar, vast time dimension that we imagine reaching back to the creation of the universe and forward to the present moment.

But, in contrast to extra spatial dimensions, new and previously unknown time dimensions would clearly require an even more profound change of our intuition. It seems to us that the intriguing possibility of new time dimensions could well play a role in future developments of our conceptions of physical reality and of natural phenomena. Starting from these mathematical objects, one might suggest a geometrical model notably of psychological time which cannot be conceived like a linear and one-dimensional concept any more, but rather as a multidimensional and polycyclical one.

Acknowledgements I wish to thank Sergio Albeverio and Philippe Blanchard for inviting me to give a lecture in the symposium at *Zentrum für interdisziplinäre Forschung* in Bielefeld, and further to write this article. I also would like to thank the anonymous referee for its useful remarks on the first version of the manuscript.

References

Cited Literature

1. Carpentier, G.A., Grossberg, S.: Neural Networks for Vision and Image Processing. MIT Press, Cambridge (1992)
2. Cowan, J.D., Ermentrout, G.B.: A mathematical theory of visual hallucination patterns. Biol. Cybern. **34**, 137–150 (1979)
3. Hawking, S., Penrose, R.: The Nature of Space and Time. Princeton University Press, Princeton (1996)
4. Horwich, P.: Asymmetries in Time. MIT Press, Cambridge (1987)
5. Isham, Ch.: Quantum gravity. In: Davies, P. (ed.) The New Physics, pp. 70–93. Cambridge University Press, Cambridge (1993)
6. Kiefer, C.: Quantum cosmology and the emergence of a classical world. In: Rudolph, E., Stamatescu, I.-O. (eds.) Philosophy, Mathematics, and Modern Physics, pp. 105–119. Springer, Heidelberg (1994)
7. Penrose, R.: Singularities and time-asymmetry. In: Hawking, S.W., Israel, W. (eds.) An Einstein Centenary Survey, pp. 581–638. Cambridge University Press, Cambridge (1979)
8. Reichenbach, H.: The Direction of Time. University of California Press, Berkeley (1956)
9. Robinson, C.: Dynamical Systems: Stability, Symbolic Dynamics, and Chaos. CRC Press, Boca Raton (1995)
10. Smale, S.: The Mathematics of Time. Essays on Dynamical Systems, Economic Processes, and Related Topics. Springer, New York (1980)
11. Whitrow, G.J.: The Natural Philosophy of Time. Oxford University Press, Oxford (1980)

Further Reading

12. Abraham, R., Marsden, J.E.: Foundations of Mechanics, 2nd edn. Addison-Wesley, Reading (1978)
13. Balian, R.: Le temps macroscopique. In: Klein, E., Spiro, M. (eds.) Le Temps et sa Flèche, pp. 155–211. Editions Frontières, Paris (1994)
14. Bergé, P., Pomeau, Y., Vidal, Ch.: L'ordre dans le chaos. Hermann, Paris (1988)
15. Boi, L.: Theories of space-time in modern physics. Synthese **139**, 429–489 (2004)
16. Boi, L.: Geometrical and topological foundations of theoretical physics: from Gauge theories to the string program. Int. J. Math. Math. Sci. **34**, 1777–1836 (2004)
17. Boltzmann, L.: Vorlesungen über Gastheorie. Barth, Leipzig (1896–1898)
18. Coleman, S.: Aspects of Symmetry, Selected Erice Lectures. Cambridge University Press, Cambridge (1985)
19. Davies, P.C.W.: The Physics of Time Asymmetries. University of California Press, Berkeley (1974)
20. Fermi, E.: Thermodynamics. Prentice-Hall, New York (1937)
21. Fraisse, P.: The Psychology of Time. Harper & Row, New York (1963)
22. Gasperini, M.: L'universo prima del Big Bang. Muzzio Editore, Milan (2002)
23. Goldstone, J., Salam, A., Weinberg, S.: Broken symmetries. Phys. Rev. **127**, 965 (1962)

24. Guckenheimer, J., Holmes, P.: Non-linear Oscillations, Dynamical Systems and Bifurcations of Vector Fields. Springer, New York (1983)
25. Guth, A.H.: The Inflationary Universe. Addison-Wesley, Reading (1997)
26. Halliwel, J.J., Perez-Mercader, J., Zurek, W.H. (eds.): The Physical Origins of Time Asymmetry. Cambridge University Press, Cambridge (1994)
27. Hartle, J.B., Hawking, S.W.: Wave function of the universe. Phys. Rev. D **28**, 2960–2992 (1983)
28. Hirsch, M.W.: The dynamical systems approach to differential equations. Bull. Am. Math. Soc. **11**(1), 1–63 (1984)
29. Husserl, E.: Vorlesungen zur Phänomenologie des inneren Zeitbewusstseins. Max Niemeyer Verlag, Tübingen (1928)
30. Lee, B.W., Zinn-Justin, J.: Spontaneously broken gauge symmetries. I. Preliminaries. Phys. Rev. D **5**, 3121 (1972)
31. Novikov, I.D.: The River of Time. Cambridge University Press, Cambridge (1998)
32. Pauri, M.: La descrizione fisica del mondo e la questione del divenire temporale. In: Boniolo, G. (ed.) Filosofia della fisica, pp. 245–333. Mondadori, Milan (1997)
33. Prigogine, I.: Introduction to Thermodynamics of Irreversible Processes. Wiley, New York (1962)
34. Rosen, J.: Symmetry in Science: An Introduction to the General Theory. Springer, New York (1995)
35. Rovelli, C.: Is there incompatibility between the ways time is treated in general relativity and in standard quantum mechanics? In: Ashtekar, A., Stachel, J. (eds.) Conceptual Problems of Quantum Gravity, pp. 126–140. Birkhäuser, Boston (1991)
36. Ruelle, D.: Irreversibility revisited. In: Fokas, A., Halliwell, J., Kibble, T., Zegarlinski, B. (eds.) Highlights of Mathematical Physics. Am. Math. Soc., Providence (2002)
37. Sachs, R.G.: The Physics of Time Reversal. University of Chicago Press, Chicago (1987)
38. Sattinger, D.: Topics in Stability and Bifurcation Theory. Springer, Berlin (1973)
39. Strominger, A.: The nature of time in string theory, 22 pages. Text of the talk presented in the String 2002 Conference, Cambridge, UK (2002)
40. Thom, R.: On the reversibility (or irreversibility) of time. In: Coechlo, Z., Shiels, E. (eds.) Workshop on Dynamical Systems, pp. 59–66. Wiley, New York (1990)
41. Veneziano, G.: The myth of the beginning of time. Sci. Am. **290**(5), 75–84 (2004)
42. Vicario, G.B.: Il tempo. Saggio di psicologia sperimentale. Il Mulino, Bologna (2005)
43. Weinberg, S.: General theory of broken local symmetries. Phys. Rev. **7**, 1068–1083 (1973)
44. Wilczek, F.: Quantum field theory. Rev. Mod. Phys. **71**, 85–95 (1999)
45. Winfree, A.: The Geometry of Biological Time. Springer, New York (1980)
46. Zeh, H.D.: The Physical Basis of the Direction of Time, 4th edn. Springer, Berlin (2001)

Chapter 12
Is Time Directed?

Michael Drieschner

Abstract This paper gives a link between the structure of time and well-known problems in the foundations of physics and probability theory: The emphasis lies on the predictive character of objective science. It is maintained that the structure of past-present-future (the "arrow of time") is presupposed in physical theory; it cannot be hoped to be derivable from physical theory. This gives a solution to the problem of irreversibility in thermodynamics. Zermelo's reversibility paradox is refuted on that ground. A definition of probability is given that allows a derivation of the rules of probability calculus. The structure of time gives a new view on the old problem of measurement in quantum mechanics as well.

To begin with, let me call your attention to a little story of Calvin and Hobbes (Fig. 1).

Why is it so clear that going to the future, the way they try it, is nonsense?—Everybody knows that we cannot move around in time at will, like in space. But we know as well that future will come by itself; we do not have to do anything for that, we could not even keep future from happening if we wanted to.—And, by the way, most of the time future does not make things better, and we will be disappointed, like Hobbes.

So this is a good place for a remark on the title of our conference, the *direction of time*. Direction is a spatial concept. The road has a direction; an arrow can have a direction as well as a river. But time?—An hour is not directed, neither is a nanosecond. Neither the present is directed nor the past nor the future.

We are used to talking about the "arrow of time", and by that we mean the difference between past and future. But we should be aware of the fact Calvin and Hobbes demonstrate us, namely that time is something entirely different from space. This is obscured by our habit as physicists of imagining time as a real parameter t that runs from $-\infty$ to ∞, just as a spatial coordinate. This comes in quite handy for the calculation of predictions that depend on the position of the hand on a clock; but it does by no means characterize what time *is*.

M. Drieschner (✉)
Institut für Philosophie, Ruhr-Universität Bochum, Bochum, Germany
e-mail: Michael.drieschner@rub.de

S. Albeverio, P. Blanchard (eds.), *Direction of Time*,
DOI 10.1007/978-3-319-02798-2_12,

Fig. 1 Calvin & Hobbes, The time warp. Calvin and Hobbes © 1987 Watterson. Reprinted by permission of Universal Press Syndicate. All rights reserved

1 Ontology of Time

Working on the question what time really is, is usually called the "ontology" of time. That is what we are going to do now, for the next few paragraphs, with the help of some considerations of physics. This kind of ontology of time will be fruitful for the physicist as well as for the philosopher.

Some of the participants of this conference quoted or even propagated the opinion that *there is no time*. This opinion is not really new. Aristotle writes in his *physics*, more than 2.300 years ago:

> That time either is not at all or scarcely and dimly is, might be suspected from the following considerations. Some of it has been and is not, some of it is to be and is not yet. From these both infinite time and any arbitrary time are composed. But it would seem to be impossible that what is composed of things that are not should participate in being.[1]

Similarly Augustine, more than 1.600 years ago, in his specific language:

> How are, then, those two times, past and future, when the past 'is' not any longer as well as the future 'is' not yet.[2]

which means about the same as Aristotle's text above.

Thus we find that time is something quite special. To call its theory "ontology" seems rather misleading, after having read those traditional observations, since we cannot just talk about the "existence" of time. Again Augustine gives a beautiful and well-known formulation of his amazement about this problem:

"So what is time?—As long as nobody asks me I know it. But when somebody asks me, and I want to explain it to him, I do not know it."[3]

Augustine himself gives an idea how time exists—or, rather, how events "in time" exist: The past "is" (present) in my memory, the future "is" (present) in my expectations, fears, hopes etc.; both modes of time are "in the soul".

Against this background let us now return to discussing time in physics. In Relativity Theory—in the Special as well as in the General Theory of Relativity—time is but another coordinate, quite analogous to the spatial coordinates. Albert Einstein seems to have believed so much in that analogy that he views past and future in some way present at once. He wrote, after his friend Besso died, and a few weeks before he died himself, to his friend's family:

---

[1] Ὅτι μὲν οὖν ἢ ὅλως οὐκ ἔστι ἢ μόλις καὶ ἀμυδρῶς, ἐκ τῶνδέ τις ἂν ὑποπτεύσειεν. Τὸ μὲν γὰρ αὐτοῦ γέγονε καὶ οὐκ ἔστι, τὸ δὲ μέλλει καὶ οὔπω ἔστιν· ἐκ δὲ τούτων καὶ ὁ ἄπειρος καὶ ὁ ἀεὶ λαμβανόμενος χρόνος σύγκειται. τὸ δ'ἐκ μὴ ὄντων συγκείμενον ἀδύνατον ἂν εἶναι δόξειε μετέχειν οὐσίας.

(Aristotle 1983; 217b32–218a3)

[2] « Duo ergo illa tempora, praeteritum er futurum, quomodo sunt, quando et praeteritum iam non 'est' et futurum nondum 'est'? » [1].

[3] « Quid est ergo tempus? Si nemo ex me quaeret, scio; si quaerenti explicare velim, nescio. » [1].

Now he has preceded me a bit even in the farewell to this strange world. This does not mean anything. For us faithful physicists the separation between past, present, and future means nothing but an, although obstinate, illusion.[4]

Here Einstein takes the framework of space-time, which is best suited for the description of measurable events, as true reality. The structure of time, on the other hand, which we know before any physics, to him looks like an illusion.

But that structure of time described above is what we know before knowing any physics. It cannot be falsified as an illusion by physics. But we have to ask ourselves what its role might be for physics, since its structure is so much mutilated in physical theory; not even the "now" has any place there.

2 The Difference Between Past and Future

One thing is true in physics as well as in everyday life: the quite simple observation that the past is actual and future is possible ("potential", according to the conventional Aristotle-translation). This characterization of the decisive difference can guide us on our search for the reasons of that mysterious "direction" of time.

Another observation can guide us as well: There are reversible theories in physics, as mechanics and electrodynamics, and there is an irreversible theory, thermodynamics. Apparently an irreversible theory is closer to the real world. We immediately see if a film is shown backwards, for it shows entirely impossible processes. On the other hand all fundamental theories of physics seem to be reversible. This is an intriguing feature, indeed, that will be worked on elsewhere. But for the moment, for our discussion of time, it is a very interesting point that in statistical mechanics the irreversible theory, thermodynamics, can be "reduced" to mechanics, the reversible theory. This seems to be the place where irreversibility comes into physics, where we might be able to tell how time begins to be "directed" within physics.

3 Statistical Thermodynamics, Part 1

Thus in our context the question of the "direction" of time boils down to the question of the difference between reversible and irreversible theories. More specifically we will ask, in the case of thermodynamics, how irreversible thermodynamics can be "derived" from reversible mechanics.

Let us recall the argument:

In the middle of the 19th century the idea began to be accepted that the thermodynamics of any system—beginning with the theory of gases, later thermodynamics in general—could be derived from the statistics of its smallest parts. For a gas e.g.

---

[4]"Nun ist er mir auch mit dem Abschied von dieser sonderbaren Welt ein wenig vorausgegangen. Dies bedeutet nichts. Für uns gläubige Physiker hat die Scheidung zwischen Vergangenheit, Gegenwart und Zukunft nur die Bedeutung einer, wenn auch hartnäckigen, Illusion" ([15, p. 537]; translation into English MD).

this is the statistics of its molecules. The great advantage of this statistical theory is that it is derivable from very general principles. So it teaches understanding thermodynamics, the science of steam engines, which looked very special in the beginning, as a general theory of approximate description of any physical system. The generality of the objections against that theory corresponded to the generality of the theory itself. Until now among these objections the "reversal objection" has been felt to be particularly serious. This objection, first formulated by Lord Kelvin and J. Loschmid in 1875, starts form the irreversibility: The second law of thermodynamics says that the entropy of a system that is closed energetically as well as materially, can increase or stay the same, but cannot decrease. This describes e.g. two systems at different temperatures. Their temperatures converge when they are brought into contact, but the temperatures will never become different by themselves. In this sense thermodynamic processes are "irreversible". This irreversibility can be derived from the mechanics of a system consisting of very many partial systems, e.g. from the mechanics of a gas that consists of very many (10^{23}!) freely moving molecules.

The argument is roughly that among all possible states of the system at one time an overwhelming majority will develop, according to mechanics, such that the entropy of the system increases in time, or it remains constant. There are states, though, that will develop in a way of decreasing entropy ("fluctuations"), but their probability is negligible. Thus with overwhelming probability, practically with certainty, the entropy will not decrease, the theory is irreversible.

Now the problem is that mechanics is a reversible theory. This means that, according to mechanics, with every process the reverse process is possible as well where velocities have reverse direction and the sequence of the positions is reversed. Statistics adds nothing to mechanics but a reduction of detail in the description of the processes in the way of only retaining average values ("coarse graining"). One cannot see how a reduction of information could possibly change anything about the basic reversibility of the theory.

This problem is stated more precisely in the "reversal objection": Regard any development of a thermodynamic system, where entropy increases. Now imagine that in the basic mechanical system all velocities are reversed such that the system passes the states it has just passed in reversed order. Then also the thermodynamic states the system has just passed will be passed in reversed order so that entropy will decrease. Mechanically the latter process is possible as well as the former, but thermodynamically it is impossible. Thus, the reversal objection says, the claim that thermodynamics follows from mechanics is not consistent.

Here the "direction" of time enters the scene.

Boltzmann gives several arguments in defense of statistical thermodynamics. At first he says that usually, in considering thermodynamic systems, we start with a state of low entropy; thus, in regarding the states of *all* mechanical systems that belong to this thermodynamic state, we find an overwhelmingly large probability for increasing entropy.

Later he explains his point of view in regard of the whole universe in a different way, namely:

> For the universe, the two directions of time are indistinguishable, just as in space there is no up or down. However, just as at a particular place on the earth's surface we call "down" the

direction toward the center of the earth, so will a living being in a particular time interval of
such a single world distinguish the direction of time toward the less probable state from the
opposite direction (the former toward the past, the beginning, the latter toward the future,
the end). By virtue of this terminology, such small isolated regions of the universe will
always find themselves "initially" in an improbable state. This method seems to me to be
the only way in which one can understand the second law—the heat death of each single
world—without a unidirectional change of the entire universe from a definite initial state to
a final state.[5]

This is a particularly obvious manifestation of the prejudice of a typical physicist.
He thinks that the really true description of the world is the one by equations of
mechanics or thermodynamics, where the real parameter t governs processes. And
on the other hand he thinks that past, present and future are "subjective" accessories
of the individual that have to be unmasked as soon as they lead into difficulties
with the true, objective description by physical equations.—That Boltzmann calls
his abstruse proposal the only method to think the "direction" of time can be read
today as admitting his failure.

4 Questions

Now we can specify, as is usually done, Augustine's question: « Quid est ergo tempus? » in a way that is adapted to modern physics:

1. Where do past, present, and future come from?
2. What is the status of past, present, and future in physic, in science?
3. How does thermodynamics (irreversibility) follow from mechanics (reversibility)?

Ad 1: Starting with the first question, my answer is probably quite different from
that of most physicists: The "direction" of time is neither derived from physics nor
can it be in any way founded on structures that are described in science. Time is
rather *presupposed* for science, it is, so to speak, systematically *before* science. For
science is empirical. Experience means that you learn from the past for the future:
this presupposes the "direction" of time. And since science presupposes (the possibility of) experience, all the more it presupposes past, present, and future. Thus

---

[5]"Für das Universum sind also beide Richtungen der Zeit ununterscheidbar, wie es im Raume
kein oben oder unten giebt. Aber wie wir an einer bestimmten Stelle der Erdoberfläche die Richtung gegen den Erdmittelpunkt als die Richtung nach unten bezeichnen, so wird ein Lebewesen,
das sich in einer bestimmten Zeitphase einer solchen Einzelwelt befindet, die Zeitrichtung gegen
die unwahrscheinlicheren Zustände anders als die entgegengesetzte (erstere als die Vergangenheit, den Anfang, letztere als die Zukunft, das Ende) bezeichnen und vermöge dieser Benennung
werden sich für dasselbe kleine aus dem Universum isolierte Gebiete, "anfangs" immer in einem
unwahrscheinlichen Zustande befinden. Diese Methode scheint mir die einzige, wonach man den
zweiten Hauptsatz, den Wärmetod jeder Einzelwelt, ohne eine einseitige Änderung des ganzen
Universums von einem bestimmten Anfangs- gegen einen schließlichen Endzustand denken kann."
(Boltzman [2, Sect. 90]).

there is no way to answer our first question within science. All we can do is state that structural dependence.

Ad 2: Our considerations within physics about the "direction" of time have the status of a consistency argument: When irreversibility, the "arrow" of time, comes in, we should see if the structure we find in its description is a contradiction to what we have presupposed in the beginning, or if it is compatible. If it is not compatible, we will be in serious trouble!

Ad 3: Even though we do not have to *derive* the "arrow" of time from our considerations about statistical thermodynamics, it is still, within a consistency argument, an important question how that "direction" of time comes in through statistical thermodynamics. For it is curious: We deal with a system of mechanical objects within a reversible mechanical model. Then we decide not to look so closely, but rather to coarse grain our view a bit, and—bingo!—the theory is irreversible! So this is our question: *Where does the arrow of time enter the theory?* We are going to deal with this question in most of the remaining part of this paper.

5 Probability

A crucial notion for our discussion is *probability*: It is a basic concept of Statistical Thermodynamics, the use of probability arguments marks the difference between Statistical Thermodynamics and Mechanics. What is probability?—In order to get some of the arguments straight that play a role in the relevant discussions, we have to indulge at least to some extent into the very complex discussion of that mysterious notion.

Until recently most experts on probability were convinced that there is no *definition* of probability. This is the result of a long discussion of that concept, rather from the beginning of probability calculus in 17th century.

6 Classical Definition

There was, historically, the so-called *classical definition of probability* by Laplace, saying: "Probability is the ratio of the number of favorable cases to the number of possible cases." [5, p. 146]. This applies mainly to the combinatorial considerations that were quite common in the beginning of probability theory, e.g. for calculating the chances in games of cards or of dice: The probability to draw a king in a deck of 52 cards is 1/13, namely 4 (the number of kings, the *favorable* cases) divided by 52 (the number of all cards, the *possible* cases). One sees quickly that this is not a proper *definition*: Laplace himself puts his "definition" under the condition that "we see no reason why one of those cases would occur more easily that any other one" [5]. He could have put it more clearly: This applies if we suppose equally probable cases. Thus Laplace reduces unequal probabilities to equal probabilities, but he does not define the *concept* of probability.

A true definition of probability, on the other hand, ran into all but insurmountable difficulties. The reason is the inherent vagueness of probability that cannot be removed by a definition, sharp as that definition might be: Apparently probability has to do with relative frequency. But it cannot simply be identified with relative frequency since the latter only roughly equals probability.

The problems about the definition of some objective concept of probability led to the introduction of *subjective* probability.[6] The latter refers to the subjective assessment for the degree of truth of a proposition, made explicit e.g. in the willingness to bet. According to this view the rules of probability theory contain nothing but the conditions for the consistency of such assessments. We can make them explicit in the condition that a bet has to be *fair*.

7 Predicted Relative Frequency

The battle between those two concepts—and a dozen more that came up in the meantime—rages until now. What is interesting for us is only the concept of probability as it is used in physics. There we see that it is intimately connected with *prediction*. Actually the solution of the problems about probability came from defining probability as a *predicted relative frequency*.

For the concept of probability, as for thermodynamics, the inclusion of the "direction" of time gives amazingly simple solutions:

At first it is clear that predictions do not have to come exactly true. Probability theory itself gives a prediction for the mean *deviation* of relative frequency from the predicted value in an actual series of experiments. In order to specify this prediction, in turn, one can calculate the deviation of those deviations from their predicted value, for series of series of experiments, etc. Thus probability has got a hierarchical structure that can be continued as far as one likes (cf. [6, 7]).

The definition of probability we gave, as a predicted relative frequency, allows us to see the systematic place of the difference between objective and subjective probability: *Prediction* always contains a *subjective* element: Somebody predicts, and predictions may turn out wrong. But predictions made in science are supposed to prove true empirically, i.e. to indicate *objective* facts. We could describe their relation in this way: The subjective interpretation of probability emphasizes the character of proposition, of knowledge: the subjective opinion about what the relative frequency will be. For the objective interpretation, on the other hand, the emphasis is on the *content* of the prediction, on the real future relative frequency that would confirm a true prediction.

Let me add a remark on the concept of probability in general. Our definition covers one of many possible meanings of probability. There could be (and are) other ways to use this word. The structure of our argument is: Science deals with relative

---

[6]Bruno de Finetti since the 1920s. Cf. in english [4].

frequencies and their prediction. I find that what traditionally is called probability usually agrees with my concept of predicted relative frequency. And I find that for this concept of predicted relative frequency some problems that are usually discussed in relation with probability find a solution. But it is quite possible that there are other concepts of probability that are not affected at all by this argument of mine.

We can ask now whether probability is objective in the sense of a measurable quantity. The fact that probability is found in a measurement only approximately is not a counter-argument. For this is true for all measurable quantities. But for probability this inaccuracy is of very fundamental nature: What we measure is a relative frequency; and above we have seen that probability is *not* the same as relative frequency. If we want to interpret probability as a property of a physical system we have to treat it, apparently, as a kind of disposition, a "propensity", as Popper calls it, to produce certain relative frequencies [13]. This propensity does not appear directly as a result of a measurement. What we measure, the phenomenon, depends on the propensity in a well-known way, but it is not identical with it.—In the discussion of quantum mechanics we will come across such structures again.

8 Axiomatic

In spite of the systematically unavoidable inaccuracies while predicting relative frequencies, probability theory allows exact calculations with real numbers. Why is that?

Probability theory has become pure mathematics since its axiomatization by A. Kolmogoroff in 1933 [12]. The crucial point in his axiomatics is banishing the problematic relation between probability and relative frequency entirely from mathematics into the "application". In his axiomatics he included only the *relations* among probabilities; that could be stated exactly and rather simply. In fact probability theory from the beginning dealt with nothing but relations among probabilities and their "consistency", as mentioned above.

Another brilliant simplification in Kolmogoroff's work is his treatment of the so-called product rule saying that the probability for events A *and* B is the product of the probability of A alone and the probability of B alone, provided the two events are *independent*. Kolmogoroff does not give a criterion for the independence of events but he introduces the product rule by a *definition*. This definition reads something like: "We call two events A and B independent if the product rule is true for them."— Looked at in this way, probability theory is pure mathematics. Mathematicians put aside the problems we mentioned above as "application problems".

We like to introduce the opposite view as well, which proves probability theory to be a *science*. This is again aided by regarding the "direction" of time. For we can see, from the structure of time, that the most general law of nature is a probability law. Since this is a very special assertion in the framework of our investigations I will explain it a bit more in detail.

What is a law of nature?—Reduced to the most general scheme every law of nature is a prescription how to get empirically testable predictions from the present state. The most general empirically testable prediction, I claim, does predict a relative frequency. Above we introduced probability as a predicted relative frequency. Thus our assertion reads: The most general empirically testable prediction is a probability statement.

Let me give a short sketch of the argument (cf. [6]): We can give, evidently, unambiguous "simple" predictions, yes or no. But one could also think of specifying predictions like: "Sometimes yes, sometimes no" in a way to make them empirically testable. This kind of prediction should be *general*, just as the simple ones. This means that they cannot apply e.g. to a definite number or a definite sequence of yes and no. For this again would be a simple prediction, only for a more complex experiment.—The only *general* prediction that specifies "Sometimes yes, sometimes no", it turns out, is the prediction of a relative frequency. This means it is a probability.

We can then derive the well-known rules of probability calculus from the definition "predicted relative frequency". In doing so we cannot, as Kolmogoroff, introduce *independence* by a definition. But we can specify the independence of events A and B by the condition that the predicted relative frequency for A is the same if either B is the case or if non-B is the case, and vice versa. With those premises we can derive Kolmogoroff's theory. The definition of probability given, using the "direction" of time, turns out to be the basis for the whole theory of probability.

9 Property of What?

It has often been argued whether one can apply probability to single events or only to series of events. Here again our approach gives us a clue.

We have seen above that we can consider at most, as a property of the system, the *propensity* to produce certain relative frequencies. A relative frequency, in turn, is a property of an actual series of measurements. This could be, e.g., a series of 14 throws of dice, and the result could be twice the "1"; whereas the corresponding probability, the propensity of the system to produce the result "1", could have been 1/6. This latter disposition, the propensity, is usually (as in our example) not exactly confirmed by the actual frequency. But the disposition is valid, by its definition, for *any* actual series of experiments.

Thus it turns out that both views are valid: One can assign a probability to the class of all possible series of experiments; but with the same right one can assign probability to one experiment as representative of that class; for it is constitutive for that class that its members are all one and the same type of experiment as far as probability is concerned.—In this description it is a problem which experiments are "of the same type" and of that same probability; it is a question of the experimentalist's skill who devises the experiment to ensure that "same type" for all experiments.

10 Statistical Thermodynamics, Part Two

We have presented so far a definition of probability that is founded on the "direction" of time. How does this go along with the traditional discussion that aims, conversely, at founding the "direction" of time on statistical thermodynamics, i.e. on a probabilistic theory?

We have already seen that it is impossible to understand how the "direction" of time could come just from coarse graining the description of a mechanical system. Is it really true that nobody has ever noticed that?

In the beginning of the 20th century J.W. Gibbs completed the theory of Statistical Thermodynamics. Naturally he encountered the same problem, but he dismissed it rather pragmatically. He writes in his standard treatise [10, pp. 150–151]

> But while the distinction of prior and subsequent events may be immaterial with respect to mathematical fictions, it is quite otherwise with respect to the events of the real world. It should not be forgotten, when our ensembles are chosen to illustrate the probabilities of events in the real world, that while the probabilities of subsequent events may often be determined from the probabilities of prior events, it is rarely the case that probabilities of prior events can be determined from those of subsequent events, for we are rarely justified in excluding the consideration of the antecedent probability of the prior events.

Here Gibbs hints, in a rather hidden way, at an idea that should later bring the solution of the problem, as we saw above: Probability is generally applied only to *predictions*, not to propositions on past events. It is possible, admittedly, to give real sense to propositions on past events, like e.g.: "Probably Napoleon was born in 1769". But the uncertainty we indicate by the word 'probably' does not refer to the past fact itself. For Napoleon was born in 1769 or he was not born that year; the fact exists, no matter we know it or not. What is uncertain is what we will possibly *know* in future. Thus, even when we assign probabilities to past facts we mean a *possibility*, namely the real knowledge we may have in future.

In 1939 C.F. von Weizsäcker picked up this thread and gave a refutation of the "reversal" objection [16]: The difference between past and future, which is characteristic for thermodynamics, does not come into the theory in a mysterious way by "coarse graining". It is rather ourselves who introduce this difference from outside, just in applying probability only to future.—This appears to be so self-evident that nobody made it explicit before 1939. In 1971 Weizsäcker himself writes, when his paper was being reprinted: "When I wrote it I felt that I had set forth something rather trivial. . . ." He calls his text nothing but an attempt at explaining Gibbs' words.

In his paper C.F. von Weizsäcker begins with stating that Boltzmann's H-theorem does not imply a difference between past and future. What Boltzmann proves is that with any state of non-maximal entropy all neighboring states have, with overwhelming probability, higher entropy, i.e. past states as well as future ones. Past and future are entirely symmetric. From Boltzmann's assumption of thermodynamic probability there rather follows "that a non-maximal value of entropy of a system we know nothing else about is, with overwhelming probability, a relative minimum of entropy", as Weizsäcker puts it [16, p. 174].

11 Thermodynamic Probability

What enters here is the very important concept of *thermodynamic probability* of a thermodynamic state. It is quite simply defined as the number of microstates in a macrostate divided by the number of all microstates. Microstates are mechanical states of the system (in its exact description), whereas macrostates are thermodynamic states, comprising many microstates, usually a huge number.

Using now the "classical definition of probability" mentioned above we call that ratio the "thermodynamic probability" of the macrostate. If this is really to be a probability, i.e. a predicted relative frequency that could be confirmed by experiment, the microstates must have equal probability—just as in the "classical definition". This is usually the case in thermodynamic equilibrium, but not in other thermodynamic states.

Once one has seen this rather trivial connection, one can explain

(a) why so often classical thermodynamics is criticized for applying only to equilibrium: this is because only in equilibrium thermodynamic probability is real probability; thus only in equilibrium probability arguments on that basis are valid.

(b) what the phrase "…of a system we know nothing else about…" means: It is supposed to imply symmetry between the states of the system, i.e. in this case, equal probability.

The latter argument, by the way, may be utterly misleading. If we really know nothing about a system then we do not know anything about the relation between its states either, and there is no way to imply any symmetry. What is usually meant by that phrase "…of a system we know nothing else about…" is rather that we have reasons—wherever they come from—to assume symmetry in the first place, and we have no reason (from more detailed knowledge, e.g.) to conclude that this symmetry does not imply in the present situation. We can assume, e.g., that a die is symmetric as long as we know nothing else about it: This is because dice are usually produced symmetric, and we usually rely on their outer appearance as long as we do not know of facts that would support the suspicion that it is otherwise.

12 Predictions

For a prediction, the original application of probability, this entails growth of entropy. For retrodictions, for the past, however, we need additional considerations.

Suppose you *know* that the system you consider is in thermodynamic equilibrium. Then Boltzmann's considerations are immediately valid, a state of non-maximal entropy is most probably an extreme of a fluctuation.

Often, however, we consider a system about the past of which we have or can infer some information. When I see, e.g., a pot of lukewarm coffee on a table I can be rather sure that the coffee was hot before and has cooled down, increasing its entropy. This conclusion seems reasonable, considering European household customs, i.e. from implied *facts* of the past. The idea, on the other hand, that lukewarm coffee could be the result of a fluctuation is absurd, considering imaginable past facts.

Thus the problem how the difference between past and future comes into Statistical Thermodynamics is resolved rather convincingly: We ourselves introduce that difference into our considerations in applying probability to predictions, but not to *retrodictions*.

Once we have drawn our attention to this structure it is not mysterious any more.—It is a pity, though, that this solution, which has been given as early as in 1939, has not yet entered the discussion within the scientific community. Recent presentations[7] still reproduce Boltzmann's discussion, which apparently is unsatisfactory. Not even Gibbs' remark (of 1902!) has brought a modification of those presentations.

Paul and Tatjana Ehrenfest have introduced in 1906 a nice model into this discussion that can be used to make many things clear within an environment that has a simpler structure than thermodynamics of e.g. an ideal gas [8, 9]. It is kind of a game with numbered balls distributed into two urns, and a lottery with cards that bear the numbers on the balls. Every step of that game consists in drawing one of the cards (with equal probability), whereupon the ball with the number drawn changes its urn.

That system has a microstate, namely the exact distribution of the individual balls in each urn. It has also a macrostate that is characterized by the number of balls in one (say, the left) urn. The analogy with thermodynamics is in approaching equilibrium (equal number of balls in both urns), where all microstates have equal probability. There are statistical fluctuations as in real thermodynamics, etc. With this system one can easily demonstrate the facts mentioned above: That probability considerations apply without restriction to the future, that they apply in the same way to the past when there is equilibrium, and that considerations about the past become much more complicated when we cannot suppose equilibrium (cf. [6, pp. 48–57; 215–219]).

13 Direction of Time?

So far we have seen that we can solve the questions we posed above only when we turn them around: It is really not from physics that the "direction" of time follows, but it is the "direction" of time we presuppose in doing physical theory. Thus it is entirely the other way round: We have no chance to understand how the "time arrow" comes into statistical thermodynamics if we do not *start* with the "direction" of time, then see that probability is understandable only as a prediction, and then find that this structure is mirrored in statistical thermodynamics.

This insight may also give us a clue to better understand what the puzzles of the interpretation of quantum mechanics have to do with the "direction of time".

---

[7]Cf. e.g. [11]; similarly [14].

14 Quantum Mechanics

Quantum mechanics can be interpreted as a generalized probability theory. To be more specific, we can show that Kolmogoroff's axioms of (classical) probability calculus allow a generalization to a quantum mechanical probability theory. Kolmogoroff bases his axioms on the set F of random events, where every random event is represented by a set of *elementary* random events. His first axiom reads:

"I. F *is a field of sets*."

A field of sets is what is today called a Boolean lattice (of sets). For quantum mechanics we use instead as a first axiom:

"I'. F *is a lattice of closed subspaces of Hilbert space*."

The difference between these two axioms contains all differences between classical physics and quantum mechanics; Kolmogoroff's other axioms remain the same in quantum mechanics. Those differences become clearer, again, when we *start* from the "direction" of time. In fact, basing the theory on a lattice of subspaces instead of a field of sets entails a fundamental *indeterminism* [6].

Indeterminism mirrors future's peculiarity as contrasted to the past: Future events are *possible*. In general with every event also alternative events are possible, the future is *open*. The quantum mechanical lattice of propositions can be understood most easily as expression of the open future, as a lattice of *predictions*. And probability, which is fundamental in quantum mechanics, involves a prediction—as we have seen above. We find that in the quantum mechanical lattice never all predictions can be made with certainty, i.e. with probability 0 or 1. There are always predictions with probability *between* those two extreme values. This is the fundamental indeterminism of quantum mechanics.

One could, in principle, treat the classical lattice of propositions (Kolmogoroff's field of sets) as a lattice of predictions as well. (In classical physics we can suppose that all predictions can be made "in principle" with certainty, i.e. with probability 0 or 1. Thus probabilities other than 0 or 1 must be due to our ignorance—as Laplace says in his classical formulation of determinism.) In this view the classical lattice of propositions is a degenerate case of the quantum mechanical lattice of propositions in such a way that it, "accidentally", contains only probabilities 0 and 1.[8]

Difficulties arise if one comes from the other side, the side of classical physics, which presumes that all predictions are certain. Such predictions can be understood as well as descriptions of properties that are there *in themselves*. If I can predict with certainty that I will find, e.g., planet X in position y, then I will be able to say as well: "Planet X is really in position y". So predicting the result of a measurement turned into stating a fact. For classical physics these are, as we can easily see, equivalent.

---

[8]Technically speaking the quantum mechanical lattice of propositions becomes a classical one when there is a complete superselection rule, i.e. when no superposition of states is possible, and therefore all observables are compatible among each other.

But in quantum mechanics, with its fundamental indeterminism, this "realism" does not work anymore.

This is apparently the source of many problems for someone who is used to the "ontology of classical physics", and this is where the dissatisfaction of "classical" physicists with quantum mechanics comes from.

It is in the same spirit that "realism" in the interpretation of quantum mechanics asks: What is the *reality* that is described by quantum mechanics?—or what is *behind* the quantum mechanical description?—We try to answer this difficulty with recourse, again, to the "direction" of time.

The primary purpose of a physical theory is generating empirically testable predictions rather than a description of existing reality. In case the predictions can be made with certainty they can be reformulated, as we saw above, as a description of reality. But this very possibility is excluded in quantum mechanics—if we exclude, for the moment, rather far-fetched variants like Bohmean mechanics. The fundamental indeterminism of quantum mechanics excludes this kind of "realism".

15 Necessitiy of Classical Concepts

In a second step we can specify the question of reality in a deeper way: Certainly every prediction presupposes a fundamental reality that allows describing the facts that form the basis for the prediction, and finally describing those facts that confirm or disprove the prediction. Niels Bohr calls this essential requirement for the interpretation of quantum mechanics the "necessity of classical concepts".

There are fundamental difficulties that result from this necessity of classical concepts. I can only sketch them here: Quantum mechanics gives nothing but probabilities for the results of possible experiments. If we want to describe unambiguously arrangements and results of measurements we need an appropriate language, appropriate concepts to describe reality, facts. Niels Bohr says we must be able to describe what we have done and what we have learnt. This is impossible in quantum mechanics alone, for this purpose we need the concepts and theories of classical physics.

But then we face a problem: Quantum mechanics was introduced because it describes phenomena classical physics was unable to describe. Whenever the results of the two theories differ, quantum mechanics is right, classical physics is wrong. How is it possible then that (true) quantum mechanics presupposes, in the end, (wrong) classical physics?

The practical physicist has a simple answer to this question: Where classical physics is needed for quantum mechanics, namely for describing arrangements and results of measurements, the two theories agree in a very good approximation. Thus we can assume the validity of quantum mechanics and still, in a good approximation, use the concepts of classical physics. For all practical proposes (FAPP as an acronym) this is quite all right.

The philosopher, though, particularly if he is mathematically and logically minded, wants to know more precisely. This "FAPP" may suffice for the practically working physicist but the logician must conclude: *Approximately correct* means, if you take it seriously, *false*. So the whole theory is apparently inconsistent!

16 Process of Measurement

The discrepancy that shows up here also formally appears in the description of the process of measurement. We cannot present here the theory of measurement in any detail (cf. [3]); but let me at least sketch a rough outline.

The process of measurement in quantum mechanics is interesting mainly because the theory is indeterministic. This means that before measurement several results are *possible*, but after measurement only one of the results has become *actual*. It is true, this occurs in classical physics as well. But there we can console ourselves with the thought that "in itself" already before measurement there existed but one possibility, and that it was only our ignorance that forced us to take more than one possibility into consideration. But in quantum mechanics even with the most exact description there remains, in general, more than one possibility for the result of a measurement. This is what we mean by its fundamental indeterminism.

The *state* of the system leaves open, before measurement, several possibilities with their corresponding probabilities. After measurement this diversity is reduced to one single, the actual, case. This case has then (for an immediately following second measurement) probability 1. That change of state is called the "reduction of the wave packet".

There are "realists" among physicists (or, still more, among philosophers of science) who look for a physical mechanism that brings about this change of the physical state. But when we take the "direction" of time seriously, as was explained above, we can see that we do not need a physical mechanism. In fact, a prediction with more than one possibility *means* nothing but that in the end one of those possibilities will be realized, the others not. That is what is meant by the predictions of quantum mechanics. The change in description after the measurement is a *decision* by the physicist. He could just as well continue with the old description and keep all prior possibilities for further predictions, including the corresponding probabilities. Then he would waive the chance of using the information from the experiment. But the latter description would be as valid as the first, "reduced" one. Thus it is obvious that there can be no physical mechanism, within the described system, for the "reduction", since the reduction depends on a decision of the one who describes that system.

There is a possibility to waive information from measurement in the formal description of the process of measurement as well. We would presuppose for that description that a measurement has actually taken place, and that, consequently, one of the possibilities has become actual; but the information, *which one* has become actual, is waived. Then our description would contain all possible outcomes with the respective probabilities, it would be a "mixture" of states.

The most interesting point in this theory is that the *mixture* mentioned above is different from the *state* that results from the initial state of system + measuring instrument by the measuring interaction according to the Schrödinger equation. A long discussion has shown that here is a fundamental problem we cannot get rid of by simple tricks. The generation of a mixture described above has been called—misleadingly—"reduction of the wave packet" as well. I recommend rather using the more precise term "disappearing of the interference terms". For the difference between the two descriptions is that the correlations between system and measuring apparatus, which are present after the measuring interaction in the *state* description, have disappeared in the *mixture*. What has disappeared are the terms that describe the "interference" between the system and the apparatus. Their disappearance is usually called the cut between system and measuring apparatus.

We can look upon this fundamental problem of the theory of measurement as the formal expression of the problem of *classical concepts* mentioned above: If we want to describe unambiguously what we have measured we have to waive the remaining correlations between system and measuring apparatus. With a good measuring apparatus this can be done quite easily. For in that case the interference terms are so small that they play no role for any practical purpose ("FAPP"); so again there is no problem for the practical physicist. But if we look closely we see that those interference terms, however small they may be, always exist, they are never exactly zero. Thus, strictly speaking, it is a mistake to neglect them. Eugene Wigner, who discussed this problem very carefully, finally could offer no other way out than putting small non-linear parts into the equations of quantum mechanics that make the interference terms disappear within a short time after the interaction [17].

To my advice Wigner's solution is wrong.

Let us again start with the "direction" of time: We are dealing here with predictions within *physics*. Physics, however, contains *approximations* in its very foundations. This is to be seen rather easily: We can do physics only if we can deal with objects independently of their environment. But "in reality" there are no separate objects; everything is related with everything. One can see this already in celestial mechanics: Conceptually isolating a planet in the solar system from the totality of celestial bodies means an approximation. Strictly speaking, according to physics itself, every celestial body, however far it may be, has influence on our planet from gravitation alone, not mentioning other types of interaction. These influences are so small that they can be practically neglected, but strictly speaking they are there. Treating planets only under the influence of the sun and the neighboring planets is an approximation and so, strictly speaking, it is wrong. If we did not use that approximation, however, we could not do any physics. And, above all, there is no way to describe a *strictly* independent object at all: At least the interaction with the measuring apparatus must exist in the way that the object can be an object *for me*.

The approximation introduced by neglecting the interference terms is of exactly the same sort: We neglect the very small interaction that relates system and measuring apparatus still after the measurement. Thus we introduce an approximation of the same sort as we have introduced in the very foundations of physics.—Translating this into the language of the "direction" of time means: We can give, fundamentally, only approximate predictions. This is true from their probability character alone, since probability propositions cannot be verified exactly. But it is true as well because of the fundamental approximation character of physics, which says: We can give predictions only about isolated objects, which, strictly speaking, do not exist.

During the last decades many proposals have been published to solve the problem of measurement, e.g. under the name of "consistent histories" or of "decoherence".[9] Those proposals amount to the same solution under a different name, namely to the old suggestion to neglect the interference terms ("FAPP"). Unfortunately, the authors of such recent proposals give the impression that they could now offer, other than the old authors, an *exact* solution of the problem. But this claim would mean more than one could make good.

A common argument against the view put forward here reads like this: "One who puts so much emphasis on *predictions* has only eyes for the possibilities of manipulation; he has an 'instrumentalist' view of nature. Genuine philosophy of nature should inquire more deeply, namely about what is *really* the basis, maybe hidden, of the outer appearances."—A suggestion like this is, not easily recognized, founded again on the ontology of classical physics. For it presupposes that "in itself", *behind* the appearances, there is something else that perhaps does not show itself easily, but whose description is the genuine goal of philosophically oriented science.

17 Objectivity

A program like this may be understandable from the point of view of classical physics. But there is nothing to justify it in this generality. For if we ask, according to empirical science, for the general structures of reality, we ask for an *objective* description according to the spirit of this science. This means we ask for a description that *everyone* could in principle verify *at any time*. But if we will be able to verify a proposition empirically this proposition must be a prediction: We must be able to look whether it is true *after* it has been made. This is what we brought out by our analysis of the "direction" of time. It is a specialty of classical physics that such predictions can also be formulated as descriptions of reality *in itself*. What makes this specialty possible is the fact that in classical physics with maximal knowledge all predictions can be made with probability 1 or 0. But now, knowing quantum mechanics, we cannot presuppose that any more. With a fundamentally indeterministic theory there is no such *reality in itself* any more.

But *objective* description is still possible, in the sense that any claim can be verified by anyone at any time: A general claim of this sort is a *prediction*.—Anyone who calls this view "instrumental" from the perspective of the ontology of classical physics spoils every chance at understanding a more generally objective description of reality.

To conclude we may say: Time *is* directed—in the sense of its structure of present, past, and future. But this structure is not implied by, nor is it derivable

---

[9]Cf. e.g. the works of Detlev Dürr or Roland Omnes and their collaborators.

from physics. To the contrary: Physics would not be possible at all if we did not start form this "direction" of time.

References

1. Augustinus, A.: Confessiones XI, 14(17)
2. Boltzmann, L.: Vorlesungen über Gastheorie, 2 vols. Barth, Leipzig (1898) (1896/98, 1910). English edition: Lectures on Gas Theory, Translated by Stephen G. Brush, Berkeley (1964), New York (1995)
3. Busch, P., Lahti, P.J., Mittelstaedt, P.: The Quantum Theory of Measurement. Springer, Berlin (1991)
4. de Finetti, B.: Probability, Induction, and Statistics. Wiley, London (1972)
5. de Laplace, P.S.: Recherches sur l'intégration des équations différentielles aux différences finies et sur leur usage dans la théorie des hasards. Mém. Acad. R. Sci. Paris (Savants Étrang.) 7, 69–197 (1773). Reprinted in: Œuvres complètes de Laplace, vol. VIII, Paris (1891)
6. Drieschner, M.: Voraussage, Wahrscheinlichkeit, Objekt. Springer, Berlin (1979)
7. Drieschner, M.: Moderne Naturphilosophie. Mentis Verlag, Paderborn (2002)
8. Ehrenfest, P., Ehrenfest, T.: Über eine Aufgabe aus der Wahrscheinlichkeitsrechnung, die mit der kinetischen Deutung der Entropievermehrung zusammenhängt. Mech.-Nat. Bl. 3(11, 12) (1906). Reprinted in: Klein, M.J. (ed.): Collected Scientific Papers. Amsterdam, New York (1959)
9. Eigen, M., Winkler, R.: Das Spiel. Piper, München (1975)
10. Gibbs, J.W.: Elementary Principles in Statistical Mechanics. Scribner's, New York (1902). Reprint: Woodbridge, CT (1981)
11. Grünbaum, A.: The anisotropy of time. In: Gold, T., Schuhmacher, D.L. (eds.) The Nature of Time. Cornell University Press, Ithaca (1967) or Grünbaum, A.: Philosophical Problems of Space and Time (ed. by Cohen, R.S., Wartofsky, M.H.). Reidel, Dordrecht (1973)
12. Kolmogoroff, A.: Grundbegriffe der Wahrscheinlichkeitsrechnung. Springer, Berlin (1933)
13. Popper, K.: The propensity interpretation of the calculus of probability and the quantum theory. In: Körner, S. (ed.) Observation and Interpretation. Butterworths, London (1957)
14. Sklar, L.: Physics and Chance. Philosophical Issues in the Foundations of Statistical Mechanics. Cambridge University Press, Cambridge (1996)
15. Speziali, P. (ed.): Albert Einstein – Michele Besso. Correspondance 1903–1955. Hermann, Paris (1972)
16. von Weizsäcker, C.F.: Der zweite Hauptsatz und der Unterschied von Vergangenheit und Zukunft. Ann. Phys. 36, 275 (1939). Reprinted in: von Weizsäcker, C.F.: Die Einheit der Natur, pp. 172–182, Hanser, München (1971)
17. Wigner, E.: Remarks on the mind-body question. In: Good, I.J. (ed.) The Scientist Speculates. p. 302. Heinemann, London (1961). New York (1962), p. 302. Reprinted in: Wigner, E.P.: Symmetries and Reflections, pp. 171–184. Bloomington and London (1967)

Chapter 13
Time in Modern Philosophy of Physics—A Survey

Holger Lyre

Abstract The *topos* of time ranges among the most puzzling and intriguing topics in our philosophical tradition—a seemingly endless source of deep and unsolved questions: What is time? What is temporal becoming? And how are we to spell out all this without using temporal notions in the first place? These questions are puzzling also in the sense that in our everyday life we seem to be quite familiar with the phenomenon of time. In a famous quote from the *Confessions*, Saint Augustine points out this discrepancy in the following way: *"What is time? If nobody asks me, I know; but if I were desirous to explain it to one that should ask me, plainly I know not."* Nevertheless, 20th century physics has seen much progress not in finally answering these questions, but in providing us with some new perspectives and perhaps also some deeper insights into the nature of time from a scientific point of view. This article is accordingly devoted to give an overview on the several aspects of the notion of time—and in particular the directedness of time—in modern physics. (A similar version has been published online as: Time in philosophy of physics: the central issues. *Phys. Phil.*, ISSN: 1863-7388, 2008, ID: 012, http://physphil. tu-dortmund.de.)

Keywords Time · Temporality · Endurantism · Perdurantism · Zeno's paradox · Arrows of time · Conventionality of simultaneity · Hole argument · Parmenides · Heraclit · McTaggart · H-theorem · Second law · Maxwell's demon · Entropy · Information · Measurement problem · Ignorance interpretation · Theory underdetermination · Bohmian mechanicstransactional interpretation

1 Philosophical Preliminaries

1.1 Time and Temporality—Being and Becoming

The notion of time has many faces. One of the most important distinctions in debates about time is the distinction between time in the sense of *being* on the one

H. Lyre (✉)
Philosophy Department, University of Magdeburg, Magdeburg, Germany
e-mail: lyre@ovgu.de

S. Albeverio, P. Blanchard (eds.), *Direction of Time*,
DOI 10.1007/978-3-319-02798-2_13,
© Springer International Publishing Switzerland 2014

hand and *temporal becoming*—tensed time–on the other. In this connection we find in Carnap's autobiographical notes the following well-known passage about his discussions with Einstein:

> Once Einstein said that the problem of the Now worried him seriously. He explained that the experience of the Now means something special for man, something essentially different from the past and the future, but that this important difference does not and cannot occur within physics. That this experience cannot be grasped by science seemed to him a matter of painful but inevitable resignation. I remarked that all that occurs objectively can be described in science; on the one hand the temporal sequence of events is described in physics; and, on the other hand, the peculiarities of man's experiences with respect to time, including his different attitude towards past, present, and future, can be described and (in principle) explained in psychology. But Einstein thought that these scientific descriptions cannot possibly satisfy our human needs; that there is something essential about the Now which is just outside the realm of science. We both agreed that this was not a question of a defect for which science could be blamed, as Bergson thought. I did not wish to press the point, because I wanted primarily to understand his personal attitude to the problem rather than to clarify the theoretical situation. But I definitely had the impression that Einstein's thinking on this point involved a lack of distinction between experience and knowledge. Since science in principle can say all that can be said, there is no unanswerable question left. But though there is no theoretical question left, there is still the common human emotional experience, which is sometimes disturbing for special psychological reasons. [11, pp. 37–38]

Quite obviously Carnap does not fully understand what Einstein really worries about. Carnap presupposes an understanding of time which coincides with the common usage of an earlier-later relation—mathematically described by a real-valued 1-dimensional parameter. Following John McTaggart [38] this one-parameter time is known as "B-series." It reflects, or at least comes very close to, the way time is treated in physical theories, especially space-time theories: time as being, positions in time as earlier-later relations.

By way of contrast, there is the strong, subjective, human experience of time in terms of the *temporal modes*, the *tenses of time*: whereas the future is open and potential, the past is actual and fixed. Possible events of the future come into being at the present, the Now, and immediately slip into the irreversible past. This represents, in McTaggart's terms, the "A-series" of time. Scientific reductionism, in its usual stance, comprises the idea of reducing the A-series to the B-series. And this was precisely what worried Einstein, since he found that the Now has no place in physics, which indeed is troublesome, if the modes of time are objective parts of the reality rather than mere subjective experiences.

1.2 The Metaphysics of Time

McTaggart's main concern was to present an argument which—purportedly—proves the unreality of time. For the sake of his argument, which we shall not pursue here, he pointed out that there is an element of permanence in the B-series, namely that once an event is earlier than another event, it is earlier at all times. In contrast to this the A-series is manifestly dynamical due to the ever-shifting of events from future to present and past. One may call this aspect of temporal becoming the "Heraclitean view" as opposed to a "Parmenidean view". According to Heraclite everything flows, nothing abides, and the present is primary. Parmenides, instead, banishes temporal changes as being illusory. Only the static "Is" exists.

The Heraclitean view asserts a diachronous existence (or *persistence*) of things in time. Any 3-dimensional spatial object is wholly present at any one time. Proponents of this view are therefore called *3-dimensionalists* or *endurantists*, and one may presumably consider it the common view of the man on the street. In contrast to this the Parmenidean view asserts the eternal existence of tenseless objects which have temporal parts as well as they have spatial parts. Proponents of this view are called *4-dimensionalists* or *perdurantists*.

Corresponding to these two different views about the existence of objects in time there are the views about the existence of time itself—the subject matter of the metaphysics or ontology of time. Here, endurantism corresponds to *presentism*, the view that only the present exists, whereas perdurantism corresponds to *eternalism*, the view that all temporal parts exist. Both ontological views about time are symmetric, which means that they do not respect the distinction between past and future. There is, moreover, *possibilism* as an intermediate view between presentism and eternalism. The possibilist asserts that the present and the actual past are real and, thus, subscribes to the asymmetry of time as attested by our experience. Accordingly, possibilism is in agreement with endurantism, but not with perdurantism.

As we will see in the sections about relativity theory there are obstacles for the views of presentism and possibilism in special as well as in general relativity theory. Another distinction related to the ontology of time, but also to the ontology of space and, hence, space-time, is expressed in the debate between *relationalism* and *substantivalism*. Whereas substantivalists consider space-time as an entity *per se*, relationalists merely think of it as a set of relations of objects. This will also be addressed in the general relativity section.

1.3 Zeno's Paradoxes

Taken at face value, the Parmenidean view seems to be absurdly wrong. Everyday experience obviously tells us that there simply is true and undeniable change in the world! Nevertheless, the Parmenidean topic of the illusory nature of change lies at the roots of western philosophy's tradition. Among the early supporters of Parmenides and his Eleatic school, Zeno of Elea was perhaps the most influential— also given the fact that both Plato and Aristotle took his arguments quite seriously.[1] He presented a host of paradoxes by using a "dialectic" method, which, following Aristotle, was his genuine methodological invention and which, apart from the arguments themselves, certainly impressed both Plato and Aristotle. The idea of the

---

[1]However, almost everything we know about Zeno and much of what we know about Parmenides is due to Plato's and Aristotle's writings.

dialectic method is to argue against a certain view by showing that it entails un-acceptable or even absurd consequences. For the particular case at hand, Zeno had argued that the denial of the Parmenidean view—the indivisibility of motion, for instance—leads to absurd consequences—namely that motion is impossible. Note that this is absurd from a non-Parmenidean point of view. What Zeno of course wanted to highlight was the cognitive inconsistency in the non-Parmenidean con-cept of motion—and, hence, the Parmenidean or Eleatic view of the illusory nature of change and multiplicity as the only viable alternative. Reality must be a single indivisible One.

Among the variety of ways Zeno presented his argument, the paradox of the race between Achilles and the tortoise is certainly the most famous one. The idea is the following: a tortoise (T) has been given a lead for her race with Achilles (A), the fastest of all the Greeks. Once A has got to the place from which T has started, T has already advanced a little farther. We may iterate this idea and come to the paradoxical conclusion that however fast A runs, he can never catch up with T! (And hence Zeno's conclusion: since this is not what we observe, our concept of motion is inconsistent and wrong.)

Another paradox, which has basically the same structure, is even simpler to grasp: Consider a runner who needs to run a finite race distance (which for sim-plicity's sake we shall normalize to 1). He first has to run the first half $x_1 = \frac{1}{2}$, next the first half of the remaining second half to reach $x_2 = \frac{3}{4}$. Then he has to got to $x_3 = \frac{7}{8}$ and so on. Again, the upshot is that the runner can never reach the end of the race track.

It is now often said that Zeno's paradoxes can easily be resolved within the mod-ern, Cantorian view of transfinites in mathematics. We simply note that the infi-nite sum $\sum_{n=1}^{\infty} \frac{1}{2^n} = 1$ indeed converges. This is also the predominant view among philosophers of science (cf. [27, 30, 50]), but with the important addendum that there is of course no *a priori* guarantee to assume that space-time has the structure of a continuum. This has to be confirmed empirically, since Zeno's problem is after all physical, not mathematical in nature.

Most certainly, however, a 'modern Aristotle' would not be very much impressed by the Cantorian resolution of the paradoxes. Aristotle's very point was to introduce and to insist on the distinction between actual and potential infinities—and he was fond of the latter (cf. Aristotle's Physics Γ, Δ, Z in Ross [47]). For him, spatial distance must be considered a whole, being only potentially divisible. A runner covering a certain race distance does therefore not actually divide this continuous whole ("synholon") into pieces. Conversely, any actual division of space unavoid-ably takes time: Achilles indeed does not catch the tortoise, if he performs a halt after each step of iteration! But only this amounts to dividing space into pieces (or, in more operational terms, to measure a certain spatial distance). It seems much likely that Aristotle would rather be gratified to hear about intuitionistic mathemat-ics as a much more appropriate tool to describe nature.

Two further remarks concerning the connection between Zeno's paradoxes and quantum mechanics should be made. The first remark is that there is an interesting analogy between Aristotle's view of the continuum and the way we describe position and motion in quantum mechanics. Suppose we have a moving particle with constant velocity, i.e. definite momentum, then due to the uncertainty relations position is indefinite! Conversely, if the particle has a definite position, its state of motion, i.e. momentum, is totally uncertain. This fits indeed quite nicely with Aristotle's views.

The second remark concerns the *quantum Zeno effect* (cf. [39]). This is not really a quantum version of any of Zeno's paradoxes, but rather a formal result with broad similarities to the original. The general idea is that in quantum theory a system "freezes up" under continuous observations or measurements. Consider, for instance, a system of radioactive, decaying atoms. The decay probability will be $p(t) \sim e^{-t}$, which for short times is proportional to t^2. Thus, after a time t_0 the probability of decay is $p(t_0) \sim t_0^2$. But now we make an observation at $\frac{t_0}{2}$, where we get $p(\frac{t_0}{2}) \sim (\frac{t_0}{2})^2$. After the observation we must reset our clock and consider the same decay rate for the second sub-period. So, effectively we get the sum $p(t_0) \sim (\frac{t_0}{2})^2 + (\frac{t_0}{2})^2 = \frac{t_0^2}{2}$. Accordingly for n observations we have $p(t_0) \sim \frac{t_0^2}{n}$, which, in the limit $n \to \infty$ of infinitely many observations leads to probability zero. Thus, for a continuous measurement the system does not change at all!

A first attempt of an experimental realization of this paradoxical prediction was made by Itano et al. [31]. The authors used trapped ions and observed certain state transitions in dependency on disturbing radiation pulses, which they considered as 'measurements.' And, indeed, the results were of the Zeno fashion in the sense that the transition rate was decreasing with increasing radiation pulse number. Surely, this particular experimental set-up raises questions about what counts as a measurement and also, more generally, whether the idea of a continuous measurement has an operational meaning (after all, any real detector has a finite responding time). The lurking discussion of the measurement process shall be postponed to Sect. 5.1.

2 Physical Preliminaries

Our considerations have already reached a technical level, but some preliminary remarks concerning the notions of time, time reversal and the arrows of time should be made before addressing the particular problems in physical theories.

2.1 Newtonian Space-Time and Time Reversal (Reversal of Motion)

Newtonian space-time is generally considered the epitome of a fixed background space-time reflecting the spatio-temporal symmetries of classical mechanics. Due to its mathematical structure $\mathbb{R}^3 \times \mathbb{T}$, Newtonian space-time allows for a unique 3-space foliation and, hence, a global cosmic time. Its 3-dimensional spatial slices can be understood as planes of absolute simultaneity, meaning that the notion and measurement of time in Newtonian space-time is independent of any reference frame. In his famous *scholium* Newton described time as an absolute entity: "*Absolute, true and mathematical time, of itself and from its own nature, flows equably without relation to anything external.*"

As well-known, Newtonian physics shows invariance under $\widehat{T} : t \rightarrow -t$ with

$$q(t) \rightarrow q(-t) \quad \text{and} \quad \dot{q}(t) \rightarrow -\dot{q}(-t), \tag{1}$$

such that the Hamiltonian transforms as $H(q, p) \rightarrow H(q, -p)$. The operation $\widehat{T}$ is usually called "time reversal." However, this should be taken with a grain of salt, since what $\widehat{T}$ really does is rather a *reversal of motion*, as should be clear from (1). Hence, physicists define temporal reversibility as reversal of motion—a reversal in the sense of the B-series.

The idea of $\widehat{T}$ is to expresses the *isotropy* of time. But of course, since $\widehat{T}$ is a *discrete* symmetry, Noether's theorem does not apply and there is no conserved quantity connected with $\widehat{T}$. Instead of isotropy, the *homogeneity* of time *is* expressed via a conserved quantity—total energy—in terms of the first law. In fact, both laws of thermodynamics can be seen as laws about the nature of time: while the first law expresses the homogeneity, the second law stresses the anisotropy of time—in contrast to the alleged isotropy of the $\widehat{T}$-symmetry. Section 4 takes up this issue.

2.2 Arrows of Time

In his 1979 paper on "Singularities and time-asymmetry," [43] Roger Penrose presented a list of seven possible arrows of time, which might be helpful to structure the following sections.[2]

1. *Weak interaction arrow*: The "decay of the K^0-meson" is a clear experimental result and as such an 'almost' direct indication that Nature at least in one manifest case distinguishes past and future. However, this is only 'almost' an indication since, first, this literally *weak* interaction effect is, as Penrose puts it, "utterly minute" (smaller than 10^{-9}) and it seems therefore highly implausible to try to establish the more apparent arrows of time on this tiny effect. Second, the K^0-decay can only be observed indirectly via *CP*-violation and under the assumption that *CPT* is conserved.

2. *Quantum mechanical arrow*: "Quantum mechanical observations," whether in terms of 'collapses of the wave function' or stated otherwise, are time-asymmetric phenomena which give rise to quantum indeterminism. The quantum measurement process is discussed in Sect. 5.1.

3. *Thermodynamical arrow*: The "general entropy increase" of isolated systems on the macro-level according to the second law clashes with $\widehat{T}$-symmetry on the micro-level. Consequences will be laid out in Sect. 4.

---

[2]The expressions in quotes are Penrose's formulations.

4. *Electrodynamical arrow*: Classical electrodynamics is time-symmetric—there are future-directed, retarded waves as well as past-directed, advanced waves possible—, but still we only observe the "retardation of radiation," as for instance the spherical emission of (point) sources into the future time direction. We touch upon this issue in Sect. 5.2.

5. *Psychological arrow*: There is our indisputable feeling that the past is fixed, whereas the future is open and mutable, and also that causation acts towards the future only. Penrose calls it the "psychological time." Here, in our subjective time perception, we clearly distinguish between A- and B-time series.

6. *Cosmological arrow*: The "expansion of the universe" favors the future direction. This arrow is often connected to the thermodynamical as well as the electrodynamical arrow. It will be mentioned in Sect. 4.

7. *Gravitational arrow*: This arrow is due to the fact that gravitational collapses result in black hole singularities, whereas white holes have not been observed so far. While Penrose is particularly concerned with it, it plays no role in this article (readers may refer to Penrose's and similar literature).

3 Relativity Theory

3.1 Special Relativity

Special relativity (SR) mainly differs from pre-relativistic, classical mechanics by the assumption of a universal and finite limiting velocity, empirically identified with the vacuum velocity of light c (we already presuppose the relativity principle for inertial reference frames, which may be reconciled with classical mechanics either). The finite c equips space-time with a causal lightcone structure and, thus, replaces Newtonian space-time by Minkowskian space-time, a united combination of space and time in the sense that, in general, Lorentz transformations mix temporal and spatial parameters. It must have been this feature of the transformations which led Minkowski in his famous 1908 Cologne lecture on "space and time" to the statement: *"Henceforth space by itself, and time by itself, are doomed to fade away into mere shadows, and only a kind of union of the two will preserve an independent reality..."* (cf. [42, p. 152]). But here we have almost obviously, from the quite contradictory nature of his quote by using "henceforth" (or, even more obviously "von Stund an"—"from this hour"—in the German original), the entire problem in a nutshell, whether time in its independency with respect to *all* its features must really be given up. Does not it seem that Minkowski did at best dispense with the independency of B-series time, while being still committed to A-series time?

Nevertheless SR's resulting *relativity of simultaneity*, that is, the frame-dependency of simultaneity and hence the denial of absolute time, poses problems for endurantism and, correspondingly, presentism or possibilism as views about the reality of temporal objects and the ontology of time. The relativity of simultaneity means that the temporal distance between two space-like separated events is not defined.

This is usually illustrated for observers with different relative velocities, which are high compared to c. But we may as well consider low velocities and far remote events instead, as Roger Penrose shows in a drastic example by considering two persons who differ in their views about the launching of a space fleet on Andromeda to invade planet Earth [44, p. 303]: *"Two people pass each other on the street; and according to one of the two people, an Andromedean space fleet has already set off on its journey, while to the other, the decision as to whether or not the journey will actually take place has not yet been made."* This is obviously an odd situation, since the existence of events itself seems to become frame-dependent. Many authors in this debate[3] are convinced that the relativity of simultaneity cannot be reconciled with presentism (or possibilism) and that we have to be eternalists instead. Parmenides strikes back!

The problem gets even worse, if we consider the further thesis of the *conventionality of simultaneity*: the view that the simultaneity relation of two inertial clocks must be chosen by convention (cf. [46, § 19] and [26]). Consider two clocks A and B in an arbitrary inertial frame of reference. To synchronize these clocks we may send a light signal at A-clock's time t_1 from A to B, where it is instantaneously reflected back to A, arriving at t_2. The standard simultaneity is then the definition that the event at $t' = t_1 + \varepsilon(t_2 - t_1)$ with $\varepsilon = 1/2$ is simultaneous with the signal's reflection at B. However, as Einstein himself has put it in his famous popular book on relativity theory: *"That light requires the same time to traverse ... [both paths] is in reality neither a* supposition nor a hypothesis *about the physical nature of light, but a* stipulation *which I can make of my own freewill in order to arrive at a definition of simultaneity"* [21, § 8]. Thus, the choice $\varepsilon = 1/2$ is a mere convention—and this, then, could be exploited to the claim that the existence of events is not only frame- but convention-dependent!

However, David Malament [36] has shown that—under some "minimal, seemingly innocuous conditions"—standard simultaneity is the only non-trivial equivalence relation in accordance with causal connectability (this assumption might be considered a version of the causal theory of time). Nevertheless, commentators have even attacked these minimal assumptions. Sarkar and Stachel [51] raised particular doubts about the fact that in Malament's proof the simultaneity relation has to be symmetric under temporal reflections $\widehat{T}$. Thus, the conventionality issue is still not settled.

3.2 General Relativity

General relativity (GR) poses even severe problems on a Heraclitean view of time than does SR. Let us start with the most prominent, recent argument concerning the

---

[3]For the more recent debate compare the contributed papers to the sections "Special Relativity and Ontology" and "The Prospects for Presentism in Spacetime Theories" (and references therein) in the Proceedings of the 1998 Biennial Meetings of the Philosophy of Science Association, Part II, *Philos. Sci.* **67**(3), Supplement (2000).

ontological status of space-time, the question, whether space-time *substantivalism*, the view that space-time has a substantial or existential status on its own, is possible at all. The question has its traditional forerunner in the famous debate between Newton and Leibniz about the status of space. Whereas Newton hold a substantivalist position, Leibniz advocated the opposing *relationalist* view according to which space is nothing but the set of possible relations of bodies (cf. [14, Chap. 6, and 9 for the following]).

When Einstein—around 1912 during his search for a relativistic gravitational theory—came to realize that the field equations must be generally covariant, i.e. invariant under all coordinate transformations, he was quite confused about the physical meaning of this requirement. He actually invented an argument saying that generally covariant field equations cannot uniquely determine the gravitational field. Part of the argument was to consider an empty region in the energy-matter distribution, and so it was dubbed the "hole argument" ("*Lochbetrachtung*" in German).[4]

In 1987, John Earman and John Norton [18] presented a new version of the hole argument focusing on its ontological implications. They considered diffeomorphic models of GR, which are usually understood to represent the same physical situation (this was Einstein's early confusion). More precisely, let $\phi : \mathcal{M} \to \mathcal{M}$ be a diffeomorphic mapping defined on the space-time manifold $\mathcal{M}$ and $M = \langle \mathcal{M}, g_{\mu\nu}, T_{\mu\nu} \rangle$ be a model of GR with metric $g_{\mu\nu}$ and stress-energy tensor $T_{\mu\nu}$, then $M' = \langle \mathcal{M}, \phi^* g_{\mu\nu}, \phi^* T_{\mu\nu} \rangle$ is also a model of the theory. The reason for this is that M and M' are empirically indistinguishable. However, under certain ontological premises, in particular under the substantivalist assumption of space-time points as entities *per se*, M and M'—despite their empirical indistinguishability—represent *different* states of reality. Since Einstein's field equations cannot uniquely determine the temporal development of different diffeomorphic models (owing to general covariance), the space-time substantivalist has to accept a radical indeterminism arising in his picture of the world. Earman and Norton chose a 'hole diffeomorphism' h with $h = id$ for $t \leq t_o$ and $h \neq id$ for $t > t_o$ (obeying usual smoothness and differentiability conditions at t_o). We then have $M = M'$ for $t \leq t_o$, but $M \neq M'$ for $t > t_o$— an apparent breakdown of determinism from the substantivalist's point of view.

The new hole argument has caused a host of debates and comments—including intriguing objections and new options for substantivalists—but the majority of philosophers of science today is convinced that such an *ad hoc* indeterminism is far too high a price to pay for space-time substantivalism. Earman has shed new light on the debate by focusing on the, as he calls it, "*ideological*" rather than ontological implications of the hole argument [16]. These implications mainly arise from the non-trivial aspect of general covariance in GR. Take, for instance, Kretschmann's famous 1917 objection against Einstein's alleged 'principle of general covariance'

---

[4]We cannot follow the original argument due to lack of space. Historians of science have wondered about the trivial nature of Einstein's hole argument (besides the fact that he could not make use of modern differential geometry), but I am inclined to follow Stachel's [57] position that it was not a trivial argument. The reader may also consult Norton [41] for a comprehensive overview on the debates about general covariance.

in GR. Indeed, general covariance as the mere requirement of covariance under coordinate transformations is physically vacuous, it should quite generally be applicable in any sensible physical theory. But in GR the situation is far more complex: we must carefully distinguish between two applications of the concept of diffeomorphisms, for they might either correspond to mere coordinate transformations, but also to transformations of reference frames in the sense of physically instantiated transformations of observers provided with measuring rods and clocks. GR is thus characterized by the fact that not only the purely mathematical requirement of general coordinate covariance holds, but also the principle of general relativity, according to which any possible reference frames are seen as physically equivalent (for non-inertial frames one has of course to take compensating gravitational fields into account).

It is possible, in fact, to reconstrue GR as a gauge theory of the diffeomorphism group. This causes, already on the level of classical GR, the infamous problem of time: motion is pure gauge, all the genuine observables (i.e. gauge invariant quantities) are constants of the motion. Taken at face value this is a dramatic result! Parmenides indeed strikes back twice as hard, since this not only means a block universe stripped of A-series change (and accordingly the problems with presentism already in SR), but no B-series change, a *truly frozen universe* as a sort of *"neo-Parmenideanism"* or *"McTaggartism,"* as Earman [17] puts it.

Physicists usually begin to pay attention to these problems on the level of quantizing gravity, since here the problem of time becomes apparent because of the timeless Wheeler–DeWitt equation. However, this equation is nothing but the quantum variant of the Hamiltonian constraint and so, strictly speaking, the problem of no B-series change already exists on the classical level. Indeed, many of the leading figures in quantum gravity, relationalists in the majority, are aware of this fact (cf. [49]). We shall not say more about quantum gravity here, but brief mention should be made about two further aspects of the concept of time as they must presumably be expected from a truly quantized space-time theory: the possibility of instants of time (e.g. "chronons," [23]), and time as a quantum operator. Another source of questions about time connected with GR is cosmology. Since the cosmological arrow also relates to the thermodynamical arrow, cosmological aspects will be touched upon in the following section.

4 Thermodynamics

Most of the arguments about time presented so far have been arguments about the ontology rather than arguments about the directedness of time. In thermodynamics, however, the general entropy increase of isolated systems according to the second law reflects an asymmetry of time: the thermodynamical arrow.[5]

---

[5]Compared to the importance of this issue the presentation in the following is far too brief. Some more elaborate references are: Ben-Menahem and Pitowsky [3], Guttmann [28], Sklar [56] and Uffink [59, 60].

4.1 The Second Theorem—A Law?

In his kinetic theory of gases, Boltzmann considered a transport equation for the distribution function $f(q, p, t)$ in phase space and was able to describe entropy as

$$S = -H\big(f(q, p, t)\big) = -\int d^3q d^3 p f(q, p, t) \log f(q, p, t). \qquad (2)$$

His aim was to arrive at a proper microscopic underpinning of macroscopic thermodynamics—and in particular to obtain a microscopic version of the second law. For this purpose he introduced the famous "Stoßzahlansatz" (a.k.a. the assumption of molecular chaos), where the two-particle distribution function is written as a simple product of one-particle functions, which amounts to the assumption of uncorrelated particles before collision. From this ansatz he was able to derive the infamous H-theorem

$$\frac{dH(f(q, p, t))}{dt} \le 0, \qquad (3)$$

which describes the tendency of a gas to evolve to the Maxwell equilibrium distribution. However, the well-known and quite general problem with this account (as expressed in the early and famous objections of Loschmidt, Poincaré and Zermelo) is the obvious contradiction between the alleged macroscopic irreversibility as opposed to the undoubtedly existing reversibility on the mirco-level of classical particle mechanics. Indeed, how should it be possible at all to infer logically from a perfectly reversible mirco-mechanics to an irreversible macro-world?

The usual stance is to consider the increase of entropy only statistically and, thus, granting the H-theorem merely the character of a statistical law. But this does not solve the problem entirely, since the main worry with Boltzmann type accounts is to understand where the incredibly low *initial* entropy state comes from. Boltzmann himself (cf. [6]) was fully aware of this problem and tried to circumvent it—in various ways. One of his ideas is known as the *fluctuation hypothesis*: our known world is a real fluctuation phenomenon in a universe of much greater spatial and temporal extension. A this point the connection between the thermodynamical arrow and the cosmological arrow comes into play.

However, there is an underlying and sometimes overlooked time-symmetry of the whole Boltzmannian approach, which becomes visible in the fluctuation hypothesis. The point is that due to (3) and starting from an initial, low entropy state at $t = t_o$ we get increasing entropy in either time direction! In other words, the H-theorem indeed establishes increasing entropy for the future direction $t > t_o$, but—from the same logic—also for the past direction $t < t_o$. One must therefore come to the conclusion that the H-theorem does not single out the future direction and, hence, is *not* equivalent to the second law (seen as a law which truly distinguishes between past and future).

An account to secure the second law and, hence, irreversibility, based on a pure epistemological consideration was proposed by Carl Friedrich von Weizsäcker. By using a transcendental argument, i.e. referring to our methodological *preconditions of experience*, Weizsäcker claims that the distinction between past and future is already a fundamental precondition of experience—as can be seen from the analysis of our usual way of defining experience:

> A possible definition of experience may be that it means to learn from the past for the future. Any experience I now possess is certainly past experience; any use I now can still hope to make of my experience is certainly a future use. In a more refined way one may say that science sets up laws which seem to agree with past experience, and which are tested by predicting future events and by comparing the prediction with the event when the event is no longer a possible future event but a present one. In this sense time is a presupposition of experience; whoever accepts experience understands the meaning of words like present, past, and future. [61]

Thus, the central argument here is that in our empirical sciences we necessarily presuppose an understanding of the tenses of time, otherwise we were not able to explain what we mean by "empirical." As a presupposition, however, we cannot expect the distinction between past and future dropping off from physics as an empirical result, since this would be circular. We rather have to make explicit the distinction as a precondition of experience, which then might help to bridge the decisive gap between the H-theorem and the second law.

4.2 Maxwell's Demon, Entropy and Information

Besides the difficulties of a microscopic underpinning of the second law, microscopic attacks on its validity, conversely, also seem to fail. The probably most famous example of this type is *Maxwell's demon*. James Clerk Maxwell's idea was the following:

> ... the second law of thermodynamics ... is undoubtedly true as long as we can deal with bodies only in mass, and have no power of perceiving or handling the separate molecules of which they are made up. But if we conceive a being whose faculties are so sharpened that he can follow every molecule in its course, such a being, whose attributes are still as essentially finite as our own, would be able to do what is at present impossible to us. For we have seen that the molecules in a vessel full of air at uniform temperature are moving with velocities by no means uniform, though the mean velocity of any great number of them, arbitrarily selected, is almost exactly uniform. Now let us suppose that such a vessel is divided into two portions, A and B, by a division in which there is a small hole, and that a being, who can see the individual molecules, opens and doses this hole, so as to allow only the swifter molecules to pass from A to B, and only the slower ones to pass from B to A. He will thus, without expenditure of work, raise the temperature of B and lower that of A, in contradiction to the second law of thermodynamics. [37]

This thought experiment of Maxwell provoked a debate which has not stopped until today[6] and from which only the most important highlights shall be mentioned: The early discussions focused on the aspect of the physical realizability of the demon and brought to light that pure technical solutions fail and that the demon must in addition be 'intelligent.' This was most clearly worked out by Leo Szilard [58],

---

[6]For a most comprehensive collection of important papers in the more than a century long debate about Maxwell's demon see Leff and Rex [35], and also Earman and Norton [19, 20].

who showed that, quite generally, any measurement produces an increase of entropy. These considerations, carried on by Brillouin, Gabor and von Neumann, led to the idea of a *thermodynamic equivalent of a bit* $\Delta E = k_B T \ln 2$, understood as the minimum energy to produce or storage 1 bit of information. The final clue, however, came with the work of Rolf Landauer [34] and Charles Bennett [4]. Landauer discovered that memory erasure in computers results in an entropy increase in the environment, and Bennett therefore argued that the demon, who has to storage and to remember the data he obtains about the molecule velocities, saves the second law by the very act of resetting his memory (which is unavoidable for any realistic demon with a finite memory).

Landauer's and Bennett's work points out the deep connection between the concepts of entropy and information, as already suggested in the thermodynamic equivalent of a bit. Indeed, their information theoretic exorcism of Maxwell's demon hints at a renewed and fundamental interpretation of entropy in pure information theoretic terms. From a mathematical point of view, the close analogy between Boltzmann's formula $S = -k_B \sum_i p_i \ln p_i$ (in different notation than (2); p_i is the probability of a system to be in a certain microstate and k_B the Boltzmann constant) and the well-known Shannon [54] *information entropy* $H = -\sum_i p_i ldp_i$ giving the expectation value of the information content of a source (where $I = -ldp$ is the information content of a sign with probability p) is already striking. A certain confusion, however, arose about the sign of both quantities. Entropy may indeed be interpreted as a specific kind of non-information—the ignorance of the particular microstate in a given macrostate. Brillouin [7], therefore, envisaged a *negentropy principle of information*. Perhaps here we have a rather verbal problem which might just be resolved by distinguishing *potential* from *actual information*, as Weizsäcker [61] has proposed. In this terminology, entropy is potential information, the possible amount of information of a given macrostate, if all the microstates were known.

Conceptual links between entropy and (potential) information have been advocated by important thinkers in the foundations of thermodynamics (cf. [32, 33, 48] and [61]). But of course, the main worry with the information theoretic view is the seemingly subjective nature of the concept of information as opposed to the alleged objective nature of entropy as a system state quantity—or, in other words, the rather epistemic nature of information as a property of the observer as opposed to the ontic nature of entropy as a property of physical systems. This is why, for instance, Earman and Norton [20] dismiss the information theoretic exorcism of Maxwell's demon altogether. On the other hand, it seems that physics in many of its modern developments uncovers the importance of the notion of information.

5 Quantum Mechanics

5.1 The Measurement Problem

As Penrose has pointed out (see Sect. 2.2), quantum mechanics gives rise to an arrow of time because of the measurement problem. To begin with, we should review the measurement problem in brief. We consider a system S and a measuring apparatus A, and split the measurement process into different steps: As a first step, S and A must couple, such that formally one has to enlarge the Hilbert space of S to the Hilbert space of the compound system $S \otimes A$, while, secondly, a measurement interaction H_{int} takes place. Next, the compound system, being still in a pure state, will be separated into subsystems S and A again. The states of the subsystems are now formally given by the reduced density operators $\hat{\rho}_S$ and $\hat{\rho}_A$. At the end of the measuring chain we may read off the measuring result—a definite pointer state of A (if all went well).

The measurement problem arises now from the fact that the operators $\hat{\rho}_S$ and $\hat{\rho}_A$, which we obtain after the formal separation of S and A, are so-called *improper mixtures*, which means that the *ignorance interpretation* is not applicable. This amounts to saying that it is not possible to attribute a definite state to S (or A, respectively)—neither of the subsystems does allow for an objectification (the assumption of a definite, i.e. observer-independent state of $\hat{\rho}_S$ leads to formal contradictions; cf. [40]). Since we do, however, expect measuring results to be definite and objective, the replacement of *improper* by *proper mixtures*, known as the reduction of the wave function, has to be put in by hand ("Heisenberg cut"). According to this *minimal instrumentalist interpretation*, as one could have it, the reduction of the wave function, which cannot be described by some unitary process, must be seen as an indeterministic element over and above the deterministic quantum dynamics.

It should particularly be emphasized that the failure of the ignorance interpretation really is the hard problem of the measurement process. This remark is in order in view of the successful and persuasive application of the various *decoherence* approaches on the market, whose importance could undoubtedly be established within the last decades: in realistic cases, the coupling of S to the environment will unavoidably destroy the typical quantum correlations (cf. [25]). However, following John Bell's classic phrasing, the vanishing of correlations FAPP ("for all practical purposes," [2]), should not be confused with the vanishing of the non-applicability of the ignorance interpretation. For even if, in a suitable pointer basis, we are left with, say, probabilities $\frac{1}{2}$ each and negligible superposition probabilities for the two outcomes of a simple binary quantum alternative (a quantum coin tossing, for instance), the failure of the ignorance interpretation implies that it is still not the case that the quantum coin does possess some definite state with corresponding probabilities as merely expressing the observer's ignorance about this very state.

This, indeed, causes a severe problem for determinism in quantum mechanics. In contrast to the classical statistical mechanics case (see Sect. 4), non-objectifiable quantum probabilities do not allow for a merely statistical indeterminism (and, hence, a hidden determinism). It has therefore become quite fashionable among 'decoherentists' to subscribe to a many worlds interpretation in order to establish an 'ontologically adequate' approach to the occurrence of quantum probabilities by asserting one real world for each measuring outcome. Those, who do not wish to enlarge reality in such a drastic manner, have to accept a radical quantum indeterminism on the bottom level—since otherwise the question, why apparently only one of the two dynamically independent components of a quantum alternative is experienced, remains entirely unexplained.

5.2 Interpretations of QM

Quantum theory—unlike other physical theories—is loaded with deep interpretational problems. The above sketched minimal instrumentalist interpretation is 'minimal' in the sense that it suffices to use the theory as a highly successful tool for applied physics. And to be sure, in this sense quantum theory is the most precise and successful physical theory mankind has ever discovered. To many and from a more concerned ontological point of view, however, the instrumentalism of the working physicist seems to be unsatisfactory. This is why we see a garden variety of competing interpretations of quantum theory—some who either deny the measurement problem or the indeterminism claim or both. In the following, we shall concentrate on two such interpretations—the Bohmian and transactional interpretation—which take different views on time-(a)symmetry and (in)determinism in quantum physics, but which are nevertheless empirically equivalent. We are therefore facing remarkable cases of *theory underdetermination by empirical evidence.*

Bohm's [5] original 1952 account of quantum mechanics is indeed basically a clever re-formulation of ordinary quantum mechanics in the sense that one extracts a term from the Schrödinger equation which formally looks like a potential—a nonlocal quantum potential, however—and which is then used in a Newton-type equation of motion. This additional equation, which does not exist in the minimalist formulation, re-introduces an ontological picture of particle trajectories into Bohmian mechanics. Bohmians consider their view as 'realistic'—without neglecting the genuine quantum non-locality (which makes the particle trajectories quite 'surrealistic'; cf. [22]).

It is an indeed remarkable fact that in Bohmian mechanics the measurement problem may be said to disappear. Given the quite general analysis in terms of the non-applicability of the ignorance interpretation in the preceding section, one might wonder how this is possible at all. So here's a first motivation: The non-applicability of the ignorance interpretation amounts to saying that an observer cannot distinguish between improper and proper mixture states of S or, in other words, that he has no means to decide whether the measuring apparatus A is still correlated to S or not. To decide this he would have to apply a suitable meta-observable on the compound system $S' = S \otimes A$, but this can obviously only be done by a meta-observer with apparatus A'. We may extend this consideration to the universe as the largest physical system possible. As inner observers we cannot distinguish between proper and improper mixtures of subsystems of the universe, such that it is logically possible to assume the initial conditions of any particle positions, as Bohmians would have it, as non-local hidden variables with determinate values fixed by a deterministic velocity equation. Hence, our usual quantum mechanical probability calculus must be interpreted as arising due to our subjective ignorance of the objective state of the universe much like the usage of probabilities in classical statistical mechanics (where we do apply an ignorance interpretation). This is why Bohmians are indeed able to circumvent the problem of the ignorance interpretation in the measurement process. We may hence conclude that *per constructionem* Bohmian mechanics is purely deterministic and time-symmetric in analogy to classical mechanics.

Let us now turn to a somewhat lesser well-known approach of quantum mechanics: Cramer's [12] *transactional interpretation*. It is mainly inspired from the Wheeler–Feynman approach [62] of electrodynamics (which has only recently attracted new interest from philosophers of physics; cf. [24, 45]). The main idea is that Wheeler and Feynman allowed for the full time-symmetric set of solutions of the Maxwell wave equations, in particular the existence of advanced solutions. Usually, these backwards-in-time radiating waves are dismissed on the basis of suitable boundary conditions as for instance the *Sommerfeldsche Ausstrahlungsbedingung*, according to which the universe must be seen a sink of radiation. Thus, the electrodynamical arrow is based in one way or the other either on the cosmological or the thermodynamical arrow.

In the same line of thinking Cramer considers both retarded and advanced wave functions. The Wheeler–Feynman absorber condition—a suitable canceling of retarded and advanced solutions—turns in Cramer's account into a *transaction* ("hand-shaking") between retarded "offer" waves from the emitter and advanced "confirmation" waves from the absorber. As an exchange between waves from the past and waves from the future the transaction as such is *atemporal*. Over and above that the approach is time-symmetric (despite, Cramer's remarks in his 1986, Sect. III.J). The situation is analogous to the underlying time-symmetry of Boltzmann's H-theorem (Sect. 4): Cramer's account cannot single out the future lightcone.

Cramer believes that his interpretation gives better explanations of non-local effects such as EPR-Bell correlations and delayed choice measurements than the standard formulation, but simultaneously emphasizes that both lead to the same experimental predictions. We are thus left with three apparent cases of theory underdetermination—the minimal interpretation, Bohmian mechanics, and transactional interpretation—which are empirically equivalent but drastically differ in ontology.[7] Thus, on the basis of pure interpretational manoeuvres one may choose between indeterminism, determinism, and partial atemporalism!

6 Conclusion

We have reached the end of our *tour de force* through questions about time and its direction in modern philosophy of physics. It goes without saying that we could only touch upon a few of a whole universe of aspects of this extensive topic. For instance, no mention was made of phenomena involving 'backwards causation,' such as time-travel (cf. [15]). Indeed, the whole issue about causation was omitted, just as counterfactuals have not been addressed (cf. [29]). Finally, some further literature shall be indicated to the interested reader: Very good physics references, for instance, are Schulman [53] and Zeh [63]. Among the philosophy of physics literature

---

[7]Some Bohmians do assert possible empirical differences to the standard approach by introducing "effective wave functions," which are completely decoupled from their environment (cf. [13]; I like to thank David Albert and Roderich Tumulka for indicating this to me).

mention should be made of Albert [1], Butterfield [8], Butterfield and Earman [9], Callender [10], Horwich [29], Savitt [52], Sklar [55], and Price [45]. Again, this little list of references is of course far from being complete, but rather provides useful entries for more elaborate studies of the fascinating issue of time and its direction in physics and philosophy.

References

1. Albert, D.: Time and Chance. Harvard University Press, Cambridge (2000)
2. Bell, J.S.: Against 'measurement'. Phys. World **8**, 33–40 (1990)
3. Ben-Menahem, Y., Pitowsky, I. (eds.): Special issue: the conceptual foundations of statistical physics. Stud. Hist. Philos. Mod. Phys. **32**(4) (2001)
4. Bennett, C.H.: The thermodynamics of computation—a review. Int. J. Theor. Phys. **21**, 905–940 (1982)
5. Bohm, D.: A suggested interpretation of the quantum theory in terms of "hidden" variables. Phys. Rev. **85**, 166–179, 180 (1952)
6. Boltzmann, L.: Vorlesungen über Gastheorie. Barth, Leipzig (1896)
7. Brillouin, L.: Science and Information Theory. Academic Press, London (1962)
8. Butterfield, J. (ed.): The Arguments of Time. Oxford University Press, Oxford (1999)
9. Butterfield, J., Earman, J. (eds.): Handbook for the Philosophy of Physics. Elsevier, Amsterdam (2006)
10. Callender, C. (ed.): Time, Reality, and Experience. Cambridge University Press, Cambridge (2002)
11. Carnap, R.: Intellectual autobiography. In: Schilpp, P.A. (ed.) The Philosophy of Rudolf Carnap, Library of Living Philosophers, vol. X. Open Court, La Salle (1963)
12. Cramer, J.G.: The transactional interpretation of quantum mechanics. Rev. Mod. Phys. **58**, 647–688 (1986)
13. Dürr, D., Goldstein, S., Zanghi, N.: Quantum equilibrium and the origin of absolute uncertainty. J. Stat. Phys. **67**, 843–907 (1992)
14. Earman, J.: World Enough and Space-Time. MIT Press, Cambridge (1989)
15. Earman, J.: Bangs, Crunches, Whimpers, and Shrieks. Oxford University Press, New York (1995)
16. Earman, J.: Gauge matters. PSA 2000 conference lecture, to appear in Philos. Sci. (2000)
17. Earman, J.: Thoroughly modern McTaggart: or what McTaggart would have said if he had learned the general theory of relativity. Philos. Impr. **2**(3) (2002). http://www.philosophersimprint.org/002003
18. Earman, J., Norton, J.: What price spacetime substantivalism? The hole story. Br. J. Philos. Sci. **83**, 515–525 (1987)
19. Earman, J., Norton, J.: Exorcist XIV: The wrath of Maxwell's demon. Part I. From Maxwell to Szilard. Stud. Hist. Philos. Mod. Phys. **29**, 435–471 (1998)
20. Earman, J., Norton, J.: Exorcist XIV: The wrath of Maxwell's Demon. Part II. From Szilard to Landauer and Beyond. Stud. Hist. Philos. Mod. Phys. **30**, 1–40 (1999)
21. Einstein, A.: Relativity: The Special and General Theory. Holt, New York (1920)
22. Englert, B.-G., Scully, M.O., Süssmann, G., Walther, H.: Surrealistic Bohm trajectories. Z. Naturforsch. **47a**, 1175–1186 (1992). Comment: Dürr et al.: Z. Naturforsch. **48a**, 1261 (1993); Reply to Comment: Englert et al.: Z. Naturforsch. **48a**, 1263 (1993)
23. Finkelstein, D.: Quantum Relativity: A Synthesis of the Ideas of Einstein and Heisenberg. Springer, New York (1996)
24. Frisch, M.: (Dis-)solving the puzzle of the arrow of radiation. Br. J. Philos. Sci. **51**, 381–410 (2000)

25. Giulini, D., Joos, E., Kiefer, C., Kupsch, J., Stamatescu, I.-O., Zeh, H.D.: Decoherence and the Appearance of a Classical World in Quantum Theory. Springer, Berlin (1996)
26. Grünbaum, A.: Philosophical Problems of Space and Time. Knopf, New York (1963). Second, enlarged edition, Reidel, Dordrecht (1973)
27. Grünbaum, A.: Modern Science and Zeno's Paradoxes. Wesleyan University Press, Middletown (1967)
28. Guttmann, Y.M.: The Concept of Probability in Statistical Physics. Cambridge University Press, Cambridge (1999)
29. Horwich, P.: Asymmetries in Time. MIT Press, Cambridge (1987)
30. Huggett, N. (ed.): Space from Zeno to Einstein: Classic Readings with a Contemporary Commentary. MIT Press, Cambridge (1999)
31. Itano, W.M., Heinzen, D.J., Bollinger, J.J., Wineland, D.J.: Quantum Zeno effect. Phys. Rev. A **41**(5), 2295–2300 (1990)
32. Jaynes, E.T.: Information theory and statistical mechanics. Phys. Rev. **106**, 620 (1957)
33. Jaynes, E.T.: Information theory and statistical mechanics II. Phys. Rev. **108**, 171 (1957)
34. Landauer, R.: Irreversibility and heat generation in the computing process. IBM J. Res. Dev. **5**, 183–191 (1961)
35. Leff, H.S., Rex, A.F. (eds.): Maxwell's Demon: Entropy, Information, Computing. Princeton University Press, Princeton (1990)
36. Malament, D.: Causal theories of time and the conventionality of simultaneity. Noûs **11**, 293–300 (1977)
37. Maxwell, J.C.: Theory of Heat. Longmans, London (1871)
38. McTaggart, J.M.E.: The unreality of time. Mind **17**(68), 457–474 (1908)
39. Misra, B., Sudarshan, E.C.G.: Zeno's paradox in quantum theory. J. Math. Phys. **18**, 756 (1977)
40. Mittelstaedt, P.: The Interpretation of Quantum Mechanics and the Measurement Process. Cambridge University Press, Cambridge (1998)
41. Norton, J.D.: General covariance and the foundations of general relativity: eight decades of dispute. Rep. Prog. Phys. **56**, 791–858 (1993)
42. Pais, A.: 'Subtle is the Lord...': The Science and the Life of Albert Einstein. Oxford University Press, Oxford (1982)
43. Penrose, R.: Singularities and time-asymmetry. In: Hawking, S.W., Israel, W. (eds.) General Relativity: An Einstein Centenary Survey. Cambridge University Press, Cambridge (1979)
44. Penrose, R.: The Emperor's New Mind. Oxford University Press, New York (1989)
45. Price, H.: Time's Arrow and Archimedes' Point. Oxford University Press, New York (1996)
46. Reichenbach, H.: Philosophie der Raum-Zeit-Lehre. de Gruyter, Berlin (1927). Engl. transl.: Philosophy of Space and Time. Dover, New York (1957)
47. Ross, W.D.: Aristotle's Physics. Clarendon Press, Oxford (1936)
48. Rothstein, J.: Informational generalization of entropy in physics. In: Bastin, T. (ed.): Quantum Theory and Beyond. Cambridge University Press, Cambridge (1971)
49. Rovelli, C.: The century of the incomplete revolution: Searching for general relativistic quantum field theory. J. Math. Phys. **41**(6), 3776–3800 (2000)
50. Salmon, W.C. (ed.): Zeno's Paradoxes. Bobbs-Merrill, Indianapolis (1970). Repr., Hackett, Indianapolis (2001)
51. Sarkar, S., Stachel, J.: Did Malament prove the non-conventionality of simultaneity in the special theory of relativity? Philos. Sci. **66**, 208–220 (1999)
52. Savitt, S.F.: Time's Arrows Today. Cambridge University Press, Cambridge (1995)
53. Schulman, L.S.: Time's Arrows and Quantum Measurement. Cambridge University Press, Cambridge (1997)
54. Shannon, C.E.: A mathematical theory of communication. Bell Syst. Tech. J. **27**(3), 379–656 (1948)
55. Sklar, L.: Space, Time, and Spacetime. University of California Press, Berkeley (1974)
56. Sklar, L.: Physics and Chance. Cambridge University Press, Cambridge (1993)

57. Stachel, J.: Einstein's search for general covariance, 1912–1915. In: Howard, D., Stachel, J. (eds.) Einstein and the History of General Relativity. Einstein Studies, vol. 1. Birkhäuser, Boston (1989)
58. Szilard, L.: Über die Entropieverminderung in einem thermodynamischen System bei Eingriffen intelligenter Wesen. Z. Phys. **53**, 840–856 (1929)
59. Uffink, J.: Bluff your way through the second law of thermodynamics. Stud. Hist. Philos. Mod. Phys. **32**(3), 305–394 (2001)
60. Uffink, J.: Issues in the foundations of classical statistical physics. In: Butterfield, J., Earman, J. (eds.): Handbook for the Philosophy of Physics. Elsevier, Amsterdam (2006)
61. Weizsäcker, C.F.v: Die Einheit der Natur. Hanser, München (1971). Engl. transl.: The Unity of Nature. Farrar, Straus and Giroux, New York (1980)
62. Wheeler, J.A., Feynman, R.P.: Interaction with the absorber as the mechanism of radiation. Rev. Mod. Phys. **17**, 157–181 (1945)
63. Zeh, H.-D.: The Physical Basis of the Direction of Time. Springer, Heidelberg (1989). Fourth edition (2001)

Chapter 14
The Direction of Time in Dynamical Systems

Interdisciplinary Perspectives from Cosmology to Brain Research

Klaus Mainzer

Abstract Dynamical systems in classical, relativistic, and quantum physics are ruled by laws with time reversibility. Dynamical systems with time-irreversibility are known from thermodynamics, biological evolution, brain research and historical processes in social sciences. They can also be simulated by computation and information systems. Thus, arrows of time and aging processes are not only subjective experiences or even contradictions to natural laws but can be explained by the dynamics of complex systems.

Keywords Symmetry of time · Reversibility · Cosmic arrow · Complex dynamical system · Time operator · Duration · Aging · Self-organization · Evolutionary time · Social time · Computational time · Unpredictability

1 Time in Classical and Relativistic Dynamics

According to Newton's laws of mechanics, a *dynamical system* is determined by a time-depending equation of motion. Newton distinguished between *relative* and *absolute time*, assuming that all clocks of relative reference systems in the Universe could be synchronized to an absolute world-time of an absolute space. The *symmetry of time* is expressed by changing the sign of the *direction of motion* in an equation of motion [2, 3]. In *classical mechanics*, mechanical laws are preserved (invariant) with respect to all inertial systems moving uniformly relative to one another (Galilean invariance). A consequence of time symmetry is the conservation of energy in a dynamical system. Newton's absolute space can actually be replaced by the class of inertial systems with Galilean invariance. But, according to the Galilean transformation of time, there is still Newton's distinguished absolute time in classical mechanics.

In 1905, Einstein assumed the *principle of special relativity* for all inertial systems satisfying the constancy c of the speed of the light ('Lorentz systems') and derived a common space-time of mechanics, electrodynamics, and optics. Their

K. Mainzer (✉)
Carl von Linde-Akademie, Technische Universität München, München, Germany
e-mail: mainzer@cvl-a.tum.de

S. Albeverio, P. Blanchard (eds.), *Direction of Time*,
DOI 10.1007/978-3-319-02798-2_14,
© Springer International Publishing Switzerland 2014

laws are invariant with respect to the Lorentz transformations. Time measurement becomes path-dependent, contrary to Newton's assumption of absolute time. Every inertial system has its *relative* (*'proper'*) *time*. An illustration delivers the Twin paradox: In a space-time system, twin brother A remains unaccelerated on his home planet, while twin brother B travels to a star at great speed. The traveling brother is still young upon his return, while the stay-at-home brother has become an old man. But, according to the *symmetry of time*, the twin brothers may also become younger. Thus, relativistic physics cannot explain the *aging of an organism* with direction of time. According to Einstein (1915), gravitational fields of masses and energies cause the *curvature of space-time*. Clocks are affected by gravitational fields: The gravitational red shift of a light beam in a gravitational field depends on its distance to the gravitational source and can be considered as dilatation of time. The effect is confirmed by atomic clocks.

Relativistic cosmology assumes an expanding universe in cosmic time. According to Hubble's law of expansion (1929), no galaxy is distinguished. The *Cosmological Principle* demands that galaxies are distributed spatially homogeneous and isotropic ('maximally symmetric') at any time in the expanding universe. In geometry, homogeneous and isotropic spaces have constant (flat, negative or positive) curvature. In two dimensions, they correspond to an Euclidean plane with flat curvature and infinite content, a negatively curved saddle, or a positively curved surface of a sphere with finite content. With the assumption of the Cosmological Principle and Einstein's theory of gravitation, H.P. Robertson and H.G. Walker derived the three standard models of an expanding universe with open cosmic time in the case of a flat or negative curvature and final collapse and end of time in the case of positive curvature. F. Hoyle's *steady state universe* (1948) without global temporal development can be excluded by overwhelming empirical confirmations of an expanding universe. K. Gödel's *travels in the past* on closed world lines in an anisotropic ('rotating') universe (1949) are excluded by the high confirmation of isotropy in the microwave background radiation.

The beginning and end of time get new impact by the theory of *Black Holes* and *cosmic singularities*. According to the theory of general relativity, a star of great mass will collapse after the consumption of its nuclear energy. During 1965–1970, R. Penrose and S.W. Hawking proved that the collapse of these stars is continued to a point of singularity with infinite density and gravity [1]. Thus, the singularity of a Black Hole is an *absolute end of temporal development*. The Schwarzschild-radius determines the event horizon of a Black Hole. Because of the *symmetry of time*, there might be also 'White Holes' with expanding world lines and exploding matter and energy, starting in a point of singularity. This idea inspired Hawking's theorem of cosmic origin (1970): Under the assumption of the theory of general relativity and the observable distribution of matter, the universe has an *initial temporal singularity* (*'Big Bang'*), even without the additional assumption of the Cosmological Principle. Time is initialized in that point.

From different philosophical points of view, theists or atheists have supported or criticized the idea of an initial point of time, because it seems to suggest a creation of the universe. The mathematical disadvantage is obvious: In singularities of zero extension and infinite densities and potentials, computations must fail. Thus, nothing can be said about the origin of time in relativistic cosmology.

2 Time in Quantum Dynamics

According to Bohr's correspondence principle, a dynamical system of quantum mechanics can be introduced by analogy to a dynamical system of classical (Hamiltonian) mechanics. Classical vectors like position or momentum are replaced by operators satisfying a non-commutative (non-classical) relation depending on Planck's constant h. The dynamics of quantum states is completely determined by time-depending equations (e.g., Schrödinger equation) with *reversibility of time*. The laws of classical physics are invariant with respect to the symmetry transformations of time reversal (T), parity inversion (P), and charge conjugation (C). According to the *PCT-theorem*, the laws of quantum mechanics are invariant with respect to the combination PCT. Thus, in spite of P-violation by weak interaction, the PCT-theorem still holds in quantum field theories. But it is an open question how the observed violations of PC-symmetry and T-symmetry (e.g., decay of kaons) can be explained [2].

An immediate consequence of the non-commutative relations in quantum mechanics is Heisenberg's principle of uncertainty which is satisfied by conjugated quantities such as time and energy: Pairs of virtual particles and antiparticles can spontaneously be generated during a tiny interval of time ('*Planck-time*'), interact and disappear, if the product of the temporal interval and the energy of particles is smaller than Planck's constant. Thus, quantum vacuum as the lowest energetic state of a quantum system is only empty of real particles, but full of virtual particles ('quantum fluctuations').

Furthermore, according to Heisenberg's uncertainty principle, there are no time-depending orbits (trajectories) of quantum systems, depending on precise values of momentum and position like in classical physics. In order to determine the temporal development of a quantum system, R. Feynman suggested to use the sum ('integral') of all its infinitely many possible paths as probability functions. In *quantum cosmology*, the whole universe is considered as quantum system. Thus, Feynman's method of path integral can be applied to the whole universe. In this case, the quantum state (wave function) of the universe is the sum (integral) of all its possible temporal developments (curved space-times). In 1983, J. Hartle and Hawking suggested a class of curved space-times without singularities, in order to avoid the failure of relativistic laws in singularities and to make the cosmic dynamics completely computational.

According to Hawking's hypothesis of an *early universe without beginning*, Feynman's path integral allows different models of temporal expansion which are more or less probable—collapsing universes, critical universes, universes with fast (inflationary) expansion. Hawking uses the (weak) Anthropic Principle to distinguish a universe like ours, enabling the evolution of galaxies, planets, and life, with an early inflation and later retarded expansion of flat curvature [1]. From his hypothesis, R. Laflamme and G. Lyons derived the forecast of tiny fluctuations of the microwave background radiation which was confirmed by the measurements of COBE in 1992. Thus, Hawking's hypothesis of an early universe without temporal beginning has been confirmed (until now), but not explained by an unified theory of quantum and relativistic physics which we still miss.

The temporal development of the universe can be considered as dynamics of phase transitions from an initial quantum state of high density to hot phase states of inflationary expansion and the generation of elementary particles, continued by the retarded expansion of galactic structures. Cosmic time is characterized by the development from a nearly uniform quantum state to more complex states of differing cosmic structures. In this way, we get a *cosmic arrow of time from simplicity to complexity*, which is characterized by a bifurcation scheme of global cosmic dynamics: An initial unified force has been separated step by step into the partial physical forces we can observe today in the universe: gravitation, strong, weak, and electromagnetic interactions with their varieties of elementary particles [2–4].

If in the early universe gravitation and quantum physical forces are assumed to be unified, then we need a *unified theory of relativity and quantum mechanics* with new objects as common building blocks of the familiar elementary particles. The string theory assumes tiny loops of 1-dimensional strings (10^{-35} m) with minimal oscillations generating the elementary particles. In a superstring theory, the unified early state corresponds to a transformation group of *supersymmetry*, which leaves the laws of the unified force invariant. During the cosmic expansion the early symmetry is broken into partial symmetries corresponding to different classes of particles and their interactions. Only three spatial dimensions of the more dimensional superstring theory are 'unfolded' and observable. Today, there are five 10-dimensional string theories and an 11-dimensional theory of supergravitation with common features ('dualities') and identical forecasts of the universe. They are assumed to be unifiable in the so-called M-theory. In this case, the *cosmic arrow of time* could be completely explained by phase transitions from simplicity to complexity.

3 Time in Thermodynamics

In physics, a direction of time was at first assumed in thermodynamical systems. According to R. Clausius, the change of the entropy S of a physical system during the time dt consists of the change $d_e S$ of the entropy in the environment and the change $d_i S$ of the intrinsic entropy in the system itself, i.e. $dS = d_e S + d_i S$. For isolated systems with $d_e S = 0$, the *second law of thermodynamics* requires $d_i S \geq 0$ with increasing entropy ($d_i S > 0$) for *irreversible thermal processes* and $d_i S = 0$ for *reversible processes* in the case of thermal equilibrium. According to L. Boltzmann, entropy S is a measure of the probable distribution of microstates of elements (e.g., molecules of a gas) of a system, generating a macrostate (e.g., temperature of a gas): $S = k_B \ln W$ with k_B Boltzmann's constant and W number of probable distributions of microstates, generating a macrostate. According to the second law, entropy is a measure of increasing disorder during the temporal development of isolated systems. The reversible process is extremely improbable. For Hawking, the cosmic arrow of the expanding universe from simplicity to complexity, from an initial uniform order to galactic diversity, is the true reason of the second law.

Nevertheless, as the second law is statistical and restricted to isolated systems, it allows the emergence of order from disorder in *complex dynamical systems* which are in energetic or material interaction with their environment (e.g., convection rolls of Bénard-experiment, oscillating patterns of the Belousov–Zhabotinsky-Reaction, weather and climate dynamics) [5]. In general, the development of dissipative systems can be characterized by *pattern formation of attractors* (e.g., fixed point attractor, oscillation, chaos) and *temporal bifurcation trees*. In a critical distance to a point of equilibrium, the thermodynamical branch of minimal production of energy ('linear thermodynamics') becomes instable and bifurcates spontaneously into new locally stable states of order ('symmetry breaking'). Then, the nonlinear thermodynamics of nonequilibrium starts [6]. If the system is driven further and further away from thermal equilibrium, a bifurcation tree with nodes of locally stable states of order is generated. Global pattern formation of complex dynamical systems can be *irreversible*, although the laws of locally interacting elements (e.g., collision laws of molecules in a fluid) are *time-reversible*.

4 Time in Evolutionary Dynamics

Life on Earth is not so special in the universe. In a prebiotic evolution, self-assembling molecular systems become capable of self-replication, metabolism, and mutation in a given set of planetary conditions. It is still a challenge of biochemistry to find the molecular programs of generating life from 'dead' matter. *Darwin's evolution of species*, as far as it is known on Earth, can also be characterized by *temporal bifurcation trees*. Mutations are random fluctuations in the bifurcating nodes of the evolutionary tree, breaking the local stability of a species. Selections are the driving forces of branches, leading to further species with local stability. The distance of sequential species is determined by the number of genetic changes. *Evolutionary time* can be measured on different scaling, e.g., by the distance of sequential species and the number of sequential generations of populations. Its *temporal direction* is given by the order of ancestors and descendants.

As conditions changed in the course of the Earth's history, *complex cellular organisms* have come into existence, while others have died out. Entire populations come to life, mature, and die, and in this they are like individual organisms. But while the sequence of generations surely represents the time arrow of life, many other distinct biological time rhythms are discernable. These rhythms are superimposed in *complex hierarchies of time scales*. They include the temporal rhythms of individual organisms, ranging from biochemical reaction times to heartbeats to jet lag, as well as the geological and cosmic rhythms of ecosystems.

Complex systems that consist of many interacting elements, such as gases and liquids, or organisms and populations, may exhibit separate temporal developments in each of their numerous component systems. The complete state of a complex system is therefore determined by statistical distribution functions of many individual states. It has been proposed by B. Misra, I. Prigogine a.o. that time can be defined as an operator which describes changes in the complete states of complex systems [6]. This *time operator* would then represent the average *age* of the different system components, each in its distinct stage of development. Accordingly, a 50-year-old could have the heart of a 40-year-old, but, as a smoker, have the lungs of a 90-year-old. Organs, arteries, bones, and muscles are in distinct states, each according to its particular condition and genetic predisposition. The time operator is thus intended to indicate the *irreversible aging of a complex system*, its *inner* or *intrinsic time*, not the *external and reversible clock time*.

The *human brain* may also be regarded as a complex system in which many neurons and different regions of the brain interact chemically and are switched among their component states by simple local rules. Our individual experience of "*duration*" and "*aging*" thus reflects the complex-system states of the brain, which are themselves dependent on different sensory stimuli, emotional states, memories, and physiological processes. Hence, our *subjective awareness* of time is not contrary to the laws of science, but is a result of the dynamics of a complex system. This in no way diminishes the intimate subjectively experienced flow of time as described in literature and poetry. Knowing the dynamical laws of the brain does not turn one into a Shakespeare or a Mozart. In this sense, the natural sciences and the humanities remain complementary.

The theory of complex systems also applies to the *temporal dynamics of socio-economic systems* [4]. A city, for example, is a complex residential region in which different districts and buildings have distinct traditions and histories. New York, Brasilia, and Rome are the result of distinct temporal development processes, which are not elucidated by external dates. The time operator of a city refers literally to the average age of many distinct stages and styles of development. Institutions, states, and cultures are similarly subject to growth and aging processes, which external dates can shed little light on. Today, there is the dramatic problem of *aging societies* in western civilization. From the point of view of complex dynamic systems, the discussion of age is not just metaphorical, but offers an explanation in terms of structural dynamics.

5 Time in Computation and Information Dynamics

Modern technical societies depend sensitively on the capacities of computers and information networks. *Computation time* is a measure of the time needed to solve a problem by a computer. As a measure of a problem's complexity, one focuses on the running time and data storage requirements of an algorithm and their dependence on the length of the input. The theory of computational complexity deals with the classification of problem into complexity classes, according to the dependence of running time on input length. It is suspected that appreciably shorter computational times were achievable with computers operating on the basis of quantum mechanics and not according to the principles of classical physics. But, as classical computers are based on classical physics, and quantum computers on quantum mechanics, both kinds of computers are based on the concept of *time reversibility*: The laws of nature under which they operate permit, in principle, their computing processes (other than the act of measurement and reading out in the case of quantum computers) to run backward in time.

This raises the question of whether it might also be possible to use computers to simulate *time-irreversible processes* that are well known from biological evolution and the self-organization of the brain. The emergence of cellular patterns was simulated for the first time in the 1950s by von Neumann's *cellular automata*. Computer experiments show the emergence of patterns that are familiar as the attractors of complex dynamic systems. There are oscillating patterns of reversible automata and irreversible developments from initial states to final patterns. For example, in the case of a fixed point attractor, all developments of a cellular automaton develop to the equilibrium state of a fixed pattern which does not change in the future. As these developments are independent of their initial states, they cannot be reconstructed from the final equilibrium state.

Further on, there are cellular automata *without long-term predictions* of their time-depending pattern formation. These are cellular automata with the property of universal computability. *Universal computation* is a remarkable concept of *computational complexity* which dates back to Alan Turing's universal machine. A universal Turing machine can by definition simulate any Turing machine. According to the Church-Turing thesis, any algorithm or effective procedure can be realized by a Turing machine. Now Turing's famous Halting problem comes in. Following his proof, there is no algorithm which can decide for an arbitrary computer program and initial condition if it will stop or not in the long run. Consequently, for a system with universal computation (in the sense of a universal Turing machine), we cannot predict if it will stop in the long run or not. Assume that we were able to do that. Then, in the case of a universal Turing machine, we could also decide whether any Turing machine (which can be simulated by the universal machine) would stop or not. That is obviously a contradiction to Turing's result of the Halting problem. Thus, systems with universal computation are unpredictable. Unpredictability is obviously a high degree of complexity. It is absolutely amazing that systems with simple rules of behavior like cellular automata which can be understood by any child lead to complex dynamics which is no longer predictable.

There are at least some few cellular automata which definitely are universal Turing machines [7]. It demonstrates a striking *analogy of natural and computational processes* that even with simple initial conditions and locally reversible rules many dynamical systems can produce globally complex processes which cannot be predicted in the long run.

The paradigms of parallelism and connectivity are of current interest to engineers engaged in the design of *neurocomputers* and *neural networks*. They also work with simple rules of neural weighting simulating local connectivity of neurons in living brains. Patterns of neural self-assemblies are correlated with cognitive states. With simple local rules neural networks can produce complex behavior, again. In principle, it cannot be excluded that this approach will result in a technically feasible neural self-organization that leads to systems with consciousness, and specifically with time awareness.

In a technical co-evolution, *global communication networks* of mankind have emerged with similarity to self-organizing neural networks of the brain [4]. Data traffic of the Internet is constructed by data packets with source and destination addresses. Local nodes of the net ('routers') determine the local path of each packet by using weighting tables with cost metrics for neighboring routers. There is no central supervisor, but only local rules of connectivity like in self-assembling neural nets. Buffering, sending, and resending activities of routers can cause high densities of data traffic spreading over the net with patterns of oscillation, congestion, and even chaos. Thus, again, simple local rules produce complex patterns of global behavior.

Global information networks store millions of human information traces. They are *information memories* of human history, reflecting the *aging process of mankind as a complex dynamical system*. What is the future of mankind and its information systems in the universe? Cosmic evolution can also be considered as the aging process of a complex dynamical system. If we are living in a flat universe according to recent measurements, then relativistic cosmology forecasts an infinite expansion into the void with increasing dilution of energy and decay in Black Holes. Does it mean the decay of all information storages and memories of the past, including mankind, an *aging universe* with 'Cosmic Alzheimer' [3]? Or may we believe in the fractal system of a bifurcating multiverse with the birth and recreation of new expanding universes? As far as we know there is a *cosmic arrow of time* in our universe, but it is still open where it is pointing at [8].

References

1. Hawking, S.W.: A Brief History of Time: From the Big Bang to the Black Holes, 10th anniversary ed. Bantam, New York (1998)
2. Mainzer, K.: Symmetries of Nature. De Gruyter, Berlin (1996)
3. Mainzer, K.: The Little Book of Time. Copernicus Books, New York (2002)
4. Mainzer, K.: Thinking in Complexity: The Complex Dynamics of Matter, Mind, and Mankind, 5th enlarged ed. Springer, Berlin (2007)
5. Mainzer, K., Müller, A., Saltzer, W.G. (eds.): From Simplicity to Complexity: Information, Interaction, and Emergence. Vieweg, Wiesbaden (1998)
6. Prigogine, I.: From Being to Becoming: Time and Complexity in Physical Sciences. Freeman, New York (1981)
7. Wolfram, S.: Cellular Automata and Complexity. Book News, Portland (1994)
8. Zeh, H.-D.: The Physical Basis of the Direction of Time, 5th edn. Springer, Berlin (2007)

Chapter 15
The Philosophical Significance of the Relativistic Conception of Time

Fabio Minazzi

We step and do not step into the same rivers, we are and are not.
Heraclitus

Abstract For relativity, time is an asymmetrical relationship is not unidirectional. Which brings us back to the profoundly anti-intuitive character of relativity for which it makes no sense to introduce a qualitative difference between the directions of the time. Also with relativity emerges a physical theory which tends to "dissolve" the physical entities in the meaning and function of integration of critical experimental dimension that they perform within a given natural horizon. From this point of view is the relativity does not destroy the concept of synthetic a priori and the transcendentalism of Immanuel Kant. Consequently, the reduction in Kant's time to causality retains all its importance heuristic and epistemological. Even so, thanks to this reduction, it is possible to assign a time to reach its full objective: the temporal dimension is objective because reducible causal order.

In classical mechanics the relativity of all movements has involved many difficulties, which James Clerk Maxwell illustrated clearly in his acute study, *Matter and Motion*, of 1876.

> As our ideas of space and motion become clearer, we come to see how the whole body of dynamical doctrine hangs together in one consistent system. Our primitive notion may have been that to know absolutely where we are, and in what direction we are going, are essential elements of our knowledge as conscious beings. But this notion, though undoubtedly held by many wise men in ancient times, has been gradually dispelled from the minds of students of physics. There are no landmarks in space; one portion of space is exactly like every other portion, so that we cannot tell where we are. We are, as it were, on an unruffled sea, without stars, compass, soundings, wind, or tide, and we cannot tell in what direction we are going.

F. Minazzi (✉)
Università degli Studi dell'Insubria, Via Ravasi 2, 21100 Varese, Italy
e-mail: Fabio.minazzi@unisubria.it

F. Minazzi
Accademia di architettura dell'Università della Svizzera italiana, Mendrisio, Switzerland

S. Albeverio, P. Blanchard (eds.), *Direction of Time*,
DOI 10.1007/978-3-319-02798-2_15,
© Springer International Publishing Switzerland 2014

> We have no log which we can cast out to take a dead reckoning by; we may compute our rate
> of motion with respect to the neighbouring bodies, but we do not know how these bodies
> may be moving in space.

Compared with the difficulty of the relativity of all movements present in mechanics, Einstein's theory brought about a revolutionary transformation of the constituent principle of the traditional limitation of classical physics. In achieving this critical rotation, relativity theory undermined the traditional physical-intuitive representation of the concepts of "time" and "space" and delineated a new epistemic scenario. Einstein's revolution, far from envisaging a new insight into physical reality, took the form of a new cognitive step that became increasingly separate from all sensorial and perceptive intuitions. In this sense the theory of relativity does not delineate a new intuitive *Weltanschauung* of the physical world or claim to resolve the ontological problem of "real" space and "real" time. Or rather: thanks to the theory of relativity, the problem of establishing *empirically* what is "real" space and "real" time loses all meaning, because relativity shows that there no longer exists any possibility of solving the question *experimentally*. For relativity, space and time no longer constitute—as was the case in classical physics—"real" entities: instead they refer to mathematical symbolism and to the elliptical and four-dimensional space of Minkowski's non-Euclidean geometry. In other words, in the theory of relativity the question of seeking to establish the "real" and "true" geometry of time loses all meaning. This conceptual upheaval grew out of the very core of relativity theory, on the basis of which, as Enrst Cassirer pointed out in his study *Zur Einstein'schen Relativitätstheorie* of 1920,

> we now limit ourselves to indicating different metric relationships within the multiple physical world, within that indissoluble correlation of space, time and physically real objects to which relativity theory adheres as the ultimate datum; and it affirms that these metric relationships find their exact mathematical expression in the language of non-Euclidean geometry.

Faced with this outcome we can, however, ask what is then the overall *philosophical and scientific* significance of the relativistic approach within the ambit of the history of modern physics and Western thought. Here we can quote Hans Reichenbach, who made important contributions to the philosophical significance of relativity, ranging from his classic *Philosophie der Raum-Zeit-Lehre* of 1928 (which later appeared in an English edition in 1958) to the posthumous *The Direction of Time* of 1956, edited by his wife Maria Reichenbach, and including a significant contribution to the volume edited by Paul Arthur Schilpp on *Albert Einstein: Philosopher-Scientist* in 1949. The following quotation comes from the last of these works:

> The close connection between space and time on the one hand and causality on the other is perhaps the outstanding feature of Einstein's theory, though the profound significance of this point has not always been recognised. ... Time is the order of causal chains; this is the highest result of Einstein's discoveries.

The absolute simultaneity presupposed by classical physics requires the existence of a world in which the speed at which signals travel is not curbed by an insuperable maximum limit. Einstein's criticism of the classical concept of simultaneity showed that the simultaneity of events can be established only by specifying the frame of reference within which a given physical phenomenon occurs. Two events that are simultaneous in a given frame of reference may not be simultaneous when the frame of reference of our measurements is modified. In this way, relativity no longer sees time as an absolute "container", a "thing-like" and physical entity, as it was traditionally conceived in classical physics, but it is increasingly transformed into a *metric measurement* that appears, rather, as an *enabling condition* for physical experimentation itself. But, as Reichenbach again wrote—this time in the *Philosophie der Raum-Zeit-Lehre*:

> The causal theory of space and time, which we came to through the epistemological study of the foundations of the theory of space-time, also constitutes the foundations of the relativity theory of gravitation.

This confirms that this causal theory of space and time undoubtedly represents, and not for Reichenbach alone, the most important and significant "philosophical result" of Einstein's theory of relativity. In the history of Western thought, an epistemology of neo-positivist derivation also sees this result as a decisive and almost irreversible contribution to the "process of dissolution of the *a priori synthetic*", a tendency that Reichenbach in 1949 considered "one of the most important characteristics of the philosophy of our time". It was not for nothing that Reichenbach, without any critical wavering, in neo-positivistic fashion enlisted Einstein among the ranks of the empiricists, because, in his judgment,

> Einstein's relativity belongs ... to the philosophy of empiricism. ... Mathematical physics always remains empiricist, because it grounds the ultimate criterion of truth on the perception of the senses. ... Despite its enormous mathematical edifice, Einstein's theory of space and time is a triumph of this radical empiricism, in a field that has always been reserved for the discoveries of pure reason.

In fact, in Einstein's relativity theory, the order of temporal succession, of the *before* and *after*, is always reducible to the causal order, on the basis of which effect always follows cause. However, for relativity theory time constitutes a relationship that is asymmetrical, but certainly not unidirectional. In contrast with Reichenbach's observations, this brings us back to the profoundly anti-intuitive character of relativity, which sees it as pointless to introduce a *qualitative* difference between the directions of time. Moreover, *pace* Reichenbach, relativity also delineates a theory of physics that increasingly tends to "dissolve" and "resolve" physical entities into the *significance* and the precise *functions* of critical integration of the experimental dimensions that they develop within a given natural horizon. From this point of view it is therefore legitimate to doubt that relativity is truly capable of decisively helping to pulverise the concept of the *synthetic a priori* and the related critical "Copernican" breakthrough introduced by Immanuel Kant with the identification of the transcendental dimension. We can criticise this interpretation on the basis of the following considerations.

In the first place, the most significant contribution made by Kant to understanding the concept of time is not found in the *Transcendental Aesthetics*, but rather in the analysis of principles, where the author of the *Critique of Pure Reason* deals with the second analogy, namely the principle of "temporal succession in accordance with the law of causality". In these pages Kant relates temporal order to causal connection, observing that a given reality can acquire a place of its own in time only on condition that in the previous state one presupposes a reality which it must always follow, in compliance with a precise rule: "So, if we experience that one thing happens, we always presuppose, in this respect, that another something precedes it, from which the first thing follows in compliance with a rule". Therefore "an event can receive its given temporal position in this relationship only by the fact that in the previous state something is presupposed which the event always follows, i.e. in accordance with a rule. This temporal series cannot be inverted: what happens cannot be placed before that which it follows." Moreover, "if the preceding state is posited, this given event will follow inevitably and necessarily". Therefore, "it is a necessary law of our sense impressions—and hence a formal condition of all perceptions—that the previous time frame necessarily determines the following one". While the imagination can always invert, *ad libitum*, the order of events, the real perception of time cannot, however, modify it: precisely this impossibility distinguishes perception from imagination and finally makes it possible to see temporal succession as the sole criterion of effect in relation to the causality of the cause. Kant therefore accepts the causal reduction of time. Naturally Kant's reduction of time to the causal order is outlined within the framework of Newtonian physics, but it should also be added that Einstein, in illustrating the relativity of measurements of time, did not modify the traditional Kantian concept of time as succession. Certainly, Einstein saw the order of succession as neither single nor absolute; but his reduction of the temporal dimension to the order of causal chains was analogous to that of Kant. So if, with Reichenbach, one maintains that the reduction of the temporal order—i.e. of the succession of *before* and *after*—to the causal order constitutes the principal result of Einstein's relativity, then we have to recognise that this is fully in harmony with Kant's epistemological analysis.

In the second place, one has to bear in mind that also in Einstein's relativity the spatio-temporal ordering of the physical world is still the result of a specific ideal construct. If it is true that the history of physics coincides with the study and the discovery of ever-new specific conceptual frameworks, used to progressively enlarge humanity's conceptual horizon, the theory of relativity can then be read as a contribution that has powerfully assisted in making fully evident the "ideal" and "objective" character of time and space. The relativity of space and time reveal, in all their purity, *the enabling conditions of knowledge*. They no longer indicate a physical entity, but on the contrary, by introducing the innovative relativistic concept of the space-time event, they introduce the theoretical premises that enable this nexus to be determined by experimental measurement. In this sense we can point out, with Cassirer, that

> the theory of general relativity has confirmed and demonstrated this union in a new way, by recognising more radically than all previous physical theories the conditional nature of every empirical measurement, of every attempt to ascertain concrete metric space-time relationships. But this result does not in the least conflict with the relationship between experience and thought as it is fixed in the criticism.

This is not to deny the important role of experimental measurements in the ambit of relativity, but to point out that all measurements presuppose given functional nexuses of connection, coordination and formation, which depend on precise theoretical principles. From this point of view, relativity is, therefore, capable of powerfully clarifying the overall theoretical nature of a physical theory, by showing not the dissolution of transcendentalism but its epistemic fertility. This is because scientific knowledge is never reduced to a "rhapsody of perceptual sensations", but requires an actual elaboration of concepts that is capable of delineating the *legitimacy* of knowledge.

In the third place, it is true that, since the speed of light is limited, in relativity theory the order of time proves to be indefinite. The relativity of simultaneous events introduced by Einstein's theory implies the possibility of an inversion of the temporal order of given events. However, consider a point of the universe P through which a timeline passes. On this timeline we can distinguish two events that are close but separate, A and B. Is it legitimate to attribute an asymmetrical relationship to these points so as to identify an oriented straight line? Can we send a telegram to Plato? The answer is obviously no, since the sending of each message is an irreversible process, entailing an increase in entropy. But this is valid only if the points A and B connected by the timeline are fairly close to each other. If, instead, they were placed at an arbitrarily great distance the answer would not be the same, above all, as Einstein pointed out, "if there exists a series of points that can be linked with time lines so that each point precedes the one before it in time, and *if the series is closed*". In the case of points of time cosmically at a great distance from each other the distinction between "before" and "after" is thus undermined and time's arrow becomes the source of paradoxes.

Moreover, the causal theory of time itself is not exempt from specific problems. Henry Mehlberg (in his *Essai sur la théorie causale du temps*, 1935, in *Durée et causalità*, 1937, and the two volumes of *Time, Causality and the Quantum Theory*, 1980) and Adolf Grünbaum (in *Philosophical Problems of Space and Time*, 1963 and 1973) observed that the temporal order is, in its turn, presupposed by the causal order; but on the other hand an author such as Michel Dummett (*Bringing About the Past*, 1964) claimed that causes could actually precede their effects or be simultaneous with them. Nor can it be forgotten, as Émile Meyerson also pointed out (in his rigorous and analytical *La déduction* relativistique, 1925), that some writings on relativity have included the typical exaggerations that push the spatialisation of time to the point where it is claimed that time is no longer distinguishable from Minkowski's four isotropic dimensions of the universe, so configuring a truly Parmenidean world. Moreover, how can the causal theory of time explain the possibility of travelling into the past, where the time traveller might modify the causes of his own existence? Nor, likewise, is the causal theory of time able to attribute to the temporal dimension unequivocal properties such as finiteness, infiniteness or unlimited finiteness. Besides, does not denying the difference between past and future, between before and after, necessarily entail a universal determinism? Nor, perhaps, is it possible to follow usefully the approach suggested by Georg Henrik von Wright in *Explanation and Understanding*, 1971, where he sought to modify restrictively the concept of cause, or that of Karl Popper (*The Arrow of Time*, 1956), which preferred to invoke probability and speak of propensities. These difficulties reveal all the intrinsically problematic nature of the reduction of time to cause, but they do not diminish its heuristic and epistemic importance. This is partly because, thanks to this reduction, it is possible to endow time with a fully *objective* significance: the temporal dimension is objective precisely because it is reducible to the causal order.

Chapter 16
Geometry of Psychological Time

Metod Saniga

Abstract The paper reviews the most illustrative cases of the "peculiar/anomalous" experiences of time (and, to a lesser extent, also space) and discusses a simple algebraic geometrical model accounting for the most pronounced of them.

Keywords Psychopathology of time · Pencils of conics · Algebraic geometry

1 Introduction

One of the most striking and persistent symptoms of so-called "altered" states of consciousness is, as we shall soon demonstrate, *distortions* in the perceptions of time and space. *Time* is frequently reported as flowing faster or slower, expanded or contracted, and may even be experienced as being severely discontinuous ("fragmented"). In extreme cases, it can stop completely or expand unlimitedly. The sense of space is likewise powerfully affected. *Space* can appear amplified or compressed, condensed or rarefied, or even changing its dimensionality; it can, for example, become just two-dimensional ("flat"), acquire another dimensions, or be reduced to a dimensionless point in consciousness.

As yet, there exists *no* mathematically rigorous and conceptually sound framework that would provide us with *satisfactory* explanations of these phenomena. Physics itself, although being the most sophisticated scientific discipline in describing the "objective" world, is not even able to account for the ordinary perception of time, let alone its other, more pronounced "peculiarities" mentioned above. Nor does it offer a plausible and convincing interpretation of the observed macroscopic dimensionality of space—giving more conceptual challenges than satisfactory answers. It was, among other things, this failure of current paradigms to accommodate

M. Saniga
International Solvay Institutes for Physics and Chemistry, Free University of Brussels (ULB), Campus Plaine, CP–231, Blvd du Triomphe, 1050 Brussels, Belgium

M. Saniga (✉)
Astronomical Institute of the Slovak Academy of Sciences, 05960 Tatranská Lomnica, Slovak Republic
e-mail: msaniga@astro.sk

S. Albeverio, P. Blanchard (eds.), *Direction of Time*,
DOI 10.1007/978-3-319-02798-2_16,
© Springer International Publishing Switzerland 2014

a vast reservoir of the phenomena described above that originally motivated our search for a rigorous and self-consistent scheme, and which ultimately led into what we call the theory of *pencil-generated* space-times [1–9].

The aim of this contribution is to demonstrate that this theory represents a cogent starting point for a deeper understanding of the altered states of consciousness in their temporo-spatial aspects. It is shown, in particular, that the three most abundant groups of "pathological" perceptions of time, namely the feeling of *timelessness* ("eternity"), time *standing still*, as well as the experience of the *dominating past*, can well be modeled by singular (space-)time configurations represented by a specific pencil of conics. Being speculative, the paper is also offered to stimulate further research into the possible links between mathematics and physics on the one side, and psychology, psychiatry, and philosophy on the other.

2 Examples of Psychopathology of Time

2.1 Near-Death Experiences

A typical near-death experience (NDE) occurs if a person is exposed suddenly to the threat of death but then survives and reports such phenomena as floating out of his/her body, moving rapidly through dark, empty space, having *the life review*, and encountering a brilliant white light. Out of these four consecutive phases it is the third one, the life review, which is of concern here. The following extract is taken from a famous book by R. Moody [10]:[1]

> After all this banging and going through this long, dark place, all of my childhood thoughts, *my whole entire life was* there at the end of this tunnel, just *flashing in front of me*. It was not exactly in terms of pictures, more in the form of thoughts, I guess. It was just *all there at once*, I mean, not one thing at a time, blinking on and off, but it was everything, *everything at one time...*

However, it is not only the *past* but—weird as it may sound—also the *future* that a subject experiencing an NDE can have access to. The first to draw attention to this fact seems to have been K. Ring [11]:

> ...the material I have collected that bears upon a remarkable and previously scarcely noted precognitive feature of the NDE I have called the personal *flashforward* (PF). If these experiences are what they purport to be, they not only have *extremely profound implications for our understanding of the nature of time* but also possibly for the future of our planet...
> Personal flashforwards usually occur within the context of an assessment of one's life during an NDE (i.e. during a life review and preview), although occasionally the PF is experienced as a subsequent vision. When it takes place

---

[1]In this and all the subsequent excerpts/quotations, italics are used to emphasize those parts of the narratives that most directly relate to the topic of the section. They are introduced by the author of the present paper, not the author(s) of the paper/book quoted.

while the individual is undergoing an NDE, it is typically described as an image vision of the future. *It is as though the individual sees something of the whole trajectory of his life, not just past events...* The understanding I have of these PFs is that to the NDEr they represent events of a *conditional future*—i.e., if he chooses to return to life, then these events *will* ensue...

A more impressive description of this fascinating phenomenon is borrowed from [12], based on Atwater's personal experience:

This time, I moved, not my environment, and I moved rapidly... My speed accelerated until I noticed a wide but thin-edged expanse of bright light ahead, like a "parting" in space or a "lip", with a brightness so brilliant it was beyond light yet I could look upon it without pain or discomfort... The closer I came the larger the parting in space appeared until... I was absorbed by it as if engulfed by a force field...

Further movement on my part ceased because of the shock of what happened next. Before me there loomed two gigantic, impossibly huge masses spinning at great speed, looking for all the world like cyclones. One was inverted over the other, forming an hourglass shape, but where the spouts should have touched there was instead incredible rays of power shooting out in all directions... I stared at the spectacle before me in disbelief...

As I stared, I came to recognize my former Phyllis self in the midupperleft of the top cyclone. Even though only a speck, *I could see my Phyllis clearly, and superimposed over her were all her past lives and all her future lives happening at the same time in the same place as her present life. Everything was happening at once! Around Phyllis was everyone else she had known and around them many others... The same phenomenon was happening to each and all. Past, present, and future were not separated but, instead, interpenetrated like a multiple hologram combined with its own reflection.*

The only physical movement anyone or anything made was to contract and expand. There was no up or down, right or left, forward or backward. There was only in and out, like breathing, like the universe and all creation were breathing—inhale/exhale, contraction/expansion, in/out, off/on.

2.2 Drug-Induced States

One of the most pronounced "distortions" in perception of time and space is encountered in the extraordinary states induced by the use of drugs. The following extract, taken from [13], illustrates this in detail:

...This and all other changes in my dreams were accompanied by deep-seated anxiety and funeral melancholy, such as *are wholly incommunicable by words*. I seemed every night to descend—not metaphorically, but literally to descend—into chasms and sunless abysses, depths below depths, from which it seemed hopeless that I could ever re-ascend. Nor did I, by waking, feel that I had re-ascended. Why should I dwell upon this? For indeed the state of gloom which attended these gorgeous spectacles... *cannot be approached by words.*

The *sense of space, and* in the end *the sense of time, were both powerfully affected.* Buildings, landscapes, etc., were exhibited in proportions so vast as the bodily eye is not fitted to receive. *Space* swelled, and was *amplified to an extent of unutterable and self-repeating infinity.* This disturbed me much less than *the vast expansion of time.* Sometimes I seemed to have lived for 70 or a 100 years in one night; nay, sometimes had feelings representative of a *duration far beyond the limits of any human experience...*

Here, one should notice that the "amplification" of space is often reported hand in hand with the "expansion" of time. Even a more dramatic and profound departure from the "consensus reality", induced by LSD, is depicted in [14]:

...I found myself in a rather unusual state of mind; I felt a mixture of serenity and bliss... It was a world where miracles were possible, acceptable, and understandable. I was *preoccupied with the problems of time and space and* the insoluble *paradoxes of infinity and eternity* that baffle our reason in the usual state of consciousness. I could not understand how I could have let myself be "brain-washed" into accepting the simple-minded concept of one-dimensional time and three-dimensional space as being mandatory and existing in objective reality. It appeared to me rather obvious that there are no limits in the realm of spirit and *that time and space are arbitrary constructs of the mind.* Any number of *spaces with different orders of infinities* could be deliberatery created and experienced. *A single second and eternity seemed* to be *freely interchangeable.* I thought about higher mathematics and saw *deep parallels between* various *mathematical concepts and altered states of consciousness...*

This description clearly indicates that the mind is not confined to the limits of conventional space and time, and what we perceive in our "normal" state of consciousness is only a tiny fraction of the world we all have potential access to. The following experience of "disordered", "chaotic" time [15], induced by the drug called mescaline, dovetails nicely with the above statement:

For half an hour nothing happened. Then I began feeling sick; and various nerves and muscles started twitching unpleasantly. Then, as this wore off, my body became more or less anaesthetized, and I became 'de-personalized', i.e., I felt completely detached from my body and the world...

This experience alone would have fully justified the entire experiment for me..., but at about 1.30 all interest in these visual phenomena was abruptly swept aside when I found that *time was behaving even more strangely than color.* Though perfectly rational and wide-awake... *I was not experiencing events in the normal sequence of time. I was experiencing the events of 3.30 before the events of 3.0; the events of 2.0 after the events of 2.45, and so on. Several events I experienced with an equal degree of reality more than once.* I am not suggesting, of course, that the events of 3.30 *happened* before the events of 3.0, or that any event *happened* more than once. All I am saying is that *I experienced them, not in the familiar sequence of clock time, but in a different, apparently capricious sequence which was outside my control.*

By 'I' in this context I mean, of course, my disembodied self, and by 'experienced' I mean learned by a special kind of awareness which seemed to comprehend yet be different from seeing, hearing, etc.... I count this experience, which occurred when, as I say, I was wide-awake and intelligent, sitting in my own armchair at home, as *the most astounding and thought-provoking* of my life...

The final experience we introduce in this section, induced and sustained by smoking *salvia divinorum*, seems to feature elements of time travel, or existing in separate realities simultaneously [16]:

> ... the salvia started to overtake me. Suddenly, I was unsure of where I was and, more specifically, when I was. I wasn't sure if I was sitting on the floor in my new apartment or on the couch of my old one the previous week. *It felt as if I were in both places at once*, smoking salvia. *I felt I became unstuck in time. It seemed I was existing simultaneously in the past week's trip, the current moment, and thousands of other times, both in the future and the past. Not only other times of my life, but of other's lives as well, all existing as a four dimensional hyperbeing linked through salvia.* My vision had a very "edged" aspect, as if everything had an extra dimension. While I was lying on the floor with eyes closed, "time tripping", I didn't exactly see anything, but *I had a definite sense of being in numerous places, a sort of mental map...*
>
> In all of my experiences, I get the impression that I am "bringing back" only a small portion of what I am experiencing. The sensations come at a breakneck pace, and it is difficult to even hang on, much less pay attention to what is actually going on. All of my experiences seem to have a somewhat consistent aspect. They feel *very real*, in a strange way...

Note a striking resemblance between this experience and Atwater's NDE described in the previous section.

2.3 Mental Psychoses

This is the domain where much is still unsettled and uncertain and which thus provides an invaluable source for scientific imagination, as we strive to decipher the laws of Nature. In the accounts sampled below we shall recognize at least four distinct types of anomalous temporal patterns reported by mentally ill patients.

The first type is what the majority of psychotics refer to as *"time standing still"*. Some spectacular examples are found in a paper by H. Tellenbach [17], namely [17, p. 13]:

> I sure do notice the passing of time but couldn't experience it. I know that tomorrow will be another day again but don't feel it approaching. I can esti-

mate the past in terms of years but I don't have any connection to it anymore. The time *standstill* is infinite, I live in a constant *eternity*. I see the clocks turn but *for me time does not flow...* Everything lies in one line, there are *no differences of depth* anymore... Everything is like a firm *plane...*

and [ibid; p. 14]:

"Everything is very different in my case, time is passing very slowly. Nights last so long, one hour is as long as usually a whole day..." Sometimes *time* had totally *stood still*, it would have been horrifying. Even space had changed: "Everything is so empty and dark, everything is so far away from me... *I don't see space as usual*, I see everything as if it were just a background. It all seems to me *like a wall*, everything is *flat*. Everything presses down, everything looks away from me and laughs..."

It is worth noticing that when time comes to a stillstand, perceived space loses one dimension, becoming only two-dimensional. We shall see later that this feature finds a very nice explanation in our model. A slightly more detailed description of this temporal mode is given in a very readable paper by Muscatello and Giovanardi Rossi [18, p. 784]:

Time is standing still for me, I believe. It is perhaps only a few moments that I have been so bad. I look at a clock and I have the impression, if I look at it again, that an *enormous period of time* has passed, as if hours would have passed instead only a few minutes. It seems to me that a duration of time is enormous. *Time does not pass any longer, I look at the clock but its hands are always at the same position, they no longer move, they no longer go on; then I check if the clock came to a halt, I see that it works, but the hands are standing still.* I do not think about my past, I remember it but I do not think about it too much. When I am so bad, *I never think about my past.* Nothing enters my mind, nothing... I did not manage to think about anything. I did *not* manage to see *anything in my future. The present does not exist for me* when I am so bad ... *the past does not exist, the future does not exist.*

The second type of temporal psychosis is what one may well call the *dominating past*. A couple of examples below, both by schizophrenics, give detailed accounts of it. The first narrative makes explicit how the temporal is devoid of the notions of both the future and the present [19, p. 563]:

I stop still, I am being thrown *back into the past* by words that are being said in the hall. But this all is self evident, it must be that way! There is *no present* anymore, there is *only* this stated *being related to the past*, which is *more than a feeling*, it goes through and through. There are all sorts of plans against me in the air of this hall. But I don't listen to them, I let my mind rest so that it doesn't corrode... Is there any future at all? Before, the future existed for me but now it is shrinking more and more. *The past is so very obtrusive, it throws itself over me; it pulls me back...* By this I want to say that there is *no future* and I am thrown back... Strange thoughts enter my mind and drive me off into the past...

The other account seems to even question the very nature of time [ibid; p. 561]:

> It pulls me back, well, where to? To where it comes from, there, where it was before. *It enters the past.* It is that kind of a feeling as if you had to fall back. This is the disappearing, the vanishing of things. *Time slips into the past,* the walls are fallen apart. Everything was so solid before. It is as if it were so close to be grabbed, as if you had to pull it back again: *Is that time?* Shifted way back!

The third characteristic type of distorted temporal dimension a psychotic often encounters is the sensation of time flowing *backward*. Of all the psycho-time-related references we have seen, no account draws a portrait of the essential properties of this mode better than that found in [19, p. 556]:

> Yesterday at noon, when the meal was being served, I looked at the clock: why did no one else? But there was something strange about it. For the clock did not help me any more and did not have anything to say to me any more. How was I going to relate to the clock? *I felt as if I had been put back, as if something of the past returned, so to speak, toward me, as if I were going on a journey. It was as if at 11:30 a.m. it was 11:00 a.m. again, but not only time repeated itself again, but all that had happened for me during that time as well.* In fact, all of this is much too profound for me to express. In the middle of all this something happened which did not seem to belong here. *Suddenly, it was not only 11:00 a.m. again, but a time which passed a long time before was there and there inside*—have I already told you about a nut in a great, hard shell? It was like that again: *in the middle of time I was coming from the past towards myself.* It was dreadful. I told myself that perhaps the clock had been set back, the orderlies wanted to play a stupid trick with the clock. I tried to envisage time as usual, but I could not do it; and then came a feeling of horrible expectation that *I could be sucked up into the past, or that the past would overcome me and flow over me.* It was disquieting that someone could play with time like that, somewhat daemonic...

A psychotic patient of Laing [20] gives a very brief and concise description:

> ...*I got the impression that time was flowing backward; I felt that time proceeded in the opposite direction,* I had just this extraordinary sensation, indeed ... the most important sensation at that moment was, *time in the opposite direction...* The perception was so real that I looked at a clock and, I do not know how, I had the impression that the clock confirmed this feeling, although I was not able to discern the motion of its hands...

A strikingly similar portrayal of time-reversal is also provided by a depressive patient of Kloos [21, p. 237]:

> As I suddenly broke down I had this feeling inside me that time had completely flown away. After those three weeks in a sick-camp, I had this feeling that the clock hands run idle, that they do not have any hold. This was my

sudden feeling. I did not find, so to speak, any hold of a clock and of life anymore, I experienced a dreadful psychological breakdown. I do not know the reason why I especially became conscious of the clock. *At the same time, I had this feeling that the clock hands run backward...* There is only one piece left, so to speak, and that stands still. I could not believe that time really did advance, and that is why I thought that the clock hands did not have any hold and ran idle... *As I worked and worked again, and worried and did not manage anything, I simply had this feeling that everything around us (including us) goes back... In my sickness I simply did not come along and then I had this delusion inside me that time runs backward...* I did not know what was what anymore, and I always thought that I was losing my mind. *I always thought that the clock hands run the wrong way round, that they are without any meaning. I just stood-up in the sick-camp and looked at the clock—and it came to me then at once: well, what is this, time runs the wrong way round?!... I saw, of course, that the hands moved forward, but, as I could not believe it, I kept thinking that in reality the clock runs backward...*

The final type of temporal psychosis can be termed the *extended present*, and is described nicely in [22, pp. 104–107]:

> The *patient elevates* herself *above normal boundaries of time* without totally surmounting them. The distinction of the present and the future is not canceled out as the patient still speaks about both dimensions, yet *the line between* the actual *present and* the only maybe-possible and unreal *future becomes swaying and possible to cross*. Both dimensions incapsulate and overlap each other without a steady transition. *The future fuses with the present* and vice versa and experiencing acquires a flickering twilight character which is radically distinguished from how a healthy person anticipates the future in day-dreams and the like... *The edge between the present and the past is swaying as well*. At the same time and in a totally different way, *the past is included in and fuses with the events of the present* as well as usually the present is part of the past. There is a kind of *condensation of time*; the present is not distinguished amidst the continuous, steady flow of the past any more, but at the same time the present is not filled with something past as it usually is with normal people; in this case it overlaps...
>
> The *three temporal levels* of past, present, and future therefore *seemed to overlap* in the psychotic experience of the patient *in an extremely peculiar simultaneousness* without really invalidating the distinction of past, present, and future.

2.4 Mystical States

In the last example the present loses its "point-like" nature and starts to expand into *both* the future *and* the past. If this expansion is not limited, the experiencer will eventually attain the state of *the all-containing present* ("eternity"), when he/she is able to see all events simultaneously, as in the following remarkably vivid narrative [23]:

> ...I get up and walk to the kitchen, thinking about what a timeless experience would be like. I direct my attention to everything that is happening at the present moment—what is happening here, locally, inside of me and near me, but non-locally as well, at ever increasing distances from me. I am imagining everything that is going on in a slice of the present—throughout the country, the planet, the universe. It's all happening at once. *I begin to collapse time, expanding the slice of the present, filling it with what has occurred in the immediate "past".* I call my attention to what I just did and experienced, what led up to this moment, locally, but keep these events within a slowly expanding present moment. *The present slice of time slowly enlarges, encompassing, still holding, what has gone just before, locally, but increasingly non-locally as well.* By now, I am standing near the kitchen sink. The present moment continues to grow, expand. *Now it expands into the "future" as well.* Events are gradually piling up in this increasingly larger moment. What began as a thin, moving slice of time, is becoming thicker and thicker, increasingly filled with events from the "present", "past", and "future". *The moving window of the present becomes wider and wider, and moves increasingly outwardly in two temporal directions at once.* It is as though things are piling up in an ever-widening present. The "now" is becoming very thick and crowded! *"Past" events do not fall away and cease to be; rather, they continue and occupy this ever-widening present. "Future" events already are, and they, too, are filling this increasingly thick and full present moment. The moment continues to grow, expand, fill, until it contains all things, all events. It is so full, so crowded, so thick, that everything begins to blend together. Distinctions blur. Boundaries melt away. Everything becomes increasingly homogeneous, like an infinite expanse of gelatine. My own boundaries dissolve. My individuality melts away. The moment is so full that there no longer are separate things. There is no-thing here. There are no distinctions.* A very strong emotion overtakes me. Tears of wonder-joy fill my eyes. This is a profoundly moving experience. Somehow, I have moved away from the sink and am now several feet away, facing in the opposite direction, standing near the dining room table. *I am out of time and in an eternal present. In this present is everything and no-thing. I, myself, am no longer here. Images fade away. Words and thoughts fade away. Awareness remains, but it is a different sort of awareness. Since distinctions have vanished, there is nothing to know and no one to do the knowing. "I" am no longer localized, but no longer "conscious" in the usual sense. There is no-thing to be witnessed, and yet there is still a witnesser.* The experience begins to fade. I am "myself" again. I am profoundly moved. I feel awe and great gratitude for this experience with which I have been blessed...

A somewhat bizarre, yet more scholarly report of an almost identical psycho-space-time pattern is found in [24]. Although the author almost exclusively focuses on the spatial fabric of existence throughout the book, it is undoubtedly the temporal aspect of this particular experience that is most fascinating:

> I woke up in a whole different world in which the puzzle of the world was solved extremely easily in a form of *a different space*. I was amazed at the wonder of this different space and this amazement concealed my judgement, *this space is totally distinct from the one we all know*. It had *different dimensions, everything contained everything else*. I was this space and this space was me. The outer space was part of this space, I was in the outer space and the outer space was in me...
>
> Anyway, I didn't experience time, time of the outer space and aeons until the second phase of this dream. In the cosmic flow of time you saw worlds coming into existence, blooming like flowers, actually existing and then disappearing. It was an endless game. *If you looked back into the past, you saw aeons, if you looked forward into the future there were aeons stretching into the eternity*, and *this eternity was contained in the point of the present*. One was situated in a state of being in which the "will-be" and the "vanishing" were already included, and this "being" was my consciousness. It contained it all. This "being-contained" was presented very vividly in a geometric way in form of *circles of different size which* again were all part of a unity since all of the circles *formed exactly one circle. The biggest circle was part of the smallest one and vice versa*. As far as the differences of size are concerned, I could not give any accurate information later on...

Note a striking similarity between this experience and the experience of Grof's subject (Sect. 2.2); in particular, both the subjects speak about the puzzling equivalence between the eternity and the moment of the present. This seems to be a very important property of a mystical state, for it is also mentioned by such famous mystics as St. Thomas and Meister Eckhart, and even by the great Dante Alighieri, as pointed out by Ananda Coomaraswamy [25, p. 110]:

> [*St. Thomas*:] Eternity is called "whole" not because it has parts, but because it is wanting in nothing... The expression "simultaneously whole" is used to remove the idea of time, and the word "perfect" to exclude the now of time... *The now that stands still is said to make eternity*...

[ibid; p. 117]:

> [*Meister Eckhart*:] God is creating the whole world now, this instant ("nu alzemale")... He makes the world and all things in this *present Now* ("gegenwuertig nu")...

[ibid; pp. 120–121]:

> Dante, when he is speaking of Eternity, makes many references to this "essential point" or "moment". *All times are present to it* ("il punto a cui tutti li tempi son presenti", Paradiso 17.17); *there every where and every when are focused* ("dove s'appunta ogni ubi ed ogni quando", Paradiso 29.12)... In it alone is every part there where it ever was, for it is not in space, nor hath it

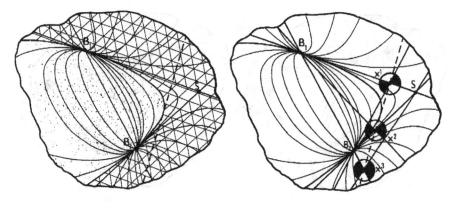

Fig. 1 A geometrical configuration representing our ordinary perception of time (*left*) and space (*right*)

poles ("in quella sola è ogni parte là ove sempr'era, perché non è in loco, e non s'impola")... Whereby it thus doth steal from thy sight (22.64).

3 An Outline of the Algebraic Geometrical Model of Time Dimension

In what follows we shall introduce the basic features of a simple geometrical model capable of mimicking remarkably well some of the most pronounced pathologies of time and space we highlighted in the preceding section. The presentation will be rather illustrative, so as to be accessible to scholars of various disciplines and diverse mathematical backgrounds. The reader wishing to go further into the details of the mathematical formalism is referred to our papers [3–5, 8].

The model in question is based on a specific *pencil* (i.e. a linear, single parametric aggregate) *of conics* in the real projective plane and its structure is illustrated in Fig. 1. We see that all the conics touch each other in two different points, B_1 and B_2, and the corresponding two common tangent lines meet at the point S. This pencil of conics is taken to generate the *time* dimension, where each conic represents a single event. The pencil, as it stands, is homogeneous in the sense that every conic has the same footing in it. Yet, from what we have just seen it is obvious that the intrinsic structure of subjective time is far from being homogeneous, being, in fact, endowed with three different kinds of event, namely the past, present, and future. Hence, the pencil has to be "de-homogenized" in order to yield the structure required.

This can be done fairly easily if, for example, we select in the plane one line (the broken line in Fig. 1) and attach to it a special status. It is clear that if this distinguished line is in a general position, it does not pass via any of the points B_1, B_2, and S. Under such an assumption, the conics of the pencil are seen to form, as far as the intersection properties are concerned, two distinct domains with respect to this line (see Fig. 1, *left*). One domain comprises those conics that have no intersection with the line (these conics are located in the dotted area and we shall call them "non-

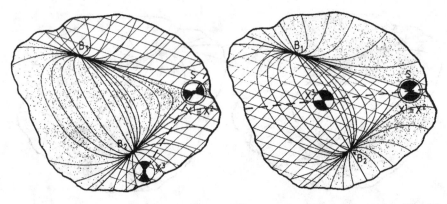

Fig. 2 A geometrical configuration representing "time standing still" (*left*) and "flat space" (*right*)

cutters"), whereas the other domain features the conics cut by the line in two distinct points (they are located in the shaded area and called "cutters"). These two domains are separated from each other by a *unique* single conic (drawn bold in Fig. 1) which has the broken line for its tangent (this conic will be referred to as the "toucher"). This is really a very remarkable pattern for it is seen to reproduce strikingly well, at least at the qualitative level, our ordinary perception of time once we postulate that the cutters represent the past events, the non-cutters the moments of the future, and that the unique toucher stands for the present, the now [1–9].

As for a *spatial* dimension, this will be modeled by a *pencil of lines*, i.e. by all the lines that pass through a *given* point (called the vertex of the pencil). Here the given point means any point which our broken line shares with each of the lines B_1B_2, B_1S, and B_2S, defined by the pencil of conics. From Fig. 1, *right*, it is evident that for a general position of the broken line there are just *three* specific pencils of lines (depicted in Fig. 1, *right*, as three half-filled circles). And there are just *three* spatial dimensions (x^1, x^2, and x^3) we perceive in our "normal/ordinary" state of consciousness! The model is thus characterized by an intricate connection between the *intrinsic* structure of time and the *number/multiplicity* of macroscopic spatial dimensions [2, 3, 8].

In order to make this link visible to the eye, let us start moving the broken line from its original, generic position of Fig. 1 towards the point S in such a way that it is eventually incident with the latter—as shown in Fig. 2. As it can easily be discerned from this figure, in this limiting case the toucher disappears and we find only the cutters (shaded area) and non-cutters (dotted area). In other words, our time dimension now *lacks* the moment of *the present*, being endowed with the past and future events only. As it is intuitively obvious that out of the three temporal levels, i.e. the past, present, and future, it is the present that seems to be fully "responsible" for what we experience as the "flow/passage" of time, its absence in the above-mentioned arrow implies that such time does *not* pass, it *stands still*. From Fig. 2 it can further be discerned that this partial "collapse" of the generic arrow of time is accompanied by a $3 \Rightarrow 2$ reduction in the dimensionality of space, because two of its coordinates (x^1 and x^2) merge with each other and form a single coordinate.

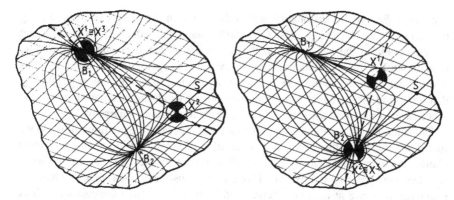

Fig. 3 A geometrical configuration representing "dominating past" (*left*) and two-dimensional space (*right*)

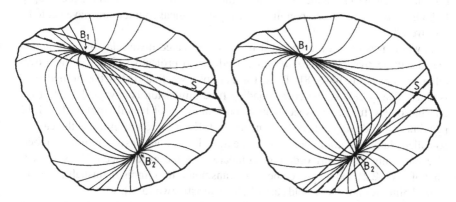

Fig. 4 Two geometrical configurations, each representing "everlasting now" and infinitely-dimensional space

This configuration thus bears a striking similarity to the space-time construct that a couple of Tellenbach's melancholic patients were trying to describe (see the first two excerpts in Sect. 2.3)!

Another kind of "degenerate" temporal arrow emerges when the broken line hits one of the points B_1, B_2, but does not incorporate the point S—the mode depicted in Fig. 3. It is obvious that the line selected in this way is a secant to *every* conic of the pencil, which means that the corresponding time dimension features *exclusively* the region of *the past*—that is, it is identical with the *dominating past* mode of F. Fischer's schizophrenic patients (see the fourth and fifth excerpts in Sect. 2.3). Note that there is again the $3 \Rightarrow 2$ drop in the number of space dimensions.

The third, and the last fundamental mode associated with this particular pencil of conics, is characterized by the broken line coinciding with one of the common tangent lines, B_1S or B_2S—as shown in Fig. 4. In this case, *every* point of the broken line is the vertex of the pencil of lines representing a spatial coordinate, thus space becomes *infinitely* dimensional (which is illustrated in Fig. 4 by two lines running parallel to the line B_1S resp. B_2S). On the other hand, the broken line is now tangent to *every* conic of the pencil, i.e. *all* the conics are its *touchers*; the corresponding time dimension thus consists *solely of* the events of *the present*, and represents thus nothing but what in the previous section was referred to as the "eternity". We see that this (kind of) space-time configuration possesses all the basic attributes of the space-time of mystics (see Sect. 2.4), as described by Huber's narrative.

The attentive reader may ask, "What kinds of (space-)time configurations do we find if we consider other types of a pencil of conics residing in the real projective plane?" As there exist as many as nine different types of these [3], we do find some new temporal patterns not exhibited by the previously discussed pencil. Thus, for example, we arrive at internally "contorted" forms of the time dimension, such as the one composed of two distinct domains of the past and two distinct moments of the present, but only *one* region of the future; or that endowed with two different domains of the past separated from each other by a single moment of the present [3]. In both the cases, the corresponding psychopathological counterparts have yet to be discovered.

A whole new class of temporal structures is revealed if we relax the assumption that the projective plane is *real* and consider also projective planes defined over other ground fields [4–6, 8]. Thus, for example, we find that if the ground field is algebraically closed, the corresponding time dimension is always devoid of the concept of the future, irrespectively of both the type of pencil employed and the position of the distinguished (broken) line [4]. Even more intriguing is the case of so-called Galois (or finite) fields: here, the time dimension consists of finite numbers of events only, lacks any ordering (compare with the experience of "disordered" time of Sect. 2.2) and may even become transmuted into (indistinguishable from) a *spatial* dimension if these fields are of characteristic two [5, 6].

Finally, a very promising generalization of the above-discussed rudimentary model is achieved if the constraint of the *linearity* of the aggregate of conics is also removed. A simple, quadratic set of conics put forward in [9] not only reproduces all the key features of the linear model, but it also leads to what we termed the *arrow-within-arrow* patterns—the structures accounting, for example, for experiences of time flowing *backward* (see Sect. 2.3). As we do not have additional space here to discuss these and other intriguing cases in further mathematical detail, the interested reader is referred to our papers [2–6, 8, 9].

4 Conclusion

The findings and results just described provide us with strong evidence that *not only* are the manifestations of mental or psychological time so diverse and unusual that they fail to conform to the generally adopted picture of the macroscopic physical world, *but* there also exists a unique mathematical framework which, at least qualitatively, underlies and unifies their seemingly bizarre properties. Hence, any

attempt to disregard these psychopathological temporal constructs as pure hallucinatory phenomena would simultaneously cast a doubtful eye on the very role of
mathematics in our understanding of Nature. To the contrary, it is just mathematics
(algebraic projective geometry here) that plainly tells us that it is far more natural
to expect all these "unusual" perceptions of time to be simply as real as our ordinary ("normal") one. We are, however, fully aware that this point of view is very
likely to meet with scepticism, and even with fierce opposition, from the side of
'hard-line' instructional scientists. Most such scientists will probably object to the
anecdotal character inherent in describing the variety of time's multifaceted phenomena. However, this inevitable anecdotal feature is necessary for research on the
qualitative aspects of time—research which profits both psychology and physics. As
very well argued by Shallis [26, p. 153]:

> *Quality* and *quantity* are somewhat like the ingredients of *descriptive* and
> *instructional* science, respectively. Because the two approaches are so differ-
> ent the sorts of evidence employed in each will also differ. *Whereas the in-
> structional approach requires, indeed demands, rigorous, quantitative, and
> reproducible evidence, the descriptive attitude, which often deals with the
> unique and individual, is mainly anecdotal.* This does not mean it is uncritical
> or sloppy, but in trying to find the whole truth everything must be taken into
> account. If some evidence turns out to be false, that too is part of the picture. In
> instructional science anecdotal evidence, *even if true*, can be dismissed as un-
> quantifiable and impossible to assess. *The techniques of instructional science
> cannot handle individual experience or admit to the quality of time. Descrip-
> tive science can. . .*

We are firmly convinced that anything that shows a definite mathematical structure,
whatever bizarre and counter-intuitive it may appear, deserves effort and ingenuity
to be thoroughly explored and examined, all the more that [ibid; pp. 153–154]:

> . . . the fact that the experience of time is *not* quantifiable puts it into arena
> of human perceptions that are at once richer and more meaningful than are
> those things that are merely quantifiable. . . *The lack of quantification of tem-
> poral experiences is not something that should stand them in low stead, to be
> dismissed as nothing more than fleeting perceptions or as merely anecdotal;
> rather that lack should be seen as their strength.* It is because the experience
> of time is not quantifiable and not subject to numerical comparison that makes
> it something of quality, something containing the essence of being. . .

Acknowledgements   I am very grateful to Mr. Pavol Bendík for painstaking drawing of the figures. I would like to express my cordial thanks to Miss Daniela Veverková and Mr. Peter Hahman
for translating into English all the excerpts taken from journals written in German. My warm thanks
are due also to Dr. Rosolino Buccheri (IASFC, Palermo) for the corresponding translation of a couple of excerpts in Italian. I am also indebted to Prof. Mark Stuckey (Elizabethtown College) for a
careful proofreading of the paper. Last, but not least, I wish to express my gratitude to my wife for
her continuous support and encouragement of my work. This work was supported in part by the
NATO Collaborative Linkage Grant PST.CLG.976850, the NATO Advanced Research Fellowship
distributed and administered by the Fonds National de la Recherche Scientifique, Belgium, and

the 2001–2003 Joint Research Project of the Italian Research Council and the Slovak Academy of Sciences "The Subjective Time and its Underlying Mathematical Structure".

References

1. Saniga, M.: Arrow of time and spatial dimensions. In: Sato, K., Suginohara, T., Sugiyama, N. (eds.) Cosmological Constant and the Evolution of the Universe, pp. 283–284. Universal Academy Press, Tokyo (1996)
2. Saniga, M.: On the transmutation and annihilation of pencil-generated space-time dimensions. In: Tifft, W.G., Cocke, W.J. (eds.) Mathematical Models of Time and Their Application to Physics and Cosmology, pp. 283–290. Kluwer Academic, Dordrecht (1996)
3. Saniga, M.: Pencils of conics: a means towards a deeper understanding of the arrow of time? Chaos, Solitons Fractals 9, 1071–1086 (1998)
4. Saniga, M.: Time arrows over ground fields of an uneven characteristic. Chaos, Solitons Fractals 9, 1087–1093 (1998)
5. Saniga, M.: Temporal dimension over Galois fields of characteristic two. Chaos, Solitons & Fractals 9, 1095–1104 (1998)
6. Saniga, M.: On a remarkable relation between future and past over quadratic Galois fields. Chaos, Solitons Fractals 9, 1769–1771 (1998)
7. Saniga, M.: Unveiling the nature of time: altered states of consciousness and pencil-generated space-times. Int. J. Transdiscipl. Stud. 2, 8–17 (1998)
8. Saniga, M.: Geometry of psycho(patho)logical space-times: a clue to resolving the enigma of time? Noetic J. 2, 265–273 (1999)
9. Saniga, M.: Algebraic geometry: a tool for resolving the enigma of time? In: Buccheri, R., Di Gesù, V., Saniga, M. (eds.) Studies on the Structure of Time: From Physics to Psycho(patho)logy, pp. 137–166. Kluwer Academic/Plenum, New York (2000)
10. Moody, R.: Life After Life, pp. 69–70. Mockingbird Books, Atlanta (1975)
11. Ring, K.: Heading Towards Omega, p. 183. Morrow, New York (1984)
12. Atwater, P.M.H.: Coming Back to Life. The After-Effects of the Near-Death Experience, Chap. 2. Dodd, Mead, New York (1988)
13. Quincey, T.: In: Confessions of an English Opium-Eater, pp. 281–282. Dent, London (1967)
14. Grof, S.: Realms of the Human Unconscious. Observations from LSD Research, p. 187. Dutton, New York (1976)
15. Ebin, D. (ed.): The Drug Experience, p. 295. Orion Press, New York (1961)
16. An anonymous account posted at http://leda.lycaeum.org/Trips/Simultaneous_Lives.5627.shtml
17. Tellenbach, H.: Die Raumlichkeit der Melancholischen. I. Mitteilung. Nervenarzt 27, 12–18 (1956)
18. Muscatello, C.F., Giovanardi Rossi, P.: Perdita della visione mentale e patologia dell'esperienza temporale. G. Psichiatr. Neuropatol. 95, 765–788 (1967)
19. Fischer, F.: Zeitstruktur und Schizophrenie. Z. Gesamte Neurol. Psychiatr. 121, 544–574 (1929)
20. Laing, R.D.: La Politica dell'Esperienza, p. 148. Feltrinelli, Milano (1968)
21. Kloos, G.: Störungen des Zeiterlebens in der endogenen Depression. Nervenarzt 11, 225–244 (1938)
22. Ciompi, L.: Über abnormes Zeiterleben bei einer Schizophrenen. Psychiatr. Neurol., Basel 142, 100–121 (1961)
23. Braud, W.G.: An experience of timelessness. Except. Hum. Exp. 13(1), 64–66 (1995)
24. Huber, G.: Akasa, der Mystische Raum, pp. 45–46. Origo, Zürich (1955)
25. Coomaraswamy, A.K.: Time and Eternity. Atribus Asiae Publishers, Ascona (1947)
26. Shallis, M.: On Time. An Investigation into Scientific Knowledge and Human Experience. Burnett Books, New York (1982)

Chapter 17
Where to Put It? The Direction of Time Between Axioms and Supplementary Conditions

Michael Stöltzner

Abstract This paper discusses the problem of implementing the unidirectionality of time into physical theory. I understand a physical theory as an axiom system instantiating some basic laws or equations that contains various models or solutions, among which the physical models form a subset. The theory expressed on these three levels must be supplemented with a set of application rules. There are, I argue, four possible ways to implement time into a theory thus conceived. They are distinguished by the systematic status of the specifically temporal concepts, that is, whether they are part of the laws, the models, or the application rules. (1) One may consider the direction of time so fundamental as to require its being expressed in the *basic axioms or basic laws* of nature. Given the fact that our present basic theories are time-reversal invariant, we have to search for new or modified basic laws. (2) According to *reductionist explanations*, the manifest arrow of time arises from a more fundamental theory, in which time does not play a role at all. (3) The direction of time is expressed in *lower level laws or supplementary conditions* that single out those models that correspond to the macroscopically observable direction of time. (4) The unidirectionality of time expresses some peculiar *non-lawlike fact about initial conditions*, perhaps of our whole Universe or the space-time region we inhabit. I will illustrate these four classes at two historical confrontations that concern Boltzmann's legacy statistical mechanics and causality-violating solutions of the general theory of relativity.

Keywords Direction of time · Axiomatic method · Basic laws and lower level laws · Physical models of an axiom system · Physical meaning criteria · Syntactic and semantic incompleteness · Time in general relativity · Time in statistical mechanics · Indeterminism · Reversibility · Second law of thermodynamics · Time travel · Gödel's rotating universe · Causality in general relativity · Boltzmann's legacy · Empiricist causality · Reductionism and emergence · Ludwig Boltzmann · Franz Serafin Exner · David Hilbert · Kurt Gödel · Max Planck · John Earman

M. Stöltzner (✉)
Department of Philosophy, University of South Carolina, Columbia, SC 29208, USA
e-mail: stoeltzn@sc.edu

S. Albeverio, P. Blanchard (eds.), *Direction of Time*,
DOI 10.1007/978-3-319-02798-2_17,
© Springer International Publishing Switzerland 2014

This paper discusses the problem of implementing the unidirectionality of time into two fundamental physical theories that are basically time-reversal invariant: Newtonian mechanics and general relativity. Rather than providing a detailed systematic treatment here, I shall use two historical encounters in order to illustrate which types of solution appear possible. For the scope of the present paper, I understand a physical theory as an axiom system instantiating the basic laws and equations that contains various models or solutions, among which the physical models form a subset. The theory expressed on these three levels must be supplemented with a set of rules how to apply the physical models to experiment or observation.

I am well aware that such an approach fails to explain the functioning of a substantial part of physical science, where models act in a much more autonomous fashion and have representative features of their own that are not provided by the theoretical framework. But rather than aspiring at generality, my point here is to show that even in cases that are close to the received view of scientific theory, implementing the direction of time poses intricate problems. There are, I argue, four possible ways to do so. They are distinguished by the systematic status of the specifically temporal concepts, that is, whether they are part of the laws (1, 2), of the models (3), or of the application rules (4).

(1) One may consider the direction of time so fundamental as to require its being expressed in the *basic axioms or basic laws* of nature. Given the fact that our present basic theories are time-reversal invariant, we have to search for new or modified basic laws that incorporate the direction of time.

(2) According to *reductionist explanations*, the manifest arrow of time arises from a more fundamental theory, in which time does not play a role at all.

(3) The direction of time is expressed in *lower level laws or supplementary conditions* that single out those models that correspond to the macroscopically observable direction of time.

(4) The unidirectionality of time expresses some peculiar *non-lawlike fact about initial conditions*, perhaps of our whole Universe or of the space-time region we inhabit.

Within this framework an answer to the ontological question whether time is 'real' will also depend upon one's general philosophical stance. Who advocates, on metaphysical grounds, an A-theory of time, that is, an objective notion of the present and the flow of time, will hardly consider (2) a viable option because in it time does not play an irreducible role. To advocates of the B-theory, who conceive time as a parameter on a par with space, all options remain possible.[1] An empiricist can easily settle for (4), while realists will find such explanation wanting. This difference in ontological attitude provides the background for the first encounter, while the second is additionally based on different intuitions about the axiomatic treatment of scientific theory. The proposed classification does not contain conceptions according to which time *tout court* is only a subjective phenomenon—even though some

---

[1]This classification going back to McTaggart [20], or elaborations thereof, still represents the shibboleth in the current philosophy of time. See, for instance, the papers in [26].

of those advocating (1) have resorted to it at least provisionally—and conceptions that ground the temporal order already in the causal order of the world [24].

I will illustrate the four classes at two historical examples. Other examples are easily found within present-day physical science. A typical argument of class (4) is to claim that causality violations in cosmology arise from atypical initial data, that is, they have measure zero in the space of solutions of Einstein's equations. But in contrast to the debates about typicality in statistical mechanics, my first example, it remains unclear how such a measure could be defined. Strategy (3) is applied when one posits a principle of self-consistency according to which the only solutions to laws of physics that can occur locally in the real universe are those, which are globally self-consistent. According to John Earman [4], this represents some sort of pre-established harmony to guarantee the universal validity of physical laws rather than new physics. New physics in the sense of class (2) is postulated by Euclidean quantum field theories, at least if one would succeed in deriving the direction of time from them. And someone following strategy (1) may hold that time-symmetry violating processes, such as the decay of neutral K-mesons, are related to the basic properties of space-time. Needless to say, a definitive unification of elementary particle physics and relativity theory is not in sight.

1 How to Understand Boltzmann's Legacy

After his unsuccessful association of Clausius' second law of thermodynamics with the principle of least action, Ludwig Boltzmann developed the statistical interpretation of the second law stepwise, in particular by countering two famous objections.[2] While he ruled out Ernst Zermelo's recurrence paradox by the unphysically large recurrence times, his rejoinder to Josef Loschmidt's reversibility objection availed itself of the typicality of initial conditions. We do not observe that broken glasses spontaneously recombine because the initial states leading to such an event are extremely improbable given the number of particles involved in macroscopic processes. This famous argument, as it stands, belongs to class (4).

When Boltzmann died in 1906, there emerged basically two readings of his legacy in statistical mechanics. His former Vienna colleagues, above all the experimentalist Franz Serafin Exner, adopted the late Boltzmann's empiricist stance and considered irreversibility as a universal phenomenon. I have called this local tradition 'Vienna Indeterminism' [28]. Looking around us, we do not discern regularities in the first place, but the fact that all natural processes are directed. Thus, so Exner declared in his inaugural address as Rector of the University of Vienna, the second law of thermodynamics becomes the basic principle in nature. Boltzmann "was the first to give a definite and clear interpretation of this direction ... showing that the world ceaselessly develops from a less probable into more probable, and hence more stable, states" [8, p. 9f]. Within Exner's empiricist and radical

---

[2]A classical historical study is [18].

probabilist approach, we only encounter "irreversible processes which can come, however, arbitrarily close to reversibility" [9, p. 710]. Not just the direction of time, but the whole lawful order of nature emerges from very many random microevents. In the molecular dynamics of a gas we "observe regularities produced *exclusively* by chance" [8, p. 13]. Accordingly, "where the random single events succeed one another too slowly there can be no talk about a law" (Ibid., p. 14), e.g., in biology or geology.

The universal validity of the second law made it possible to ponder about its cosmological implications. How could the notorious heat death of the Universe be reconciled with the apparent fact that in our region we are witnessing the emergence of large and complex structures including ourselves? Boltzmann's [2] solution—for which he credited his former assistant Ignaz Schütz—was to argue that in a sufficiently old universe there had been enough time for an improbable configuration to emerge in a finite region of it. But the Boltzmann-Schütz argument is not without problems because given our present entropy state it is highly unlikely that earlier states had even lower entropy. As Huw Price has argued, this permits fake historical records because a random emergence of Shakespeare's works seems more probably than that they were written, in an epoch of lower entropy, by William Shakespeare. For this reason, Price holds that the "appropriate attitude is a kind of healthy skepticism about the universality of the second law of thermodynamics" [23, p. 269].

But this universality was precisely the main point of the Vienna Indeterminists, even if one had to pay the price that the macroscopic laws emerged from random events on the microscopic level, that the laws of nature possibly underwent changes on the cosmological scale, and that the laws of energy conservation and gravitation were presumably of statistical validity only. In his lectures on the philosophy of nature, Boltzmann additionally contemplated an atomistic nature of time that lapses like the pictures in a cinematograph. (Cf. [10, p. 105].) In a letter to Brentano,[3] Boltzmann estimated the number of atoms in a second as $10^{10^{10^{10}}}$ — a number which grossly exceeds the $6 \cdot 10^{23}$ material atoms in a gram molecule (Loschmidt–Avogadro number). "The number of points of time can be made so great that the probability becomes great that a very improbable condition can occur in the whole world." [10, p. 282f]. Thus, in the (presumably finite) Universe there could be regions in which the entropy decreases and time flows backward. And thus the "force law must differ in time depending on whether one proceeds in time in one or another direction." (Ibid., p. 283). Boltzmann has made his argument not sufficiently precise to judge whether it falls victim of Price's criticism as well. For, the temporal atoms need not be uniformly distributed throughout the universe—now understood as a $3 + 1$ dimensional entity in the sense of B-theorists of time—and might cluster around certain events as do slow motion passages in a movie.

Must we, at bottom, simply accept that we are living in a universe where entropy increases and which provides an arrow of time corresponding to ours? What here once again comes as an explanation of class (4), may also be seen as an early

---

[3] See [17, II, p. 384].

instance of anthropic reasoning in physical cosmology, i.e., an explanation of the highly special nature of our present constitution of the Universe that appeals to our existence—or the emergence of carbon-based life—being possible within in. (Cf. [1].)

It is important not to understand Vienna Indeterminism as a metaphysical stance. On Exner's account, a deterministic micro-theory remains a possible justification of the statistical macro-laws we observe, but it is not the only possible derivation of them. Taking a thoroughly indeterminist tack can count among its virtues ontological parsimony and greater methodological coherence. When quantum mechanics entered the scene in 1926, the Vienna Indeterminists, in particular Exner's former assistant Erwin Schrödinger, felt themselves confirmed not to have insisted on a categorical presupposition of causality. Let me now move to the opposite side in the debate about Boltzmann's legacy in the decade following his death.

In his seminal 1900 theory of black-body radiation, Max Planck availed himself of Boltzmann's atomism without subscribing to the indeterminism prevailing among the Vienna physicists. Such a view contradicted his Kantian outlook in which deterministic ("dynamical") laws enjoyed a preferred categorical status. In his 1908 Leiden speech, which opened the polemics with Mach, he held that Boltzmann's lifework was "the emancipation of the concept of entropy from the human art of experimentation", [21, p. 14] that is, from the impossibility of a perpetuum mobile of the second kind. The price was to render the second law a merely probabilistic regularity that admitted 'exceptions', among them spontaneously recombining glasses.

> Boltzmann has drawn therefrom the conclusion that such strange events contradicting the [classical] second law of thermodynamics could well occur in nature. ... To my mind, this is, however, a matter in which one does not have to comply with him. For a nature in which such events happen ... would no longer be our nature. ... Boltzmann himself has formulated that condition for gas theory [which excludes these events], it is generally speaking the "hypothesis of elementary disorder". ... By introducing this condition the necessity of all natural events is restored; for if this condition is realized the increase of entropy directly follows in virtue of the calculus of probability, so that one can nearly call *the principle of elementary disorder* the essence of the second law of thermodynamics. (Ibid., p. 15)

What Planck requires is thus the introduction of a lawlike supplementary condition in the sense of (3) which rules out those strange events that appear to us as if our familiar nature was running backward in time. On the cosmological level, to be sure, such a supplementary condition would have to be implemented globally. By securing causal order, it would also preserve the universality and unidirectionality of time.

Planck rejected Exner's claim to the universality of the second law. Already in the Leiden speech he emphasized the importance of a rigid distinction between reversible and irreversible processes, even though he was well aware that the former were only idealizations. Being "much deeper than, for instance, the opposition between mechanical and electrical processes, this distinction accordingly ... will become the most distinguished explanatory foundation for classifying all physical processes and finally play the lead in the physical world view of the future."

(Ibid., p. 11.) While the principle of least action was enthroned over the whole domain of reversible physics, "the principle of entropy increase introduces an entirely novel element into the physical world view that is in itself extraneous to the action principle." (Ibid., p. 11)

This dualism carried over into the one between statistical and dynamical, strictly causal, regularities. Planck emphasized that, whereas for practical investigations statistical methods are unavoidable, the theorist must insist on the distinction between necessity and probability. And, in a rectorial address of 1914, he attacked Exner's indeterminism.

> This dualism ... to some may appear unsatisfactory, and one has already attempted to remove it—as it does not work out otherwise—by denying absolute certainty and impossibility at all and admitting only higher or lower degrees of probability. Accordingly, there would no longer be any dynamical laws in nature, but only statistical ones; the concept of absolute necessity would be abrogated in physics at all. But such a view should very soon turn out to be a fatal and shortsighted mistake. [22, p. 63]

The disagreement between Exner and Planck included their conceptions of probability. While Exner advocated the relative frequency interpretation, Planck—as a Kantian—remained committed to the idea that any physical probability required a deterministic foundation. To conclude, although Planck treated the specific nature of time in thermodynamics in the sense of class (3), the basic distinction between reversible and irreversible physics pointed to a solution, to be found by the physics of the future, in terms of (1).

2 Hilbert Versus Gödel on General Relativity and Cosmology

With his Sixth Problem of 1900, David Hilbert [13] had become the main advocate of the axiomatization of physics. Also Boltzmann's statistical mechanics was among his main targets, and Hilbert criticized the mathematical shortcomings of Boltzmann's way of performing the thermodynamic limit. More important for the present paper is, however, Hilbert's 1915 axiomatization of general relativity. It contained two levels of axioms and two supplementary conditions.[4]

The first two axioms were of primarily geometrical nature. (I) Mie's axiom of the world function H demands that the variation of $\int H \sqrt{g}\, d\omega$ vanishes for each gravitational potential $g_{\mu\nu}$ and each electromagnetic potential q_s, where g is the determinant of $g_{\mu\nu}$ and $d\omega = d\omega_1\, d\omega_2\, d\omega_3\, d\omega_4$ is the differential of the world parameters ω_k uniquely fixing the world points. H contains gravitational arguments, the $g_{\mu\nu}$ and their first and second partial derivatives with respect to the ω_k, and electromagnetic arguments, the q_s and their first partial derivatives with respect to the ω_k. Axiom (II) states that H be invariant with respect to an arbitrary transformation of the world parameters ω_k. Hilbert considers this axiom as "the simplest mathemati-

---

[4]There exists a rich literature on the character of Hilbert's independent derivation of Einstein's equations and its relationship to Einstein's work, among them [3] and [25]. I am quoting here from the 1924 reprint of Hilbert's 1915/16 papers, where he tacitly skipped some unwarranted claims of the original publication. All translations are mine.

cal expression of the requirement that the coordinates in themselves do not possess any physical significance, but only represent an enumeration of the world points" [14, p. 50].

Two further axioms were required to fix H uniquely. (III) Demanding the additivity of pure gravity and electromagnetism $H = R + L$, with R being the usual Riemann scalar curvature and L not containing second derivatives of the $g_{\mu\nu}$, guarantees that no higher than second order derivatives of the $g_{\mu\nu}$ appear in the field equations, such that one obtains a reasonable dynamics. Axiom (IV) "further elucidates the connection of the theory with experience" [14, p. 57] by specifying the signature of the metric in order to obtain the required $3 + 1$ pseudo-geometry for space-time.

Hilbert added two supplementary conditions that singled out the physical solutions. (V) protected the causal order by a meaning criterion. "From knowing the state variables at present, all future statements about them follow necessarily and uniquely, *provided they are physically meaningful.*" [14, p. 64]. (VI) Hilbert insisted on the *regularity* of the physical solutions, although non-regular ones could be used in approximations. Condition (VI) was probably motivated by an important feature of variational calculus, in virtue of which the actual solutions are typically smother than the class of candidates originally assumed. Nonetheless Hilbert's definition of singularity was too restrictive to encompass black holes and other phenomena that would stand in the center of research from the 1960s onward. As regards condition (V), Hilbert hoped that

> only a sharper comprehension of the idea basic to the principle of general relativity is needed in order to maintain the principle of causality also within the new physics. That is, in accordance with the essence of the new principle of relativity we have to require not only the invariance of the general laws of physics, but also attribute an invariant character to each single statement in physics, in case it shall be physically meaningful—chiming with that, after all, every physical fact must be established by light clocks, i.e., by instruments of an *invariant* character. [14, p. 63]

In a footnote, Hilbert [14] discussed a simple invariant electromagnetic Lagrangian that fulfilled this causality condition and, accordingly, protected temporal order. Thus Hilbert believed that factual conditions of class (4) could be made sufficiently precise and placed on solid physical foundations to act—within a general axiomatic framework based on concepts as deep as invariance—as model selection criteria of class (3).

Hilbert's view on time in statistical thermodynamics initially followed Boltzmann's rejoinder to Loschmidt. However, in his 1919/20 lectures on "Nature and Mathematical Knowledge" he continued in a fashion that was substantially different from condition (V) above.

> Summing up, we can say that in the physical happenings one cannot furnish a proof that one direction of time is distinguished. There are irreversible processes; but at closer look one recognizes that in the respective cases this irreversibility is just apparent and rests upon the particular conditions of experimentation, hence upon anthropological reasons. [16, p. 85]

Notice that Hilbert here turns the insight that the direction of time is implemented by a condition of type (4) not only into a consideration about an experimenter's intervention, but into the claim of the subjectivity of time. Apparently, the direction of time was less fundamental than causal order and invariance. But one may also view Hilbert's claim as adherence to Kant's philosophy of time, an adherence he shared with the most prominent critic of the logical side of his axiomatic method. This was not Hilbert's only reference to Kant. In his 1930 Königsberg [15] address to the *Naturforscherversammlung*, he considered the finite character of his meta-mathematics, the formal analysis of axiomatic systems, as the only legitimate heir of the Kantian synthetic a priori.

On the same meeting, the young Viennese mathematician Kurt Gödel presented an incompleteness theorem that ultimately dashed Hilbert's hopes to arrive at an ab-solute foundation of classical mathematics. Two decades later, in 1949, Gödel ob-tained a new solution to Einstein's equations that had surprising properties.[5] While all solutions known until then admitted the definition of a global cosmological time, Gödel's globally rotating universe allowed one to travel into one's own past and, as an illustration for the resulting inconsistencies, prevent one's grandfather from marrying one's grandmother. Gödel's original solution contradicted red-shift obser-vations, and time travelers in the Gödel universe would have to burn large part of their galaxy. (Cf. [19].) But Gödel was not content that time travel was just empiri-cally unfeasible.

> The mere compatibility with the laws of nature of worlds in which there is no distinguished absolute time ... throws some light on the meaning of time also in those worlds in which an absolute time *can* be defined. For ... whether or not an objective lapse of time exists ... depends on the particular way in which matter and its motion are arranged in the world. [11, p. 206]

At first glance one would think that Gödel here settles for a criterion of class (4), the distribution of matter. But from a strictly axiomatic point of view, he could not disregard unintended models, the time-travel worlds, without providing a condition that renders them unphysical in terms of a well-defined supplementary condition of class (3). It seems that Gödel could agree with Hilbert's axiomatic approach. But in an unpublished manuscript of those years Gödel rejected such a strategy for genuinely philosophical reasons. "A lapse of time ... would have to be founded, one should think, in the laws of nature." [12, p. 238].

So Gödel was left with the alternatives (1) and (2). As time already figured in the basic axioms of general relativity but failed to yield an objective lapse of time, (1) was not convincing either. Especially since Gödel rejected the option to amend the laws of nature by a separate cosmological time. "If ... such a world time were to be introduced in these [rotating] worlds as a new entity, independent of all observable magnitudes, it would violate the principle of sufficient reason." (Ibid., p. 237.) Given the fundamental and unique nature of time, Gödel's rotating universe shows as se-mantic poverty of the axiomatic concept of time figuring in the theory of relativity. If one views producing a solution to Einstein's equations—although this cannot be achieved by an algorithm—in analogy to a formal proof, there exists an interest-ing parallel between Gödel's rotational universe and his incompleteness theorems.

---

[5]For the important role Gödel's solution played as a motivation for subsequent research in general relativity, see [7] and my [27].

In bothcases, as it were, the mathematical machinery does not succeed within the limits set by the axiom system. While in the case of arithmetic—or a comparably rich axiom system—the incompleteness is syntactic, the rotating universe reveals a semantic incompleteness of general relativity with respect to the concept of time. (Cf. [29].) As a consequence, Gödel was essentially left with alternative (2).

> In the present imperfect state of physics, however, it cannot be maintained with any reasonable degree of certainty that the space-time scheme of relativity theory really describes the objective structure of the material world. Perhaps it is to be considered only as one step beyond the appearances and towards the things. Quantum physics in particular seems to indicate that physical reality is something still more different from the appearances than even the four-dimensional Einstein–Minkowski world. [In a footnote Gödel cites the Einstein-Podolsky-Rosen paper and shows a certain sympathy for non-local action-at-a-distance.] T. Kaluza's fifth dimension points in the same direction. [12, p. 240]

Gödel's expectation that a more fundamental theory was lying ahead corresponded to his ideal of a stepwise approach to the things-in-themselves, be it part of the empirical world or Platonic mathematical truths. (Cf. [27].)

But without such a more fundamental theory actually in place, the philosophical situation was of a different kind. As time does not 'lapse' or 'pass by' globally, there is no objective change either. "The concept of existence, however, cannot be relativized [to observers] without destroying its meaning completely." [11, p. 203, fn. 5]. Thus time, to Gödel's light, can only be a subjective concept as Kant had argued. "Kant says that for beings with other forms of cognition 'those modifications which we represent to ourselves as changes would give rise to a perception in which the idea of time, and therefore also of change, would not occur at all.'" [12, p. 235]. Thus at bottom Gödel resorted to a transcendental argument.

Einstein [6] brushed Gödel's reasoning aside and contemplated a physicality condition in the sense of (3). Earman even calls Gödel's train of thought "a pretty piece of ordinary language philosophizing" [4, p. 195] that "add[s] up to something less than an argument." (Ibid., p. 197). To my mind, Gödel has nevertheless spotted an important point as regards the problem of rigorously implementing the direction of time even though I fully agree with Earman that his philosophical way out, the subjectivity of time, is unsatisfactory. This can be best seen comparing Gödel's view to Hilbert's.

First, Hilbert had deliberately operated with axioms and supplementary conditions at different levels; Gödel's semantic incompleteness suggests that a layered structure is typically unavoidable in a theory as fundamental as general relativity. Second, Hilbert's dashed hopes about securing causality show that the physicality argument has to come in the right form, that is, we cannot just single out solutions by way of their empirical adequacy because we do not overlook the set of all models to a sufficient extent. This is a rather generic feature of partial differential equations. As for general relativity, there exists a long ladder of causality conditions of different strength, only some of which allow the introduction of a global temporal order. (See [5].) Third, supplementary conditions can be considered as (secondary) laws of nature. An axiomatic perspective might cure some philosophical itches of such an account. But not all of them. Earman's conclusion in this regard is quite pessimistic: "I do not see any prospect for proving that time travel is impossible in an interesting sense" [4, p. 194]. The therapy for time-travel malaise he proposes instead, is based upon distinguishing two senses of physical possibility.

possible$_1$ = *local* solution of the laws
possible$_2$ = *global* solution of the laws
possible$_2$ = possible$_1$ + *consistency constraints*

The consistency constraints will typically be global and induce new physics. Earman treats them as new supplementary laws in the sense of strategy (3) and discusses a suitable philosophical notion of law. According to the empiricist Mill–Ramsey–Lewis (MRL) account,

> a law for a logically possible world W is an axiom or theorem of the best overall deductive system for W. . . . A deductive system for W is a deductively closed, axiomatizable, set (of non-modal) sentences each of which is true in W. Deductive systems are ranked by how well they achieve a compromise between strength or information content on the one hand and simplicity on the other. (Ibid., p. 178)

Yet there are problems in comparing worlds with closed timelike curves and worlds without them. In particular, "it could turn out that although (by construction) the MRL laws of this world are all *true of* a time-travel world W, they are not all *laws of W*, except in a very tenuous sense. . . . this possibility is realized in cases where the consistency constraints are so severe as to supplant the laws of this world. In such cases the time travel is arguably such a remote possibility that it loses much of its interest." (Ibid., p. 179). Moreover, in space-times with low symmetry the consistency constraints could be so complicated that "they will not appear as axioms or theorems of any theory that achieves a good compromise between strength and simplicity." (Ibid., p. 182). "In some time travel worlds it is plausible that the MRL laws include the consistency constraints; in these the grandfather paradox has a satisfying resolution. In other cases the status of the consistency constraints remains obscure; in these cases the grandfather paradox leaves a residual itch." (Ibid., p. 194)

Earman's bottom-up approach, in the end, arrives at a layering between basic laws and consistency constraints, some of which—relative to the MRL account—can be considered as supplementary laws in the sense of (3) while some cannot and accordingly remain of class (4). This shows that even though much depends upon philosophical presuppositions, even elaborated contemporary accounts do not uniquely pick one of the classes (1)–(4).

References

1. Bettini, S.: Anthropic reasoning in cosmology. A historical perspective. In: Stöltzner, M., Weingartner, P. (eds.) Formale Teleologie und Kausalität, pp. 35–76. Mentis, Paderborn (2005)
2. Boltzmann, L.: On certain questions on the theory of gases. Nature **51**, 413–415 (1895)
3. Corry, L.: From Mie's electromagnetic theory of matter to Hilbert's unified foundations of physics. Stud. Hist. Philos. Mod. Phys. **30B**, 159–183 (1999)
4. Earman, J.: Bangs, Crunches, Whimpers, and Shrieks. Oxford University Press, New York (1995)

5. Ehrlich, P.A., Emch, G.G.: Gravitational waves and causality. Rev. Math. Phys. **4**, 163–221 (1992)
6. Einstein, A.: Remarks to the essays appearing in the collective volume. In: Schilpp, P.A. (ed.) Albert Einstein: Philosopher-Scientist, pp. 687–688. Open Court, Evanston (1949)
7. Ellis, G.F.R.: Contributions of K. Gödel to relativity and cosmology. In: Hájek, P. (ed.): Gödel'96. Logical Foundations of Mathematics, Computer Science and Physics – Kurt Gödel's Legacy, pp. 34–49. Springer, Berlin (1996)
8. Exner, F.S.: Über Gesetze in Naturwissenschaft und Humanistik. Alfred Hölder, Wien & Leipzig (1909)
9. Exner, F.S.: Vorlesungen über die physikalischen Grundlagen der Naturwissenschaften, 2nd edn. Franz Deuticke, Leipzig-Wien (1922)
10. Fasol-Boltzmann, I.M. (ed.): Ludwig Boltzmann. Principien der Naturfilosofi. Lectures on Natural Philosophy. Springer, Berlin (1990)
11. Gödel, K.: (1949) A remark about the relationship between relativity theory and idealistic philosophy. In: Feferman, S., et al. (eds.) Collected Works, vol. 2, pp. 202–207. Oxford University Press, Oxford (1990)
12. Gödel, K.: Some observations about the theory of relativity and Kantian philosophy (manuscript B2). In: Feferman, S., et al. (eds.) Collected Works, vol. 3, pp. 230–246. Oxford University Press, Oxford (1995)
13. Hilbert, D.: Mathematische Probleme. Nachrichten von der Königlichen Gesellschaft der Wissenschaften zu Göttingen. Mathematisch-Physikalische Klasse, pp. 253–297 (1900). English translation reprinted in Bull. Math. Soc. **37**, 407–436 (2000)
14. Hilbert, D.: Die Grundlagen der Physik (1924). Second version, reprinted in Hilbertiana – Fünf Aufsätze von David Hilbert, WBG, Darmstadt, pp. 47–78 (1964)
15. Hilbert, D.: Naturerkennen und Logik. Naturwissenschaften **18**, 959–963 (1930)
16. Hilbert, D.: Natur und Mathematisches Erkennen. Vorlesungen gehalten in Göttingen 1919–1920, ed. by David E. Rowe, Birkhäuser, Basel (1992)
17. Höflechner, W. (ed.): Ludwig Boltzmann. Leben und Briefe. Akademische Druck- und Verlagsanstalt, Graz (1994)
18. Klein, M.J.: The development of Boltzmann's statistical ideas. In: Cohen, E.G.D., Thirring, W. (eds.) The Boltzmann Equation. Theory and Applications. Acta Physica Austriaca, Suppl. X, pp. 53–106 (1973)
19. Malament, D.B.: Minimal acceleration requirements for 'Time travel' in Gödel spacetime. J. Math. Phys. **26**, 774–777 (1985)
20. McTaggart, J.E.: The unreality of time. Mind **18**, 457–474 (1908)
21. Planck, M.: Die Einheit des physikalischen Weltbildes. In: Wege zur Physikalischen Erkenntnis. Reden und Vorträge, pp. 1–24. S. Hirzel, Leipzig (1944)
22. Planck, M.: Dynamische und statistische Gesetzmäßigkeit. In: Wege zur Physikalischen Erkenntnis. Reden und Vorträge, pp. 54–67. S. Hirzel, Leipzig (1944)
23. Price, H.: Time's arrow, time's fly-bottle. In: Stadler, F., Stöltzner, M. (eds.): Time and History. Proceedings of the 28th International Ludwig Wittgenstein Symposium, pp. 253–273. Ontos, Frankfurt am Main (2006)
24. Reichenbach, H.: The Direction of Time. University of California Press, Berkeley (1956)
25. Sauer, T.: The relativity of discovery: Hilbert's first note on the foundations of physics. Arch. Hist. Exact Sci. **53**, 529–575 (1999)
26. Stadler, F., Stöltzner, M. (eds.): Time and History. Proceedings of the 28th International Ludwig Wittgenstein Symposium. Ontos, Frankfurt am Main (2006)
27. Stöltzner, M.: Gödel and the theory of everything. In: Hájek, P. (ed.): Gödel'96. Logical Foundations of Mathematics, Computer Science and Physics – Kurt Gödel's Legacy, pp. 291–306. Springer, Berlin (1996)
28. Stöltzner, M.: Vienna Indeterminism: Mach, Boltzmann, Exner. Synthese **119**, 85–111 (1999)
29. Yourgrau, P.: The Disappearance of Time. Kurt Gödel and the Idealistic Tradition in Philosophy. Cambridge University Press, Cambridge (1991)

Chapter 18
Models of Time

Luigi Accardi

Abstract The first part of the chapter describes, in a qualitative way, a scheme of axiomatic approach to the notion of time. It is shown that, even restricting the physical requirements to a minimum, a multiplicity of inequivalent models are possible. In particular the idea of *topological relativity* suggests the impossibility to distinguish, on experimental bases, between linear and circular time. The second part of the chapter is framed within the usual quantum mechanical context and is focused on the notion of *statistical reversibility* and its possible extensions to non-equilibrium situations.

Keywords Stochastic limit · Statistical reversibility · Topological relativity · Quantum Markov processes · Detailed balance

1 How Old is an Electron?

The notion of time has been baffling mankind since many centuries. We attribute an age to many things: living beings, archaeological findings, rocks, stars, and even the universe. But we are less confident when we try to attribute an age to a single atom, or to an electron or to a photon. We speak of lifetimes of elementary particles, but we cannot distinguish an "old" electron from a "young" one. So, to answer questions like: "does time pass for an electron?", we have to clarify our ideas on time.

Philosophical knowledge is based on descriptions. This has some advantages but it is not easy to decide if the meaning attributed by different people to a word is the same or not. For example suppose one asks you to interpret the following definition of time:

> ...The parts of time have their being from the coupling or continuation through the indivisible present instant, given that it be always other and other, from its parts other and other succeed each other and always exist...

L. Accardi (✉)
Centro Vito Volterra, Università di Roma "Tor Vergata", Via Columbia, 2, 00133 Rome, Italy
e-mail: accardi@volterra.mat.uniroma2.it
url: http://volterra.mat.uniroma2.it

S. Albeverio, P. Blanchard (eds.), *Direction of Time*,
DOI 10.1007/978-3-319-02798-2_18,

and to frame it into an historical context by attributing it to either:

(i) Heidegger or
(ii) Lacan or
(iii) Henry of Gent.

I wonder how many would guess that the correct attribution is to Henry of Gent (1279) (the original statement being: "...Partes temporis habere esse ex copulatione seu continuatione ad instans indivisible praesens, licet illud semper sit aliud et aliud ex partes eius succedunt et semper sunt aliae et aliae...").

Saint Augustine (in the *Confessioni*) expresses more clearly the same idea:

> ...Past no longer exists, future not yet exists. But if present would remain always present and never fade away in the past, there would not be time, but eternity...

In classical physics of Galilei and Newton space and time are containers with an ontological autonomy:

> ...Time, absolute, true, mathematical, in itself and by its own nature without any relation with anything external, flows uniformly... (*Principia mathematica*).

According to Kant space and time are not objects but forms of human knowledge: we do not "know" space and time, but our knowledge is organized in a space-time manner:

> ...The idea of time does not originate from senses, but is presupposed by them... Time is not something objective and real. It is neither substance, nor accident, nor relation, but a necessary subjective condition, due to the nature of human mind, to coordinate within itself all perceptible things according to a fixed law... (from: *De mundis sensibilis atque intelligibilis forma et principiis*)

Lucrezio had a similar point of view: *tempus item per se non est* (time, in itself, is not an object).

This is an appealing point of view but leaves the following old question open: how to speak of space and time before the emergence of human consciousness in the universe?

The attempt to reconcile the time of physics with the psychological time is usually attributed to Bergson (time as interior duration), but his known statement: *time is an invention or nothing* is (surprisingly, due to the fact the two thinkers considered their ideas on time in mutual disagreement) close to the well known and widely quoted passage from Einstein's condolence letter to Michele Besso's sister (March 1955): "...For us practicing physicists, the distinction between past, present and future is only an illusion, even if a tenacious one..."

The artistic arbitrariness in Borges' statement:

> ...time is a river that sweeps me away, but I am the river; it is a tiger that tears me to pieces, but I am the tiger; it is a fire that devours me, but I am the fire.

should be compared with the puny attempt to disguise a technical expedient as a deep concept, in Hawking's statement:

> ...The concept of imaginary time is the fundamental concept on the basis of which the mathematical model has to be formulated; ordinary time would be in this case a derived model that we invent—as a part of a mathematical model—with the goal of describing subjective impressions about the universe (*Halley Lectures*, 1989)

The list of metaphors used to describe time could continue indefinitely (see [13] for a survey of the multiple definitions of time elaborated by mankind in the course of history) and in some sense they all confirm the famous saying of St. Augustine according to which time belongs to that class of concepts that everybody believes are clear but nobody can explicitly define. This impossibility has been even theorized Paul Ricoeur in the three volume work "Temps et récit" [27] whose main thesis is that the nature of time cannot be object of rational thinking, but only of a "poetic resolution" through the production of histories, tales, and novels, through which we acquire an indirect comprehension of the notion of "time" and of our existence in it.

A similar purely esthetic attitude, even if not so explicit, but with a slightly ency-clopedic character and with original combinations of text and images can be found in [25]. The philosophical and theological aspects of the debate on time are de-veloped in parallel with the scientific aspects in the book by Castagnino and San-guineti [17]. The sociological aspects of time are discussed in [14] and there is a huge literature on the historical evolution of the different ways to measure time (see for example [13, 18, 20, 25]).

In this paper we will look at time from the point of view of mathematics. Scien-tific knowledge is based on different kinds of metaphors, called definitions, models, and procedures (protocols). Scientists do not illude themselves to overcome com-pletely the intrinsic ambiguity of language, but they try to limit it with the help of models. Thus models act as intermediaries between our mind and reality, whatever the latter is.

A model is a simple example of an axiomatic theory: usually the term "model" is referred to a very specific situation and the term "axiomatic theory" to a wider en-terprise such as the unification of different contexts, however, the logical structures of the two are the same: to define a context using axioms and to draw consequences from it according to the rules of logic.

The maturity of a science is measured by the degree in which it succeeds in condensing its knowledge in mathematical models, or equivalently axioms, and in deducing its procedures from them. The multiplicity of possible models is an healthy antidote to the illusions of certitude.

Scientific activity oscillates among the three poles of:

 (i) inventing new models;
 (ii) deducing observable consequences from them;
(iii) verifying experimentally these consequences.

In the present paper we will play this game with the notion of time. Emphasis will be on incompleteness (in the sense explained below): even on an extremely basic and fundamental level our ideas on time are not sufficiently precise to fix a unique class of mathematical models. In the first part of this paper this idea will be illustrated with simple geometrical models; in the second part we will concentrate on time reflections and discuss the various attempts which have been made in the physical and mathematical literature to substantiate this notion.

The goal of an axiomatic theory of time should be that of challenging clever physicists to discover empirically observable differences between different mathematical models of time. The examples discussed below show that this challenge might not be a trivial one. For example, according to the principle of "topological relativity", discussed in Sect. 4 below it is not possible to distinguish experimentally between the western tradition of linear time and the Indian tradition of circular time. In particular, as shown in Sect. 4, circular time does not necessarily imply eternal recurrence of history.

In Greek philosophy change and movement was related to imperfection. Perfection was related to immobility. Thus perfect objects, like fixed stars, just as planetary motions inspired the idea of cyclicity of time, biological and historical experiences inspired the idea of direction (arrow) of time (a notion to which Prigogine and its has dedicated several books, see e.g. [26]). Contemporary science has shifted its idea of perfection from immobility (like fixed stars) to elementarity. When we try to describe our idea of the flow of time, i.e. of irreversibility, independently of mathematical formulas, unavoidably we end up in describing a situation in which a multiplicity of interacting systems (i.e. a complex system) are separated into non-interacting systems (thus decreasing the complexity). To reconstruct the complex system from the simple ones may not be logically impossible but is surely much more complicated than the converse operation. In fact in order to break many different interactions, it is simply required to create a single interaction which dominates them all, but to reconstruct them one has to act individually on each broken tie. For example, to put and keep a gas in a box is a relatively simple operation if one can act collectively on the gas as a whole, but if, after opening the box and allowing expansion, the gas is mixed with another volume of gas made of the same molecules, to reproduce exactly the original configuration is a physically impossible task because it requires the possibility to distinguish those molecules that belonged to the former volume of gas from those who did not.

Is it a logically possible task? A thumb rule evaluation of the order of magnitude of information bits one should elaborate to achieve this goal suggests a negative answer.

In this sense the difference between life and death is the same as the difference between a set of individuals and an organization: dis-integration is different from annihilation.

The above example shows that the notions of existence, identity, system, motion, space, time, signal..., are strictly related: if some of them are assumed as primitive, the other ones can be introduced as derived.

The question whether some of them have to be considered as "objectively primitive" with respect to the other ones is interesting and has a long history. For example, Aristotle takes motion as primitive notion and defines time as the measure of motion. In a world without consciousness, hence without knowledge, motion is still conceivable. From this point of view Aristotle is closer to the scientific mentality than Kant in the sense that his approach allows a more objective, human independent, definition of time. For Newton, time and space (more generally: state) are primitive notions and "motion" is defined as change of state in time.

In special relativity one assumes as primitive much more complex notions such as:

(i) the notion of signal;
(ii) the notion of event;
(iii) the finiteness of the speed of light

and, having paid this price, one can achieve a complete "space-time democracy". These notions are used to distinguish between "space differences" and "time differences" among events. Two events which cannot be connected by a signal are defined to be simultaneous: if they are different, their difference is of space type. Two events that can be connected by a signal are always different and their difference is of time type.

Thus in relativity we can distinguish between space and time diversity, hence motion can be introduced as a measure of time diversity. However, since time reflection is admitted as a *physical transformation*, in relativity the distinction between past and future is relative to the reference frame.

Any model of time depends on the notion of "system" but, from an holistic point of view the notion of system itself is a quite anthropomorphic notion: if two systems interact, what does it mean to distinguish between them? The boundary is necessarily arbitrary and the separation depends on the choice of some scales of magnitude ("large" distances, "weak" interactions, ...) which might be quite natural for human beings but from a non-anthropocentric point of view are not privileged.

If we accept the existence of elementary particles, then we can give an (approximate) definition of decay as decomposition into (approximately) non-interacting elementary constituents. But from a more sophisticated, field theoretical point of view, which includes self-interactions, the "elementary" particles are simply manifestations of the field and they too can decay. In fact, from this point of view, one should not speak of "decays" but simply of "transformations" from one manifestation of the field to another one.

Transformations among elementary particles may be reversible, but what does this precisely mean? This question is nontrivial because of the necessity to distinguish between the reversibility of the time evolution and the "time reversal symmetry", which in many theories such as nonrelativistic classical and quantum mechanics can be defined independently of any specific time evolution (cf. Sect. 12).

2 Time as a One-Dimensional Connected Continuum

Any model translates one's intuition of a physical object or phenomenon. Abstraction of properties leads to the construction of a mathematical model. Then one tries to go back to the original intuition, i.e. to check to what extent this is reflected by the model.

The obstruction to this procedure is that there may be, and in general there are, many inequivalent models. In the case of time this leads to the following alternative:

(i) our intuition is incomplete: physical time has more properties than those specified by the model;
(ii) all models are present in nature.

Here there is an historical asymmetry between space and time: in fact it is now commonly accepted that there exist many simultaneous models of space, while many models of time can exist but not "simultaneously". The postulate of homogeneity is an extrapolation from local to global, but since all our perceptions are local both in space and in time, the way to match together a multiplicity of local perceptions into a single global picture necessarily introduces some elements which go beyond experimental evidence.

In the following of this section we shall discuss some consequences of the axiom which underlies most of contemporary scientific models of time:

(A1) Time is a 1-dimensional connected continuum.

This axiom excludes the existence of time quanta (chronons, ... [15]). Discrete models of time have been investigated in the mathematical, physical, psychological, ... literature. We emphasize that all the mathematical models described in the present section continue to be meaningful also in the case of discrete time. However, our point in this paper is that, even keeping a conservative view of time as a 1-dimensional connected continuum, still there is a lot of space for non-trivial inequivalent possibilities.

The continuum, like the infinite, is not accessible to human experiments in actual form, but only in its potential form. If one believes in the time-energy indeterminacy principle, even this potentiality is questionable from a physical point of view because the measure of extremely small periods of time would imply extremely large fluctuations in energy.

According to Hilbert the goal of an axiomatization of a physical theory is: ... *to formulate the physical requirements so that the mathematical model is uniquely determined*... (at least up to isomorphism). This requirement is usually called *completeness* in logic.

It is clear that axiom (A1) is far from complete, in fact there exist many 1-dimensional connected continua! For example the real line $\mathbb{R}$ (linear time) and the circle (circular time). The Peano curve or any other fractal curve provides additional models which satisfy axiom (A1) but which suggest different intuitive images of time (fat time, fractal time, ... (see Fig. 1)).

The introduction of a unit of time is equivalent to the introduction of an action, on this continuum, of the positive rational numbers (multiplication).

The archimedean nature of this action is implicitly postulated when one identifies time with the real line, but this is clearly an additional axiom and non-archimedean models of time are certainly possible from a mathematical point of view and not

Fig. 1 Time: linear, circular,
fat (fractal), ...

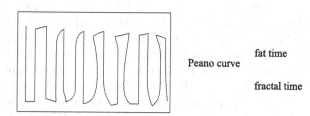

——————————————— linear time

circular time

Peano curve

fat time

fractal time

necessarily implausible from the physical point of view, as shown by the example in the following section.

3 A Non-archimedean Model of Time

The following construction was inspired by a discussion with Professor E. Brieskorn in which he explained to me a model of the extended real line due to Hausdorff ($\sim$1903). The model that follows is less sophisticated than Hausdorff's but it helps getting an intuition of how incomplete axiom (A1) is. The construction goes as follows: for each $n \in \mathbb{Z}$ fix an homeomorphism u_n between the real line $\mathbb{R}$ and the open interval $(n, n+1)$. Now fix an arbitrary, strictly increasing, sequence (α_n) of ordinals and associate to each $n \in \mathbb{Z}$ the corresponding α_n. This construction gives a one-to-one correspondence (see Fig. 2)

$$ u : \widehat{\mathbb{R}} \equiv \bigcup_n \{(n, n+1)\} \cup \left(\bigcup_n \{\alpha_n\} \right) \equiv \bigcup_n (\mathbb{R})_n \cup \left(\bigcup_n \{\alpha_n\} \right) \to \mathbb{R} $$

and we can define a topology on $\widehat{\mathbb{R}}$ so that the map u is an homeomorphism. Moreover we can use this map to transport on $\widehat{\mathbb{R}}$ the usual order structure on $\mathbb{R}$. The usual

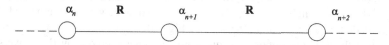

Fig. 2 Non-archimedean time

multiplication on $\mathbb{R}$ can be extended to $\widehat{\mathbb{R}}$ by the prescriptions:

$$\begin{aligned}
\lambda \cdot x &= (\lambda x) && \text{if } x \in (\mathbb{R})_n \quad \text{for some } n \\
\lambda \alpha_n &= \alpha_n && \text{if } \lambda > 0 \\
\lambda \alpha_n &= \alpha_{n+1} && \text{if } \lambda < 0 \\
\lambda \alpha_n &= 0 && \text{if } \lambda = 0
\end{aligned}$$

This extension, which does not preserve all the elementary properties of multiplication among numbers, is such that the multiplication by positive numbers is continuous but the time reflection $t \to -t$ is not. Moreover $\widehat{\mathbb{R}}$ is connected but the action of $\mathbb{R}$ on $\widehat{\mathbb{R}}$ is non-archimedean.

Recently, in the attempt to model an intrinsic asymmetry in the time evolution of physical systems, S. Wickramasekara [28] introduced the topology on $\mathbb{R}$, generated by the basic open sets $[a, b)$. In this topology multiplication by positive real numbers is continuous, but time reflection $(t \to -t)$ is discontinuous (e.g. $a + 1/n$ converges to a in this topology but $-a - 1/n$ cannot be in any interval of the form $[-a, -a + \varepsilon)$ with $\varepsilon > 0$ and therefore it does not converge to $-a$ in this topology). In this topology past and future are not symmetric because the basic open intervals include the left point but not the right one and the discontinuity of time reflection reflects this asymmetry.

In the above discussed non-archimedean model, past and future are symmetric and the discontinuity of time reflection comes from the fact that the "special points" α_n represent "singularities of time" in the sense that each of them plays the role of "infinite past" for one interval and of "infinite future" for another one. This interpretation reflects the idea of the simultaneous existence of infinitely many time flows (parallel universes). A more traditional interpretation may regard each interval as an "era" and the time singularity α_n represents the "big crunch" for the nth era and the "big bang" for the $n + 1$th. The non-archimedean character of the model reflects the incommunicability between different epochs.

4 Time in Classical Physics

For classical Hamiltonian mechanics time is an external parameter, i.e. strictly speaking it is not an observable of the theory, while observables are sections in the cotangent bundle. For Einstein they are sections in the tangent bundle so that the basic observables are restricted to the kinematical observables of classical physics: positions (space), time, their conjugate observables, velocities and energy, and functions of them. In the concrete models one has to further specify which class of functions are allowed (measurable, smooth, analytic, compact support, rapidly decaying, . . .): different choices lead to different theories. One can conceive more general bundles whose fiber includes the tangent space and some new, non-kinematical degree of freedom (such as spin, color, . . . (see Fig. 3)).

Classical theories (as opposed to quantum) are characterized by the universal compatibility of all the observables. A maximal family of independent observables

Fig. 3 Relativity: space-time
democracy

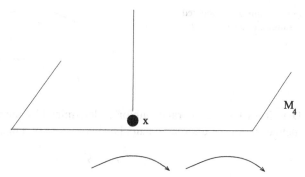

Fig. 4 Time as an oriented
manifold: $(r, s)(s, t) = (r, t)$

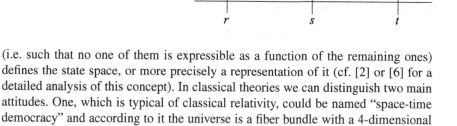

(i.e. such that no one of them is expressible as a function of the remaining ones)
defines the state space, or more precisely a representation of it (cf. [2] or [6] for a
detailed analysis of this concept). In classical theories we can distinguish two main
attitudes. One, which is typical of classical relativity, could be named "space-time
democracy" and according to it the universe is a fiber bundle with a 4-dimensional
manifold M_4 as a basis. Another one, which is at the basis of non-relativistic physics,
can be named "time supremacy" and according to it:

- the universe is a fiber bundle with a 1-dimensional oriented manifold (see Figs. 4
  and 5) as a basis;
- the fiber is the state space S;
- the time evolution $T(s, t)$ is parallel transport from one fiber to another.

The mathematical model of this scenario should be a fiber bundle with basis
$\mathbb{R}$ and fiber a space S. The curves on $\mathbb{R}$ are the ordered pairs (s, t). They form a
groupoid for the multiplication

$$(s, t)(r, s) = (r, t)$$

The dynamical evolution is a parallel transport, i.e. a representation of this groupoid

$$(s, t) \to T(s, t) : S_s \to S_t$$

The existence of two time orientations gives rise to two multiplications corresponding to two orientations

$$(s, t) : s \to t; \qquad (s, t) : t \to s$$
$$(r, s)(s, t) = (r, t); \qquad (t, s)(s, r) = (t, r)$$

To use motion as a measure of time only means that the parallel transport is given
a priori. To use time as order of motion means that we have a priori decided the
direction of time.

Fig. 5 Time as an oriented
manifold: $(t, s)(s, r) = (t, r)$

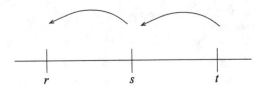

Definition 1 A temporal evolution is determined by the assignment, for each element of family of unordered pairs $\{s, t\} \subseteq T \times T$, of a map (see Fig. 10)

$$T_{s,t} : S_s \to S_t$$

Moreover it is required that, if the pairs $\{s, t\}$ and $\{t, u\}$ are in the family, then the map $\{s, u\}$ is in the family and satisfies

$$T_{t,u} T_{s,t} = T_{s,u} : S_s \to S_u \tag{1}$$

The idea that time orders motion is expressed by the requirement that:

(i) the base manifold is ordered and parallel transport (evolution) respects the order

$$r < s < t \quad \Rightarrow \quad T(s, t)T(r, s) = T(r, t)$$

Suppose we want to take seriously Aristotle: "... This is time in reality: the number of motion according to the before and the after..." i.e. it is motion that gives the direction of time, then we might argue as follows. A "curve" on the time manifold is given by an unordered pair $\{s, t\}$. However, the associated parallel transport must go from one fiber to another. In other terms: the motion must have a direction. Suppose that

$$T_{\{s,t\}} : S_s \to S_t \tag{2}$$

then we say by definition that

$$s < t$$

In other words, if the dynamics (motion) $T(s, t)$ is physically realized, then we say that $s < t$ (s is in the past of t). Thus the orientation of the curve $\{s, t\}$ is given by the fact that the parallel transport $T_{\{s,t\}}$ maps the fiber at s to the fiber at t and not conversely (see Fig. 6). In conclusion: the Aristotelian view that motion orders time is expressed by the requirement that we use parallel transport to induce an order on the base manifold

$$T(s, t)T(r, s) = T(r, t) \quad \Rightarrow \quad r < s < t$$

This leads to the following.

Definition 2 The Aristotelian future of $s \in T$ is the set of all $t \in T$ such that the parallel transport (2) is well defined. Similarly we say that $s \in T$ is in the past of $t \in T$.

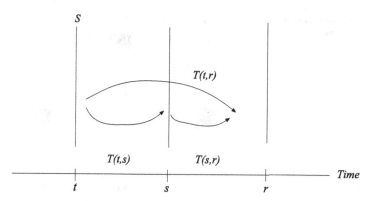

Fig. 6 Time evolution as parallel transport: motion orders time

Definition 3 $\{T, \{T_{s,t}\}\}$ is called linearly ordered if no $s \in T$ belongs to its proper future. It is called without origin if the proper past of each $s \in T$ is non-empty. It is called connected if each $s \in T$ is in the past of some $t \in T$ and in the future of some $t' \in T$. It is called circular if every $s \in T$ belongs to its proper future.

In a circular time, recurrence (of a state with respect to a given evolution) can be defined canonically because $T_{s,s}$ is well defined for each $s \in T$ and one can say that a state $x \in S_s$ is s-recurrent if $T_{s,s}x = x$.

In a linearly ordered time, in order to define recurrence (of a state with respect to a given evolution) we need to fix an identification of the different fibers of the basic time manifold. Without it would be meaningless to speak of "the same state at different times". The minimal requirement on such an identification is that, for any $t \in \mathbb{R}$, there is given a one-to-one map $j_t : S_t \to S$, where S_t is the fiber at t and S is a fixed space. For human beings this identification map is given by the "memory", but from a logical point of view this is non-canonical and therefore, as already noticed by Heraclitus, the notion of "the same state at different times" is problematic.

Given such an identification, we say that a state $x_t \in S_t$ is "the same" as the state $x_s \in S_s$ if

$$j_t(x_t) = j_s(x_s) \in S$$

and that a state $x_s \in S_s$ is "recurrent" with respect to the dynamics $\{T_{s,t}\}$ if there exists a $t \in T$ such that x_s is "the same" as the state $T_{s,t}x_s \in S_t$ (see Fig. 7) (notice that this definition is independent on time orientation). Furthermore, given such an identification, one can use the structure of additive (semi)group on $\mathbb{R}$ to define time homogeneity of an evolution by

$$T(r, s) = T(r + \tau, s + \tau); \quad \forall \tau$$

where the $=$ symbol means that "the same states", on the fibers S_r and $S_{r+\tau}$ are mapped into "the same states", on the fibers S_s and $S_{s+\tau}$.

Fig. 7 Recurrence of a state
in a time evolution

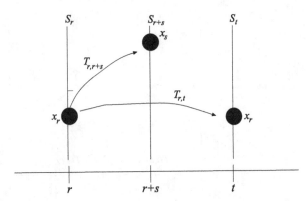

Fig. 8 Possibility of
recurrence of states in a
circular time

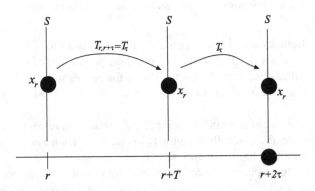

One easily sees that recurrence of a state x_s in the interval $[s, s + \tau]$ plus time homogeneity implies cyclicity of this state with period τ because (see Fig. 8)

$$T\left(s + n\tau, s + (n + 1)\tau\right)x_s = T(s, s + \tau)x_s = x_s$$

conversely: suppose time is circular with period τ

$$t + \tau \equiv t \quad \text{on the time manifold}$$

but that the initial state x_r is "wandering", i.e. its orbit has no loops, then after a time cycle the state (say of the universe) cannot be the original one,

$$x_\tau = T_\tau x_0 \neq x_0$$

hence it is not possible to prove that a time cycle has been closed. Summing up we can now formulate the "topological relativity" of time as follows: *it is not possible to distinguish experimentally between:*

(i) *circular time and a wandering state* (see Fig. 9(1));
(ii) *linear time and no recurrence* (see Fig. 9(2)).

Fig. 9 Cyclicity and
recurrence

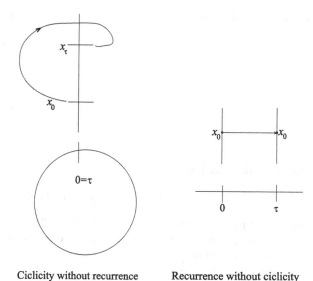

Ciclicity without recurrence   Recurrence without ciclicity

Fig. 10 Forward evolution

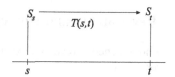

The point of view (i) would lead to a cylinder as the basic model of "flat" (i.e. without matter) space-time rather than the $\mathbb{R}^4$ of special relativity. The theory of Lorentz transformations can be extended to such a space but, again, if we insist on the usual topology for the circle, time reflection will introduce a discontinuity. In any case from both the mathematical and the conceptual point of view, this is an interesting possibility to investigate. An intriguing discussion of cyclic and linear time in physics and chemistry is in Di Meo's monograph [19].

Finally let us consider the connection between time reversibility and invertibility of the evolution. Time reversibility means that the pair (s, t) belongs to the family of admissible *time curves* if and only if also the pair (t, s) belongs to the same family. Invertibility of $T(s, t)$ means that $T(s, t)^{-1}$ exists but it can be of the form $T(s', t')$ only if time is circular. Thus we conclude that, if motion measures time and if $T(s, t) : S_s \to S_t$ is physically realized, then if time is not circular,

$$T(t, s) : S_t \not\to S_s$$

it is not a physical object even if it exists as a mathematical object.

If one weakens the definition of *time evolution,* dropping the requirement that $(t, s) \mapsto T(t, s)$ is a representation of the groupoid of admissible curves, then one can say that system is called weakly reversible if it has both a forward and a back-

Fig. 11 Backward evolution

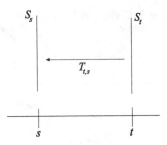

ward evolution (see Fig. 11):

$$T_b(t,s); \qquad T_f(s,t) \quad (s < t)$$

but without requiring that $T_b(t,s) = (T_f(s,t))^{-1}$. This means that evolution backward in time is admitted, but it is not necessary that, starting from an endpoint of a trajectory, the system goes through the same states in reversed order. With such a definition *Reversibility $\neq$ Recurrence*.

5 Reversibility and Homogeneity

An homogeneous reversible system is one whose dynamics is homogeneous and invertible, i.e.

$$\left(T^{-t}\right) \quad T^{-1} \text{ exists}$$

Reversible homogeneous discrete systems cannot go to any form of (dynamical or thermodynamical) equilibrium in finite time because

$$T^{n+1}x = T^n x \quad \Leftrightarrow \quad Tx = x$$

The same is true for continuous systems because

$$T^{t+\varepsilon}x = T^t x \quad \Leftrightarrow \quad T^\varepsilon x = x$$

However, non-time-homogeneous systems can go to equilibrium in a finite time because relations such as

$$T(n, n+1) \cdot T(n-1, n) \ldots T(0, 1)x = T(n-1, n) \ldots T(0, 1)x$$

$$T(s, t)T(0, s)x = T(0, s)x$$

mean only that there exists x_s such that

$$T(s, t)x_s = x_s; \quad \forall s > t$$

As already discussed above, this identity is problematic because $T(s,t): S_s \to S_t$ so we have to give a meaning to the notion of "being the same state on different time fibers".

6 Arrows of Time

Definition If there is an observable A and an expectation (mean) value (i.e. a probability measure) $\langle A \rangle$ such that the map

$$t \mapsto \langle T(s,t)[A] \rangle$$

is monotonic, then we speak of an "arrow of time".

In the cosmological arrow A is the distance of galaxies: the expansion of the universe (if it exists) gives an arrow of time. Also contraction would be an arrow of time. Thus, in agreement with the point of view of Aristotle, the real arrow comes from change, i.e. from motion.

In the thermodynamical arrow A is entropy and time flows in the direction of degrading energy. The thermodynamical arrow creates conflicting intuitions because suggests that complex structures should degrade to simpler ones, which is apparently in contrast with the biological or historical evolution where we see creation of more and more complex structures. This contrast might be only apparent because a global increase of entropy is not in contradiction with a local decrease.

A more subtle arrow of time emerged from the stochastic limit of quantum theory and is discussed in Sect. 1.18 of [11]. This has to do with the fact that the quantum transport coefficients (or generalized susceptivities), deduced from the stochastic limit of an Hamiltonian evolution in the forward time direction, is the complex conjugate of the coefficient deduced from the backward time evolution. Since the imaginary part of the quantum transport coefficient describes an energy shift, it follows that a red shift in the forward direction of time should become a blue shift in the backward direction. A possible astrophysical interpretation of this fact, based on the analysis of the Pioneer 6 data, was discussed in [8].

7 Time Reversal: Axiomatic Approach

The usual mathematical model of the set of states (in the statistical mechanics sense) of a physical system is given by the convex set

$$\mathcal{S} := \mathcal{S}(\mathcal{A})$$

of normal states of a von Neumann algebra $\mathcal{A}$. If $\mathcal{A} = \mathcal{B}(\mathcal{H})$ is the algebra of all bounded operators on a Hilbert space, we will use the notation $\mathcal{S}(\mathcal{H})$ instead of $\mathcal{S}(\mathcal{B}(\mathcal{H}))$.

The following definition is often adopted.

Definition 4 A symmetry of a quantum system is a weakly continuous affine bijection of the set $\mathcal{S}$ of states.

A time reversal is an involutory symmetry $\mathcal{T}$ of $\mathcal{S} = \mathcal{S}(\mathcal{A})$ such that

$$\mathcal{T}^2 = \mathcal{T} \tag{3}$$

If $\mathcal{A} = \mathcal{B}(\mathcal{H})$, the structure of the symmetries of a quantum system is described by the following theorem due to Wigner (cf. [16] for an elegant proof).

Theorem 1 *A map u of $\mathcal{S}(\mathcal{H})$ into itself is an affine bijection if and only if it is of the form*

$$u(\rho) = U\rho U^{-1} \quad \text{for all } \rho \text{ in } \mathcal{S}(\mathcal{H}) \tag{4}$$

where U is a unitary or antiunitary operator in $\mathcal{H}$, determined by u up to a phase factor. Any operator U satisfying (4) will be said to implement the symmetry u.

Theorem 2 *Let $\mathcal{T}$ be a time reversal on $\mathcal{S} = \mathcal{K}(\mathcal{H}_S)$. Then there exists a unitary or antiunitary operator T on $\mathcal{H} = \mathcal{H}_S$ such that*

$$\mathcal{T}(\rho) = T\rho T^{-1} \quad \text{for all } \rho \text{ in } \mathcal{K}(\mathcal{H}) \tag{5}$$

Proof It follows from Theorem 1 that a symmetry $\mathcal{T}$ is involutory if and only if any unitary or antiunitary operator T implementing it in the sense of (4), i.e.

$$\mathcal{T}(\rho) = T\rho T^{-1} \quad \text{for all } \rho \text{ in } \mathcal{S}(\mathcal{H}) \tag{6}$$

satisfies

$$\rho = \mathcal{T}^2(\rho) = T^2\rho T^{-2} \quad \text{for all } \rho \text{ in } \mathcal{S}(\mathcal{H}) \tag{7}$$

and therefore there must exist $\lambda \in \mathbb{C}$ such that $T^2 = \lambda$. For a unitary operator T this implies that $T^2 = \lambda T_0^2$ where $T_0^2 = \pm 1$ and λ is a complex number of modulus 1. $\square$

Remark The following lemma shows that the phase ambiguity is reduced for antiunitary operators.

Lemma 1 *If an antiunitary operator T is such that $T^2 = \lambda$ with $\lambda \in \mathbb{C}$, then $\lambda = \pm 1$.*

Proof Since $T^2 = \lambda$ and T^2 is unitary, λ must have modulus 1. Moreover, $\forall \psi \in \mathcal{H}$, the identity $T^2\psi = \lambda\psi$ and antiunitarity imply that

$$TT^2\psi = T\lambda\psi = \bar{\lambda}T\psi$$

On the other hand

$$TT^2\psi = T^2T\psi = \lambda T\psi$$

It follows that $\lambda = \bar{\lambda}$ and therefore $T^2 = \pm 1$. $\square$

8 Examples

Example 1 Complex conjugation in L^2-spaces is the basic example of an antiunitary operator.

Lemma 2 *The complex conjugation in $L^2(S, \lambda)$ is defined by*

$$C\psi := \bar{\psi}$$

C is an antiunitary operator satisfying $C^2 = 1$.

Proof

$$\forall \psi \in \mathcal{S}(\mathcal{A}), \quad \forall \lambda \in \mathbb{C}$$
$$C(\lambda\psi) = (\lambda\psi)^- = \bar{\lambda}\bar{\psi} = \bar{\lambda}C\psi$$
$$\langle C\psi_1, C\psi_2 \rangle = \int (C\psi_1)^-(C\psi_2)\,d\lambda = \int \psi_1\bar{\psi}_2\,d\lambda = \langle \psi_2, \psi \rangle$$

Define

$$\mathcal{T}\rho = C\rho C$$

The dual map of $\mathcal{T}$ is defined by

$$\operatorname{Tr} \mathcal{T}(\rho)X = \operatorname{Tr}\rho\mathcal{T}'(x); \quad x \in \mathcal{A}$$

Since

$$\operatorname{Tr} \mathcal{T}(\rho)X = \operatorname{Tr} C\rho Cx = \operatorname{Tr}\rho CXC$$

it follows that

$$\mathcal{T}'(x) = CXC \tag{8}$$

$\square$

Lemma 3

$$C^* = C$$

Proof Since C is antiunitary $C^*C = 1$. But also $C^2 = 1$. Hence $C^* = C$. $\square$

Lemma 4 $\mathcal{T}'$, *defined by* (8) *is an antilinear $*$-automorphism.*

Proof $\forall x \in \mathcal{A}$

$$\mathcal{T}'(\lambda x) = C\lambda XC = \bar{\lambda}CXC = \bar{\lambda}\mathcal{T}'(x)$$
$$\mathcal{T}'(x)\mathcal{T}'(y) = CXC^2 yC = CXYC = \mathcal{T}'(x)$$
$$\mathcal{T}'(x)^* = (CXC)^* = CX^*C = \mathcal{T}'(X^*)$$

$\square$

Remark For integral spin time reversal is implemented by complex conjugation.

Example 2 Let $\mathcal{A} = M(2, \mathbf{C})$ and

$$\tau : \begin{pmatrix} a_{11} & a_{12} \\ a_{21} & a_{22} \end{pmatrix} \longrightarrow \begin{pmatrix} a_{11} & a_{21} \\ a_{12} & a_{22} \end{pmatrix}$$

denote the usual transposition. Notice that

$$\tau x = J x^* J; \qquad J \begin{pmatrix} z_1 \\ z_2 \end{pmatrix} = \begin{pmatrix} \bar{z}_1 \\ \bar{z}_2 \end{pmatrix}$$

Example 3 Let

$$\tau_0 : \begin{pmatrix} a_{11} & a_{12} \\ a_{21} & a_{22} \end{pmatrix} \longrightarrow \begin{pmatrix} a_{22} & -a_{12} \\ -a_{21} & a_{11} \end{pmatrix}$$

Then

$$\tau_0 x = K x^* K$$

$$K \begin{pmatrix} z_1 \\ z_2 \end{pmatrix} = \begin{pmatrix} -\bar{z}_1 \\ \bar{z}_2 \end{pmatrix}$$

Remark For Pauli matrices

$$\tau : \sigma_1, \sigma_2, \sigma_3 \longrightarrow -\sigma_1, -\sigma_2, -\sigma_3$$

thus for spin $1/2$ (in general any half integer spin), time reversal is implemented by K because, in analogy with classical angular momentum, you want spin (and orbital angular momentum) to change sign under time reversal.

Example 4 Let

$$\tau_2 : \begin{pmatrix} a_{11} & a_{12} \\ a_{21} & a_{22} \end{pmatrix} \longrightarrow \begin{pmatrix} -a_{22} & a_{12} \\ a_{21} & -a_{11} \end{pmatrix}$$

Then

$$\tau x = L x^* L$$

$$L \begin{pmatrix} z_1 \\ z_2 \end{pmatrix} = \begin{pmatrix} -\bar{z}_1 \\ \bar{z}_2 \end{pmatrix}$$

Notice that now $L^2 = -1$.

9 Time Reflections in von Neumann Algebras

Wigner's theorem, discussed in Sect. 7 has been extended to more general von Neumann algebras according to the following lines.

Definition 5 A Jordan *-automorphism of $\mathcal{A}$ is a linear map τ of $\mathcal{A}$ in itself satisfying

$$\tau(x^*) = \tau(x)^*; \quad \forall x \in \mathcal{A} \tag{9}$$

$$\tau^2 = id \tag{10}$$

$$\tau(xy + yx) = \tau(x)\tau(y) + \tau(y)\tau(x) \tag{11}$$

One first proves that the dual of a weakly continuous affine bijection of the set $\mathcal{S}$ of normal states of a von Neumann algebra $\mathcal{A}$ is a Jordan *-automorphism.

Definition 6 A *time reflection* in $\mathcal{A}$ is an involutive Jordan *-automorphism of $\mathcal{A}$.

The following extension of Wigner's theorem is due to Kadison.

Theorem 3 *Let α be a Jordan *-automorphism of $\mathcal{A}$ then there exists a maximal central projection $z \in A$ such that*

$$\alpha(xy) = \alpha(x)\alpha(y)z + \alpha(y)\alpha(x)(1 - z) \tag{12}$$

Moreover if

$$\alpha^2 = id$$

then

$$\alpha(z) = z$$

(*this is not true in general*).

In particular,

Corollary 1 *If $\mathcal{A}$ is a factor, then a Jordan *-automorphism is either a *-automorphism or a *-anti-automorphism.*

In particular (Stormer), if $\mathcal{A}$ is a factor and τ is implemented, then it is implemented by a unitary or antiunitary hermitian operator.

Theorem 4 *If $A = \mathcal{B}(\mathcal{H})$ every involutive Jordan *-automorphism is given by*

$$\tau x = J x^* J$$

where J is antilinear isometric and $J^2 = 1$, or

$$\tau x = -K x^* K$$

where K is antilinear isometric $K^2 = -1$.

The possible use of general involutory automorphisms in von Neumann algebras to describe time reflections has been extensively discussed in [22–24]. A discussion more oriented towards the time reversibility of the Markov property is in [5].

10 Time Reversibility and Time Reversal Invariance

In general it is assumed that for any system S there exists an involutory symmetry $\mathcal{T}$ of S which can be interpreted as the time reversal operation.

The problem to distinguish which additional conditions should be satisfied by an involutive symmetry of a given state space to guarantee the uniqueness of the time reversal requires a deep investigation. Usually these additional conditions are expressed:

(i) either in terms of privileged observables such as positions, momenta, . . .
(ii) or in terms of a given, privileged, dynamics.

The basic example of the attitude (i) is given by the following:

Definition 7 The *time reversed* state $\tilde{\rho}$ of a state ρ of a system S of n structureless particles is defined by

$$\mathrm{Tr}\big[F(\{X_j\}, \{P_j\})\tilde{\rho}\big] = \mathrm{Tr}\big[F(\{X_j\}, \{-P_j\})\rho\big] \tag{13}$$

Corollary *The time reversed $\tilde{\rho}$ of a state ρ is unique whenever the set of functions F is large enough to separate the states (basic example: the Weyl algebra of the CCR).*

Proof Clear.   □

Definition 8 The map

$$\mathcal{T} : \rho \to \tilde{\rho}$$

is called *time reversal*.

Corollary $\mathcal{T}$ *is an affine map such that*

$$\mathcal{T}^2 = id \quad \Leftrightarrow \quad \mathcal{T}^{-1} = \mathcal{T}$$

Proof Clear.   □

Definition 9 A *dynamics*, or time evolution of a physical system, is a 2-parameter family of symmetries u_{t,t_0} of its set of states.

(i) For each fixed t, t_0, u_{t,t_0} is an affine map of the convex set S into itself, i.e.:

$$u_{t,t_0}\big(\alpha\rho' + (1-\alpha)\rho''\big) = \alpha u_{t,t_0}\big(\rho'\big) + (1-\alpha)u_{t,t_0}\big(\rho''\big)$$

for all ρ', ρ'' in S and for all α in $(0, 1)$;

(ii) For all states ρ in $\mathcal{S}$, all observables A, all Borel subsets I of $\mathbb{R}$ and all initial times t_0, the functions

$$t \mapsto \mathrm{Tr}\big[E_A(I)u_{t,t_0}(\rho)\big]$$

are continuous;

(iii) For all $t_0 < t_1 < t_2$, we have

$$u_{t_2,t_0} = u_{t_2,t_1}u_{t_1,t_0}$$

The dynamics u_{t,t_0} is called homogeneous if

(iv) For all $t > t_0$ and all real s,

$$u_{t,t_0} = u_{t+s,t_0+s}$$

The dynamics u_{t,t_0} is called reversible if

(v) u_{t,t_0} is a one-to-one map of $\mathcal{S}$ *onto* itself.

The dynamics of conservative systems is usually reversible. The notion of time reversal invariance is different from that of reversibility.

Definition 10 The time evolution u_t is called *time reversal invariant* if

$$\mathcal{T}u_t\mathcal{T}u_t\rho = \rho \quad \text{for all } \rho \text{ and for all } t \tag{14}$$

Note that time reversal invariance implies that u_t has the everywhere defined inverse $\mathcal{T}u_t\mathcal{T}$, even if the time evolution had been originally defined for positive time only. So, time reversal invariance implies reversibility (but not the other way around).

11   Evolutions

Let H be the Hamiltonian of a system. We will always suppose that

$$H \geq 0$$

The Schrödinger evolution of state ψ is

$$\psi_t = e^{-itH}\psi_0 = |\psi_t\rangle$$

hence the evolution of the corresponding density matrix is given by

$$|\psi_t\rangle\langle\psi_t| = e^{-itH}|\psi_0\rangle\langle\psi_0|e^{itH}$$

Therefore an arbitrary density matrix evolves according to the law

$$\rho \to e^{-itH}\rho e^{itH}$$

The dual evolution is defined by

$$\mathrm{Tr}\,\rho_t A = \mathrm{Tr}\big|e^{-itH}\psi_0\big\rangle\big\langle\psi_0|e^{itH}A\big| = \big\langle\psi_0, e^{itH}Ae^{-itH}\psi_0\big\rangle$$

Thus the Heisenberg evolution of an observable A is

$$A \to e^{itH}Ae^{-itH}$$

Let A be a reflection invariant observable:

$$\mathcal{T}A = A; \quad t \le 0$$

$$\mathcal{T}u_{-t}\mathcal{T}A = \mathcal{T}u_{-t}A = \mathcal{T}e^{itH}Ae^{-itH} = Te^{itH}Ae^{-itH}T = e^{TitHT}Ae^{-TitHt}$$

If T is antiunitary this is equal to

$$e^{-itTHT}Ae^{itTHT}$$

and, if T is unitary, this is equal to

$$e^{itH}Ae^{-itH}$$

12 Time Reversal in the Schrödinger Representation

In the representation of $\mathcal{H}$ as $L^2(\mathbb{R}^{3n}, dx)$, it is immediately seen that (13) holds if we set

$$\mathcal{T}\rho := \tilde{\rho} = C\rho C \quad (= C\rho C^{-1}), \quad \rho \in \mathcal{K}(\mathcal{H}) \tag{15}$$

where C is the natural complex conjugation on $L^2(\mathbb{R}^{3n}, dx)$:

$$C\psi := \bar{\psi}$$

indeed, we have

$$CX_jC = X_j, \qquad C(-i\hbar\nabla_j)C = i\hbar\nabla_j \tag{16}$$

It follows that the map $\mathcal{T}$ transforming ρ into $\tilde{\rho}$ is a *symmetry*, implemented by an antiunitary operator, and satisfying $\mathcal{T}^2 = 1$ (the identity map). If we define the time reversed evolution $\tilde{u}_t$ by

$$\tilde{u}_t = \mathcal{T}u_{-t}\mathcal{T}; \quad \forall t \le 0 \tag{17}$$

we obtain

$$\tilde{u}_t\rho = \exp\left[-\frac{i}{\hbar}\tilde{T}H\right]\rho\exp\left[\frac{i}{\hbar}\tilde{T}H\right] \tag{18}$$

where

$$\tilde{H} = CHC \tag{19}$$

Note that a Hamiltonian of the form

$$H = -\Delta + V \tag{20}$$

satisfies $CHC = H$, so that $\tilde{u}_t = u_t$; in words, the evolution is time reversal invariant. In fact, let us consider the Schrödinger equation

$$i \frac{\partial \psi(q,t)}{\partial t} = H\psi(q,t) \tag{21}$$

where

$$H = -\frac{\Delta}{2} + V(q) \tag{22}$$

$q = (q_1, \ldots, q_n)$ and $V(q)$ is a real-valued function.

Theorem 5 *If $\psi(q,t)$ is a solution of the Schrödinger equation (1) then its time reciprocal wave function*

$$\psi'(q,t) = \bar{\psi}(q,-t) \tag{23}$$

is also a solution of the same Schrödinger equation

$$i \frac{\partial \psi'(q,t)}{\partial t} = H\psi'(q,t) \tag{24}$$

Proof One gets from (23)

$$-i \frac{\partial \psi(q,-t)}{\partial t} = i \frac{\partial \psi(q,-t)}{\partial(-t)} = H\psi(q,-t) \tag{25}$$

Taking the complex conjugate of (25) we get (24).

If we define $\tilde{u}_t$ as in (17), we have, using also $\mathcal{T}^{-1} = \mathcal{T}$,

$$\tilde{u}_t \rho = \exp\left[-\frac{i}{\hbar}\tilde{H}t\right] \rho \exp\left[\frac{i}{\hbar}\tilde{H}t\right] \tag{26}$$

where

$$\tilde{H} = \sigma T^{-1} H T \tag{27}$$

and where

$$\sigma = -1 \quad \text{if } T \text{ is unitary} \tag{28}$$

$$\sigma = +1 \quad \text{if } T \text{ is antiunitary} \tag{29}$$

If we require that both H and $\tilde{H}$ are bounded from below and unbounded from above, we are forced to assume that T is antiunitary, in agreement with the example (15). We shall assume in general that T is antiunitary.   □

13 Time Reflection and Positivity of the Spectrum

Suppose that, in an algebraic set up, time reflection is implemented by a self-adjoint involution

$$\tau(A) = K A^* K; \qquad K^2 = 1$$

Since

$$K U_t K = U_{-t}$$

it follows that, if K is unitary then, deriving the evolution, one has

$$K H K = -H$$

Therefore if τ is an automorphism, then the spectrum of H cannot be positive. Hence τ, and therefore K, must be antilinear.

14 Antilinear Anti-automorphisms

Definition 11 Let $\mathcal{A}$ be a $*$-algebra. An antilinear anti-automorphism of $\mathcal{A}$ is a map

$$\rho_0 : \mathcal{A} \to \mathcal{A}$$

satisfying

$$\rho_0(a + b) = \rho_0(a) + \rho_0(b) \quad \text{(additivity)}$$

$$\rho_0(\lambda a) = \bar{\lambda} \rho_0(b) \quad \text{(anti-linearity)}$$

$$\rho_0(ab) = \rho_0(b) \rho_0(a)$$

The following simple remark shows that there is a one-to-one correspondence between linear $*$-automorphisms and antilinear $*$-anti-automorphisms.

Lemma 5 *Let $u \in \text{Aut}(\mathcal{A})$ be a linear $*$-automorphism. Then*

$$u^*(x) := u(x^*); \quad x \in \mathcal{A}$$

is an anti-linear $*$-anti-automorphism of $\mathcal{A}$. Conversely, if u^* is an antilinear $*$-anti-automorphism of $\mathcal{A}$, then

$$u(x) := u^*(x^*)$$

is a linear $*$-automorphism of $\mathcal{A}$.

Proof u^* is clearly antilinear

$$u^*(xy) = u(y^* x^*) = u(y^*) u(x^*) = u^*(y) u^*(x)$$

$$(u^*(x))^* = (u(x^*))^* = u(x) = u^*(x^*)$$

Conversely, if u^* is an antilinear $*$-anti-automorphism, then u is linear and

$$u(xy) = u^*\left((xy)^*\right) = u^*\left(y^*x^*\right) = u^*\left(x^*\right)u^*\left(y^*\right) = u(x)u(y)$$
$$\left(u(x)\right)^* = \left(u^*\left(x^*\right)\right)^* = u^*(x) = u\left(x^*\right) \qquad \qquad \square$$

15 Anti-states

An anti-linear map, ρ_0, does not map states into states. This justifies the following.

Definition 12 Let $\rho_0 : \mathcal{A} \to \mathcal{A}$ be an antilinear map. A state φ on $\mathcal{A}$ is called ρ_0-invariant (or anti-invariant) if

$$\varphi \circ \rho_0 = \bar{\varphi} \qquad (30)$$

where, by definition,

$$\bar{\varphi}(a) := \overline{\varphi(a)} = \varphi\left(a^*\right) \qquad (31)$$

We have seen that, for anti-automorphisms ρ_0, the natural notion of invariance is

$$\varphi \circ \rho_0 = \bar{\varphi}$$

If φ is a state on $\mathcal{A}$, $\bar{\varphi}$ is defined by

$$\bar{\varphi}(a) = \overline{\varphi(a)} = \varphi\left(a^*\right)$$

hence it is not a state.

Definition 13 An anti-linear positive functional φ on $\mathcal{A}$ such that

$$\varphi(1) = 1$$

is called an anti-state.

Since φ is an anti-state if and only if $\bar{\varphi}$ is a state, all the notions and constructions for states are extended to anti states.

16 Automorphisms with Anti-invariant States

Lemma 6 Let R be an anti-unitary such that $R^2 = 1$ (so that $R^* = R$) and define

$$RaR =: \rho_0(a) \qquad (32)$$

Then ρ_0 is an antilinear $*$-automorphism. Suppose moreover that

$$R\Phi = \Phi \qquad (33)$$

and define

$$\langle \Phi, a\Phi \rangle =: \varphi(a)$$

Then

$$\varphi \circ \rho_0 = \bar{\varphi}$$

Proof

$$\varphi\big(\rho_0(a)\big) = \langle \Phi, RaR\Phi \rangle = \langle RR\Phi, RaR\Phi \rangle$$
$$= \langle aR\Phi, R\Phi \rangle = \langle a\Phi, \Phi \rangle = \big\langle \Phi, a^*\Phi \big\rangle = \varphi\big(a^*\big) = \bar{\varphi}(a)$$

That ρ_0 is an automorphism is clear because

$$\rho_0(a)\rho_0(b) = (RaR)(RbR) = RaR^2bR = RabR = \rho_0(ab)$$

Finally ρ_0 is a $*$-automorphism because $R^* = R$.   □

17 Motivations for the Definition of Time Reversal

Let (u_t^0) be a free evolution. Solving the Schrödinger equation in interaction representation

$$\partial_t U_t = -i H_I(t) U_t \tag{34}$$

for the interaction H_I

$$u_t^0(H_I) = H_I(t)$$

with initial condition $U_0 = 1$ and for *positive times* t, we obtain a cocycle for the free evolution (u_t^0). Following [4] (Sect. 1.1.29), let us show that there is a unique way to extend this cocycle to negative times so that, composing this cocycle with the free evolution, one obtains a 1-parameter automorphism group on all the real line. This extension is called *the time reflected cocycle*.

Theorem 6 *Let $\mathcal{A}$ be an algebra, (u_t^0) a 1-parameter automorphism group on $\mathcal{A}$ and $(U_t)_{t \geq 0}$ a 1-parameter family of unitary operators in $\mathcal{A}$ such that the 1-parameter family*

$$u_t := U_t^* u_t^0(\cdot) U_t =: j_{0,t} u_t^0; \quad t \geq 0 \tag{35}$$

is a 1-parameter semigroup of automorphisms of $\mathcal{A}$. Then there exists a unique 1-parameter automorphism group (u_t) $(t \in \mathbb{R})$ on $\mathcal{A}$ coinciding with (35) for positive values of t.

Moreover u_t has the form, for $a \in \mathcal{A}$,

$$u_t(a) = \begin{cases} j_{0,t} u_t^0(a) = U_t^* \cdot u_t^0(a) \cdot U_t; & t \geq 0 \\ u_t^0 j_{0,-t}^{-1}(a) = u_t^0(U_{-t}) \cdot u_t^0(a) \cdot u_t^0(U_{-t})^* = u_t^0(U_{-t} a U_{-t}^*); & t \leq 0 \end{cases} \tag{36}$$

Proof By assumption each u_t, with $t \geq 0$, is invertible hence, if an (u_t) as in the thesis exists, it is uniquely defined by

$$u_{-t} = (u_t)^{-1}; \quad t \geq 0 \tag{37}$$

Now let us check that $(u_t)^{-1}$ is given by the right hand side of (36). For each $a \in \mathcal{A}$ and $t \geq 0$, one has

$$u_{-t} u_t(a) = u_{-t}^0 \left(U_t [U_t^* u_t^0(a) U_t] U_t^* \right) = u_{-t}^0 u_t^0(a) = a$$

This implies that, if $0 \leq s < t$,

$$u_{-s} u_t = u_{-s} u_s u_{t-s} = u_{t-s} \tag{38}$$

and similarly if $s > t$. Since (u_t) is a semigroup for $t \geq 0$ (hence also for $t \leq 0$, due to (37)), (38) implies that u_t is a 1-parameter automorphism group and this ends the proof. $\square$

Notice that the solution of equation (34) for $t \geq 0$ is

$$\vec{T} e^{-i \int_0^t H_I(s)\, ds} = U_t \tag{39}$$

Now suppose that there exists an anti-automorphism satisfying

$$\rho_0 \big(H_I(s) \big) = H_I(-s)$$

Then since ρ_0 is an antilinear anti-automorphism, we have, with $t \geq 0$,

$$\rho_0(U_t) = \vec{T} \exp i \int_0^t H_I(-s)\, ds = \overleftarrow{T} e^{-i \int_0^{-t} H_I(\sigma)\, d\sigma} = \overleftarrow{T} e^{i \int_{-t}^0 H_I(s)\, ds} = (U_{-t})^* \tag{40}$$

Proceeding more constructively let, for $t \leq 0$, u_t^0 be the backward free evolution. Then if U_t is the evolution in interacting representation, we must have, because of Theorem 6,

$$U_t = u_t^0(U_{-t})^* = u_t^0 \big(\vec{T} e^{-i \int_0^{-t} H_I(s)\, ds} \big)^* = u_t^0 \big(\overleftarrow{T} e^{+i \int_0^{-t} H_I(s)\, ds} \big)$$
$$= \overleftarrow{T} e^{i \int_0^{-t} u_t^0(H_I(s))\, ds} = \overleftarrow{T} e^{i \int_0^{-t} H_I(s+t)\, ds} = \overleftarrow{T} e^{i \int_t^0 H_I(\sigma)\, d\sigma}$$

With the change of variables $\sigma := s + t$, we obtain

$$U_t = \overleftarrow{T} e^{i \int_t^0 H_I(\sigma)\,ds}; \quad t \leq 0$$

comparing this with (40) we conclude that, for $t \leq 0$, one has

$$\rho_0(U_{-t}) = u_t^0(U_{-t})^* \tag{41}$$

This identity motivates the definition introduced in the following section.

18 Time Reflections in Local Algebras

Let there be given a von Neumann algebra $\mathcal{A}$ with a time localization. This means a triple $\{\mathcal{A}, \mathcal{A}_0, (u_t^0)\}$ such that

$$u_t^0 \in \mathrm{Aut}(\mathcal{A}) \tag{42}$$

$$\mathcal{A}_t := u_t^0(\mathcal{A}_0) \tag{43}$$

$$\mathcal{A} = \bigvee_{t \in \mathbb{R}} \mathcal{A}_t \tag{44}$$

There exists at most one additive map ρ_0 satisfying

$$\rho_0\big(A(t_1)\dots A(t_n)\big) = A(-t_n)^* \dots A(-t_1)^* \tag{45}$$

for any choice of $t_1, \dots, t_n$ (not necessarily ordered). Any map with this property satisfies, for any $A, B \in \mathcal{A}$:

$$\rho_0(\lambda A) = \bar{\lambda}\rho_0(A)$$

$$\rho_0\big(A^*\big) = \rho_0(A)^*$$

$$\rho_0(AB) = \rho_0(B)\rho_0(A)$$

The first two properties are clear. The third one follows from

$$\rho_0\big(A(t_1)\dots A(t_n) \cdot B(s_1)\dots B(s_m)\big) = B(-s_m)^* \dots B(-s_1)^* A(-t_n)^* \dots A(-t_1)^*$$

Definition 14 An anti-automorphism $\rho_0 : \mathcal{A} \to \mathcal{A}$ satisfying (45) will be called a time reflection with respect to the time localization $\{\mathcal{A}, \mathcal{A}_0, (u_t^0)\}$. If $\mathcal{A}$ is generated by the $\mathcal{A}_t$'s topologically, then ρ_0 is required to be continuous.

Lemma 7 Let ρ_0 and u_t^0 be as above and let φ be an u_t^0-invariant state. Then $\varphi \circ \rho_0 = \bar{\varphi}$.

Proof For any

$$F(\{A(s)\}) \in \mathcal{A}_{[0,t]}$$

one has

$$\varphi(\rho_0(F(\{A(s)\}))) = \varphi(F(\{A(-s)\})^*) = \varphi(u_s^0[F(\{A(-s)\})^*])$$
$$= \varphi(F(\{A(s)\})^*) = \bar{\varphi}(F(\{A(s)\}))$$

which is the thesis.   □

In the following we study the existence of time reflections on special algebras. In Sect. 22 we will prove that such an anti-automorphism can be explicitly constructed for any mean zero gauge invariant Gaussian field with standard free evolution.

19 The Adjoint and the Time Reversed of a Markov Semigroup

The following discussion abstracts a general scenario which is realized in several concrete examples of physical interest in the stochastic limit of quantum theory [3, 4, 9–11].

Let there be given:

(i) an algebra $\mathcal{A}$, a state φ on $\mathcal{A}$, and a unitary in the algebra $U_t \in Un(\mathcal{A})$;
(ii) an anti-automorphism $\rho_0 \in \text{Antiaut}(\mathcal{A})$ of $\mathcal{A}$ leaving φ anti-invariant

$$\rho_a(ab) = \rho_0(b)\rho_0(a); \qquad \rho_0(a^*) = \rho_0(a)^*; \qquad \rho_0\text{—antilinear}$$
$$\varphi \circ \rho_0 = \bar{\varphi}$$
$$(46)$$

(iii) let $\mathcal{A}$ be realized on a Hilbert space $\mathcal{H}$ and

$$\varphi(\cdot) = \langle \Phi, (\cdot)\Phi \rangle$$

Then we have

$$\langle X\Phi, U_t^* Y U_t \Phi \rangle = \varphi(X^* U_t^* Y U_t) = \bar{\varphi}(\rho_0(X^*[U_t^* Y U_t]))$$
$$= \bar{\varphi}(\rho_0(U_t)\rho(Y)\rho_0(U_t^*) \cdot \rho_0(X^*))$$
$$= \varphi(\rho_0(X)\rho_0(U_t)\rho_0(Y^*)\rho_0(U_t^*))$$
$$= \langle \rho_0(X^*)\Phi, \rho_0(U_t)\rho_0(Y^*)\rho_0(U_t^*)\Phi \rangle$$
$$= \varphi(\rho_0(X)\rho_0(U_t)\rho_0(Y^*)\rho_0(U_t))$$
$$(47)$$

Suppose moreover that the following conditions are satisfied:

(iv) X, Y are such that

$$\rho_0(X) = X^*; \qquad \rho_0(Y) = Y^*$$

(recall that ρ_0 is antilinear);

(v) for some 1-parameter group (u_t^0) of automorphisms of $\mathcal{A}$ leaving φ invariant

$$\rho_0(U_t) = u_{-t}^0 (U_t)^*; \quad t \geq 0$$

(vi) $u_t^0(X) = X; u_t^0(Y) = Y; \forall t.$

Then the identity (47) becomes, for $t \geq 0$:

$$\langle X\Phi, U_t^* Y U_t \Phi \rangle = \langle X\Phi, u_{-t}^0(U_t) Y u_{-t}^0 (U_t^*)\Phi \rangle$$

equivalently, introducing the notations

$$U_{[0,t]} := U_t; \qquad U_{[-t,0]} := u_{-t}^0(U_t^*) = u_{-t}^0(U_t)^*$$

$$\varphi\big(X^* U_{[0,t]}^* Y U_{[0,t]}\big) = \varphi\big(X^* U_{[-t,0]}^* Y U_{[-t,0]}\big)$$

In the case of a Markovian structure compatible with φ, i.e. X, Y are in the time zero algebra and

$$\varphi \circ E_{0]} = \varphi \circ E_{[0} = \varphi$$

we obtain

$$\varphi_0\big(X^* E_{0]}\big(U_{[0,t]}^* Y U_{[0,t]}\big)\big) = \varphi_0\big(X^* E_{[0}\big(U_{[-t,0]}^* Y U_{[-t,0]}\big)\big)$$

Thus, introducing the notations

$$P^t(Y) := E_{0]}\big(U_{[0,t]}^* Y U_{[0,t]}\big)$$

$$P_{\text{rev}}^t(Y) := E_{[0}\big(U_{[-t,0]}^* Y U_{[-t,0]}\big)$$

we obtain the duality

$$\varphi_0\big(X^* P^t(Y)\big) = \varphi_0\big(X^* P_{\text{rev}}^t(Y)\big) \tag{48}$$

which, since the identity takes place for all $X, Y \in \mathcal{A}_0$, is equivalent to

$$P^t = P_{\text{rev}}^t \tag{49}$$

Notice the difference between the duality (48) and the usual duality for Markov semigroups,

$$\varphi_0\big(X^* P^t(Y)\big) = \varphi_0\big(P_+^t(X)^* Y\big) \tag{50}$$

It is well known that, while in a classical (commutative) context the adjoint semi-group P_+^t always exists, existence in a quantum context undergoes severely restrictive conditions. In fact in the above construction we have also used the invariance of the "time zero algebra" under time reflection. This is automatically satisfied in many concrete models, however, it is not difficult to modify the above construction so to include also a non-trivial action of the time reflection of the "time zero algebra".

In classical probability the condition (49) is equivalent to *statistical reversibility* of the associated Markov process, i.e. to the existence of a time reversal transformation (or a 1-parameter family of such transformations, see [1]) leaving invariant the measure of the process. For a proof of the fact that the existence of the adjoint Markov semigroup is a characteristic of equilibrium situations cf. Theorem 1.41 in [4] and the discussion following it where it is emphasized that the above conclusion strongly depends on the existence of "sufficiently many allowed transitions among the atomic levels". In the presence of forbidden transitions the physical situation is much richer and the mathematical situation is much more complex and should be discussed case by case. Under these conditions the correct quantum generalization of the symmetry condition (49) was introduced in the paper [21] and, in terms of the generators $\mathcal{L}$ and $\mathcal{L}_*$ of P^t and P_+^t, respectively, is expressed by the identity

$$\mathcal{L}^* = \mathcal{L} - 2i\Delta$$

where Δ is a derivation on $\mathcal{A}_0$ (or a sub-space of $\mathcal{A}_0$ in the unbounded case). In absence of time reversal invariance of the initial state for the composite system (which is the typical case), concrete examples of time reversed semigroups obtained by comparing the stochastic limit of quantum systems in the forward and in the backward direction of time are discussed in [3, 10]. These examples led to an extension, to the non-equilibrium case, of the Frigerio, Kossakowski, Gorini, Verri equivalence between detailed balance and KMS (equilibrium) condition, which has been generalized recently by Accardi, Fagnola and Quezada [12]. These recent results give a first mathematical expression to the intuitive connection between lack of statistical reversibility and non-equilibrium. However, they constitute only the first steps of a huge staircase which has to be climbed.

20 Field Algebras

Theorem 7 *Let a_k, a_k^+, be a Boson field and $\langle \cdot \rangle_0$ a Gaussian mean zero gauge invariant state on it. Let $\omega : \mathbb{R}^d \to \mathbb{R}$ be a sufficiently good function.*

Then there exist a unique Boson field $a(t, k)$, $a^+(t, k)$ and a state $\langle \cdot \rangle$ on it such that the map

$$a(t, k) \mapsto e^{-it\omega_k} a_k = u_t^0(a_k)$$

extends to a 1-parameter automorphism group u_t^0 of the $a(t, k)$-field algebra.

Definition 15 The algebra generated (in the operator valued distribution sense) by the fields $a(t, k)$ in Theorem 7 is called the free field algebra.

21 Existence of a Time Reflection on the Field Algebra

We will prove that on the field algebra there exists a unique map ρ_0 such that

(i) ρ_0 is an anti-automorphism, i.e.

$$\rho_0(AB) = \rho_0(B)\rho_0(A)$$
$$\rho_0(A)^* = \rho_0(A)^*$$

(ii) if $A_0 \in \mathcal{A}_0$ (time zero algebra) and u_s^0 is the free evolution, then $\forall s \in \mathbb{R}$

$$\rho_0\left(u_s^0(A_0)\right) = u_{-s}^0(A_0)^*$$

(in particular the time zero algebra is left invariant).

Let there be given a boson field with a free evolution $a(t, k) = e^{-it\omega_k} a_k$ and let $A(S_t g)$ and $A^+(S_t f)$ denote the corresponding evolutions of the smeared fields. We have

$$a(-t, k)^* = \left[e^{-i(-t)\omega_k} a_k\right]^* = \left[e^{it\omega_k} a_k\right]^* = e^{-it\omega_k} a_k^+$$

Then an anti-automorphism ρ_0, as in Sect. 18, exists and it is characterized by

$$\rho_0\left(a(t, k)\right) = a(-t, k)^* = e^{-it\omega_k} a_k^+$$

Equivalently the map ρ_0 can be obtained in two steps:

(i) exchange of a and a^+;
(ii) replacement of ω_k by $-\omega_k$.

22 ρ_0-Anti-invariant Gaussian States

For Gaussian states φ everything is reduced to the pair correlations

$$\varphi\left(A(g)A^+(f)\right) = \langle g, Qf \rangle$$

Assuming that the one-particle free evolution commutes with the covariance:

$$S_t Q = Q S_t$$

one has

$$\varphi\left(A_f(s)A_g^+(t)\right) = \langle f, Q S_{t-s} g \rangle$$
$$\varphi\left(A^+(t)A(s)\right) = \varphi\left([A^+(t), A(s)]\right) + \varphi\left(A(s)A^+(t)\right)$$
$$= \langle g, S_{t-s} f \rangle + \langle g, Q S_{t-s} f \rangle = \langle g, (1+Q)S_{t-s} f \rangle$$

Therefore

$$\varphi\big(A_g(-t)A_f^+(-s)\big) = \langle g, QS_{(-s)-(-t)}f\rangle = \langle g, QS_{t-s}f\rangle$$
$$= \overline{\langle f, QS_{t-s}g\rangle} = \varphi\big(A_f(s)A_g^+(t)\big)$$

Therefore

$$\varphi \circ \rho_0 = \bar{\varphi}$$

In particular any gauge invariant Gaussian state is ρ_0-anti-invariant.

References

1. Accardi, L.: On the quantum Feynman–Kac formula. Rend. Semin. Mat. Fis. Milano **48**, 135–179 (1978)
2. Accardi, L.: Stato Fisico. Enciclopedia Einaudi, vol. 13, pp. 514–548 (1981)
3. Accardi, L., Imafuku, K.: Dynamical detailed balance condition and local KMS condition for non-equilibrium state. Preprint Volterra N. 532, October 2002
4. Accardi, L., Kozyrev, S.V.: Quantum interacting particle systems. Lectures given at the Volterra–CIRM International School, Levico Terme, 23–29 September 2000. In: Accardi, L., Fagnola, F. (eds.) Quantum Interacting Particle Systems. World Scientific, Singapore (2002). Preprint Volterra, N. 431, September 2000
5. Accardi, L., Mohari, A.: Time reflected Markov processes. Infin. Dimens. Anal. Quantum Probab. Relat. Top. (IDA–QP) **2**(3), 397–425 (1999). Preprint Volterra N. 366, 1999
6. Accardi, L., Reviglio, E.: The concept of physical state and the foundations of physics. Preprint Volterra N. 198, 1996
7. Accardi, L., Frigerio, A., Lewis, J.: Quantum stochastic processes. Publ. Res. Inst. Math. Sci., Kyoto Univ. **18**, 97–133 (1982)
8. Accardi, L., Laio, A., Lu, Y.G., Rizzi, G.: A third hypothesis on the origins of the red shift: applications to the Pioneer 6 data. Phys. Lett. A **209**, 277–284 (1995). Preprint Volterra N. 208, 1995
9. Accardi, L., Imafuku, K., Kozyrev, S.V.: Stimulated emission with non-equilibrium state of radiation. In: Proceedings of XXII Solvay Conference in Physics, 24–29 November 2001. Springer, Berlin (2002). Preprint Volterra N. 521, July 2002
10. Accardi, L., Imafuku, K., Lu, Y.G.: Onsager relation with the "slow" degrees of the field in the white noise equation based on stochastic limit. In: Proceedings of the Japan–Italy Joint Waseda Workshop on "Fundamental Problems in Quantum Mechanics", Tokyo, Japan, 27–29 September 2001. World Scientific, Singapore (2002). Preprint Volterra N. 522, July 2002
11. Accardi, L., Lu, Y.G., Volovich, I.: Quantum Theory and Its Stochastic Limit. Springer, Berlin (2002)
12. Accardi, L., Fagnola, F., Quezada, R.: Weighted detailed balance and local KMS condition for non-equilibrium stationary states. Bussei Kenkyu **97**(3), 318–356 (2011/2012). Perspectives of Nonequilibrium Statistical Physics, Dedicated to the memory of Shuichi Tasaki
13. Aveni, A.: Gli imperi del tempo: Calendari, orologi e culture (1999)
14. Bardis, P.D.: Cronus in the eternal city. Sociol. Int. **16**(1–2), 5–54 (1978)
15. Caldirola, P., Recami, E.: The concept of time in physics (considerations on physical time). Epistemologia **I**, 263–304 (1978)
16. Cassinelli, G., De Vito, E., Levrero, A.: Symmetry groups in quantum mechanics and the theorem of Wigner on the symmetry transformations Rev. Math. Phys. **9**(8), 921–941 (1997)
17. Castagnino, M., Sanguineti, J.J: Tempo e Universo, Un approccio filosofico e scientifico. Armando Editore, Roma (2000)

18. Davies, P.: I misteri del tempo. L'Universo dopo Einstein. Arnoldo Mondadori Editore, Milan (1996)
19. Di Meo, A.: Circulus Aeterni Motus: Tempo ciclico e tempo lineare nella filosofia chimica della natura. Piccola Biblioteca Einaudi (1996)
20. Dominici, P.: Evoluzione delle misure orarie in Italia. Istituto dell'Enciclopedia Italiana, Dizionario della Fisica (1993)
21. Frigerio, A., Kossakowski, A., Gorini, V., Verri, M.: Quantum detailed balance and KMS condition. Commun. Math. Phys. **57**, 97–110 (1977). Erratum: Commun. Math. Phys. **60**, 96 (1978)
22. Majewski, W.A.: On the relationship between the reversibility of dynamics and detailed balance conditions. Ann. Inst. Henri Poincaré **XXXIX**(1), 45–54 (1983)
23. Majewski, W.A.: The detailed balance condition in quantum statistical mechanics. J. Math. Phys. **25**(3), 614–616 (1984)
24. Majewski, W.A.: Dynamical semigroups in the algebraic formulation of statistical mechanics. Fortsch. Phys. **32**(1), 89–133 (1984)
25. Priestley, J.B.: L'Uomo e il Tempo. Sansoni, Firenze (1974)
26. Prigogine, I.: From Being to Becoming. Freeman, New York (1980)
27. Ricoeur, P.: Tempo e racconto. Jaca Book, Milano (1999), Italian translation. Temps et récit. Cerf, Paris
28. Wickramasekara, S.: A note on the topology of space-time in special relativity. Class. Quantum Gravity **18**, 5353–5358 (2001)

Chapter 19
The Arrow of Time and Information Theory

Vieri Benci

Abstract In this paper first we present the notions of Boltzmann entropy and Shannon entropy and some notions from information theory. Also we define some new concepts such as Combinatorial Entropy and Computable Information Content. In the second part, we argue that the mechanisms which determine the two arrows of time (the thermodynamic arrow and the evolution arrow) can be modeled and better understood using these concepts.

Keywords Shannon entropy · Boltzmann entropy · Information theory · Algorithmic Information Content · Irreversible systems · Maxwell devil

1 Introduction

1.1 The Arrows of Time

The problems related to time are as old as human thinking. One of the most fascinating and unsettling problem is the arrow of time. In modern science the most meaningful indicators of flowing of time and of its direction are essentially two:

- (*II law*) The second law of thermodynamics: the passing of time destroys information. Time is Kronos who devours his offspring. Everything is consumed by the flow of time, and, at the very end, the universe will be an undifferentiated mass where even light and darkness will be hopelessly mixed together.
- (*Evolution*) Historical, biological, cosmological evolution: the passing of time creates information. In cosmological evolution, light is separated from darkness; galaxies, stars, and planets take their form; in biological evolution life arise from mud and bacteria, protista, fungi, plants, and animals evolve in always more complex forms; and then intelligence appears and evolution continues in history and gives origin to more and more complex civilizations.

V. Benci (✉)
Dipartimento di Matematica Applicata "U. Dini", Università degli Studi di Pisa, Largo Bruno Pontecorvo 1/c, 56127 Pisa, Italy
e-mail: benci@dma.unipi.it

S. Albeverio, P. Blanchard (eds.), *Direction of Time*,
DOI 10.1007/978-3-319-02798-2_19,
© Springer International Publishing Switzerland 2014

These two aspects of the arrow of time are apparently contradictory with each other. But there is more to say: both of them are in contradiction with the fundamental laws of physics.

In fact the fundamental laws of physics are reversible: they do not distinguish past and future. From a mathematical point of view, the law of physics are expressed by differential equations (e.g. the equations of Hamilton, Maxwell, Schrödinger, Einstein, etc.). The "state" of any physical system at time t is described by a function $u(t)$ which solves the equations involved. A peculiarity of all these equations lies in the fact that if $u(t)$ is a solution, also $u(-t)$ is a solution. This is the meaning of the word "reversibility", at least in this paper. This fact, in the physical world has the following meaning: if $u(t)$ describes the evolution of a physical system, also $u(-t)$ represents a possible evolution of the same system (with different initial conditions). This fact is in evident contradiction with experience. Nobody is born old and gets younger until becoming a child and finally disappearing in an egg.

1.2 The Aim of This Paper

Today, these apparent contradictions are understood reasonably well. The relation between the second law and the reversibility of the fundamental law of physics has been explained by Boltzmann. His theory has received and still receives many objections, but it is essentially correct (at least in my opinion and in the opinion of most scientists). The evolution arrow and its apparent contradiction with the second law, in recent times, has been object of a lot of attention and the study of dissipative systems explains reasonably well the underlying mechanisms. Nevertheless, still there are many subtle questions to be settled from many points of view: philosophical, physical and mathematical. One of these questions is related to the meaning of "information" and its relation with the many notions of "entropy". In this paper, we propose new and more precise definitions of these concepts which help, we hope, to clarify some delicate points on this matter. My point of view arises from the consideration that these concepts are universal and should be applicable to many contexts and not only to physical systems. Moreover, in any model in which the existence and coexistence of the two arrows of time are present, the basic definitions must be relatively simple. These ideas are supported by the empirical experience which we get from computer simulations. If we simulate a dynamical system of mixing type (such as the cat map) and we start with an ordered distribution of the initial points, we experience a growth of disorder which we can assimilate to the growth of entropy in a physical systems. On the other hand, if we simulate an irreversible dynamical system (such as the Conway "game of life") we see complex structures to appear. We experience a sort of creation of order and information. If creation and destruction of information can be simulated so easily, the right mathematical definitions which can describe ad eventually explain these phenomena, must be relatively simple.

2 Review of the Main Notions

2.1 Clausius and Boltzmann Entropy

One of the classic formulations of the second law of thermodynamics is the following:

> "The entropy of an isolated system grows until the thermodynamical
> equilibrium which corresponds to the maximum of entropy"

In this case, by entropy, we mean the thermodynamic entropy defined (by Clausius in 1850) as follows:

$$S = \int \frac{dQ}{T}$$

According to the theory of Boltzmann the thermodynamic entropy can be defined from the law of classical dynamics. Boltzmann's theory is based on the distinction between macrostate and microstate.

The macrostate is defined by the property accessible to the observer, namely by the quantity which can be experimentally measured; the microstate is given by its complete description, namely the position and the velocity of all the elementary components of the system. For example, if we are dealing with a perfect gas, the macrostate is described by volume, pressure and temperature; the microstate is described by the position and the velocity of each molecule with a given accuracy.

According to Boltzmann, the entropy of the system can be defined by the following equation:

$$S = k \ln W \tag{1}$$

where k is the Boltzmann constant and W is the number of microstates compatible with the given macrostate.

This theory is consistent with many theoretical and experimental facts and it is accepted by the majority of scientists.

2.2 Shannon Information and Entropy

Let $\mathcal{A}$ be an alphabet, namely a finite collection of symbols (letters).

Given a finite string σ (namely a finite sequence of symbols taken in our alphabet), the intuitive meaning of *quantity of information* $I(\sigma)$ contained in σ is the following one:

> $I(\sigma)$ *is the length of the smallest binary message from which you can*
> *reconstruct* σ.

Thus, formally

$$I : \Sigma(\mathcal{A}) \to \mathbf{N}$$

I is a function from the set of finite strings in a finite alphabet $\mathcal{A}$ which takes values in the set of natural numbers. There are different notions of information and some of them will be discussed here. The first one is due to Shannon.

In his pioneering work, Shannon defined the quantity of information as a statistical notion using the tools of probability theory. Thus in Shannon framework, the quantity of information which is contained in a string depends on its context. For example the string $'pane'$ contains a certain information when it is considered as a string coming from a given language. For example this world contains a certain amount of information in English; the same string $'pane'$ contains much less Shannon information when it is considered as a string coming from the Italian language because it is much more common (in fact it means "bread"). Roughly speaking, the Shannon information of a string σ is given by

$$I(\sigma) = \log_2 \frac{1}{p(\sigma)}$$

where $p(\sigma)$ denotes the probability of σ in a given context. The logarithm is taken in base two so that the information can be measured in binary digits (bits).[1]

If in a language the occurrences of the letters are independent of each other, the information carried by each letter is given by

$$I(a_i) = \log \frac{1}{p_i}$$

where p_i is the probability of the letter a_i. Then the average information of each letter is given by

$$H = \sum_i p_i \log \frac{1}{p_i} \qquad (2)$$

Shannon called the quantity H entropy for its formal similarity with the Boltzmann's entropy. Now, we will discuss the reason of this similarity.

2.2.1 Shannon Versus Boltzmann Entropy

Consider a set of n particles and suppose that the phase space X of each particle is divided in L small cells. We can label any cell by a letter a_i of an ideal alphabet. Then the microstate of the system (with the accuracy given by the grain of our

---

[1]From now on, we will use the symbol "log" just for the base 2 logarithm "$\log_2$" and we will denote the natural logarithm by "ln".

partition of the phase pace) can be represented by a string of letters. Namely the string

$$a_1 a_2 a_3 \ldots$$

represents the state in which the particle 1 is in the cell a_1, the particle 2 is in the cell a_2, and so on. At this point it is possible to compute the Boltzmann entropy of our microstate; it is given by $\log_2 W$ where W is the set of all the configurations of our system. Clearly two configurations must be considered as belonging to the same macrostate if they have the same number of particles np_i in the cell a_j. Then the number W takes the following form:

$$W = \frac{n!}{\prod_{i=1}^{N} n_i!} \tag{3}$$

where $n := \sum_{i=1}^{L} n_i$ is the number of particles, n_i is the number of particles in the cell a_i.

We can give a nice form to the number W using the following approximation given by the Stirling formula:

$$\log n! = n \log n + n \log e + O(\log n) \tag{4}$$

Using this formula we get

$$\log W = \log \frac{n!}{\prod_{i=1}^{N} n_i!} = \log n! - \sum_{i=1}^{L} \log n_i!$$

$$= n \log n + n \log e - \sum_{i=1}^{L} n_i \log n_i - e \sum_{i=1}^{L} n_i + O(\log n)$$

$$= n \log n - \sum_{i=1}^{L} n_i \log n_i + O(\log n) \simeq n \log n - \sum_{i=1}^{L} n_i \log n_i$$

Now, setting $p_i = n_i/n$, we get the equation

$$S = \log W \cong n \sum_{i=1}^{L} p_i \log \frac{1}{p_i} \tag{5}$$

where the number p_i can be interpreted as the probability that a particle lies in the cell a_i. The formal similarity between (2) and (5) is evident. The main difference consists of the factor n. This is because the entropy (5) represents the "information" necessary to describe the full microsystem, while the (2) represents average the information of each letter. The full information of a typical message of n letter is given by nH. In this comparison we can identify a language $\mathcal{L}$ with a macrostate, provided that we define a language as the set of all messages (strings) of length n which contain exactly np_j times the letter a_j. A more appropriate definition of language (or, information source) will be given in the next section. Anyhow, in this heuristic description, the Shannon entropy could be defined as

$$H = \frac{\log W}{n}$$

Thus we have the following scheme:

	Boltzmann	**Shannon**
$\mathcal{A}$	*set of the cell*	*alphabet*
n	*number of particles*	*length of the message*
$\mathcal{L}$	*macrostate*	*language*
σ	*microstate*	*string*
W	*microstates in $\mathcal{L}$*	*strings in $\mathcal{L}$*
$S(\mathcal{L})$	$\log W$	$-$
$H(\mathcal{L})$	$-$	$\frac{1}{n}\log W$

(6)

Notice that the factor n which distinguish $S(\mathcal{L})$ from $H(\mathcal{L})$ makes $S(\mathcal{L})$ an "extensive" measure while $H(\mathcal{L})$ is an average measure. For example, in a particles gas, $H(\mathcal{L})$ is the average Boltzmann entropy for particle. This point is source of many misunderstanding when we use the word "entropy" in an interdisciplinary context. In the following we will define different "kind" of entropies; in order to avoid these misleading facts, we will use the letters S and H as extensive and average "measures", respectively.

If we assume that the strings are very long, the statistical properties of a language can be studied letting $n \to \infty$. This fact, in our comparison correspond in taking the thermodynamic limit.

2.2.2 A Mathematical Definition of Shannon Entropy

In this section we will give the exact mathematical definition of Shannon entropy. We will define Shannon entropy in a new way, which emphasizes its similarity with Boltzmann entropy and which will be useful later when we will introduce the notion of Computable Information Content. We refer to [1] for more details on this point.

Let σ be a finite string of length n. We set

$$S_0(\sigma) = \log W(\sigma) \tag{7}$$

where $W(\sigma)$ is the number of strings which can be obtained by σ permuting its letters. Notice that

$$S_0(\sigma) \leq |\sigma| \cdot \log |\mathcal{A}|$$

where $\mathcal{A}$ is the alphabet of σ, namely the set of letters which appear in σ. Moreover $S_0(\sigma) = 0$ iff σ is constant.

We will call a parsing of σ a partition of σ in shorter strings w which we will call words. For example if

$$\sigma = \text{"betubetube"}$$

two parsings of σ are given by

$$\alpha_1 = (be, tube, tube)$$

$$\alpha_2 = (bet, u, bet, u, e)$$

Given a parsing α, we will denote by $W(\alpha)$ the number of strings which can be obtained permuting the words of α. In our example we have

$$W(\alpha_1) = 3;$$

$$W(\alpha_2) = \frac{5!}{2! \cdot 2! \cdot 1!} = 30$$

Given a parsing α, we will call dictionary of α the set $V(\alpha)$ of words w (with $|w| > 1$) which appear in α. In our example we have

$$V(\alpha_1) = \{be, tube\}$$

$$V(\alpha_2) = \{bet\}$$

We define the *combinatorial entropy of* σ as follows:

$$S_{\text{com}}(\sigma) = \min_{\alpha}\left[\log W(\alpha) + \sum_{w \in V(\alpha)} S_0(w)\right] \qquad (8)$$

Notice that $S_{\text{com}}(\sigma) \leq S_0(\sigma)$. In fact, if α contains only one-letter words, we have $\log W(\alpha) = \log W(\sigma)$ and $V(\alpha) = \emptyset$. Since $S_{\text{com}}(\sigma)$ is obtained taking the minimum over all the partitions, it turns out that $S_{\text{com}}(\sigma) \leq S_0(\sigma)$.

Given any string σ we denote by $\alpha(\sigma)$ the partition which gives the minimum in (8) and set $V(\sigma) = V(\alpha(\sigma))$; if two partitions give the same minimum value, we take the one which corresponds to the smaller dictionary.

In our example, we have $\alpha(\sigma) = \alpha_1$, $V(\sigma) = V(\alpha_1) = \{be, tube\}$

$$S_{\text{com}}(\sigma) = \log W(\alpha(\sigma)) + S_0(be) + S_0(tube)$$

$$= \log 3 + \log 2 + \log 4! \cong 7.169$$

We define the *average combinatorial entropy* of σ in the following way:

$$H(\sigma) = \frac{S_{\text{com}}(\sigma)}{|\sigma|}$$

Now let ω be an infinite string and let $\omega^n \in \mathcal{A}^n$ be the finite string obtained taking the first n digits of ω and set

$$H(\omega) = \max \lim_{n \to \infty} H(\omega^n)$$

Since $H(\sigma) \leq \frac{S_{\text{com}}(\sigma)}{|\sigma|} \leq \log|\mathcal{A}|$ the maximum limit is finite.

Let $\mathcal{A}^{\mathbf{N}}$ be the set of all the infinite string in the alphabet $\mathcal{A}$, let μ be a probability measure on $\mathcal{A}^{\mathbf{N}}$ and let $T : \mathcal{A}^{\mathbf{N}} \to \mathcal{A}^{\mathbf{N}}$ be the shift map (defined as follows: $(T\sigma)_i = \sigma_{i+1}$). μ is called invariant (or stationary) if for every $A \subset \mathcal{A}^{\mathbf{N}}$, $\mu(A) = \mu(T^{-1}(A))$. If μ is invariant the couple $(\mathcal{A}^{\mathbf{N}}, \mu)$ is called an information source.

Now, we can give a definition of Shannon entropy which can be proved to be equivalent to the usual one.

Definition 1 The Shannon entropy of $(\mathcal{A}^{\mathbf{N}}, \mu)$ is defined by

$$h_\mu = \int_{\mathcal{A}^{\mathbf{N}}} H(\omega) \, d\mu$$

It is also possible to prove that for μ-almost every string $\omega \in \mathcal{A}^{\mathbf{N}}$ the limit

$$H(\omega) = \lim_{n \to \infty} H(\omega^n) \tag{9}$$

exists (see [1]). Clearly, if μ is ergodic, $H(\omega) = h_\mu$ for μ-almost every string $\omega \in \mathcal{A}^{\mathbf{N}}$.

2.3 Information Content

As we have seen the Shannon notion of information relies strongly on the notion of probability and this is very disappointing for the aims of this paper for the following reasons:

- we think that from an epistemological point of view the definition of probability presents many problems and does not help to clarify the nature of notion such as "entropy" and irreversibility
- we think that the notion of information is primitive and that the notion of probability should be derived by it
- our goal is to give definition which can be applied also to cellular automata and to computer simulations and this objects are strictly deterministic; thus the notion of probability should be avoided at least as primitive concept.

Moreover the Shannon information is context dependent and also this fact is in contrast with our aims. However, there are measures of information which depend intrinsically on the string and not on its probability within a given context. We give a general definition of information content which apply to many different contexts.

Definition 2 Let

$$U : \Sigma(\mathcal{A}) \to \Sigma(\{0, 1\})$$

be an injective map and set

$$I_U(\sigma) = |U(\sigma)|$$

The function

$$I_U : \Sigma(\mathcal{A}) \to \mathbf{N}$$

is called information function relative to U if, for any infinitely long string ω for which the limit (9) exists, we have

$$H_U(\omega) \leq H(\omega) \tag{10}$$

where

$$H_U(\omega) = \min \lim_{n \to \infty} \frac{I_U(\omega^n)}{n} \tag{11}$$

The number $I_U(\sigma)$ will be called U-information content of σ.

$U(\sigma)$ can be thought as a coding of the string σ in binary alphabet. (10) relates the information content to the Shannon entropy. Actually, (10) represents a kind of *optimality* of the coding U. The U-information content I_U allows to define the U-entropy, $H_U(\omega)$ of a single infinite string ω by (11). $H_U(\omega)$ represents the average information content of the string ω and it des not depend on any probability measure.

$H_U(\omega)$ allows to give an exact relation between the Shannon entropy h_μ and the information content I_U:

Theorem 3 *Let $(\mathcal{A}^{\mathbf{N}}, \mu)$ be an information source with entropy h_μ; then*

$$h_\mu = \int H_U(\omega) \, d\mu$$

Proof See [1]. □

Of course there are many functions U and I_U which satisfy Definition 2; for infinitely long strings they are equivalent in the sense of Theorem 3. However, they can be very different from each other when we consider finite strings, particularly when these strings are generated by a non-stationary information source. In the following we will discuss some of them.

2.4 Algorithmic Information Content

One of the most important of the information functions is the Algorithmic Information Content (*AIC*). In order to define it, it is necessary to define the notion of partial recursive function. We limit ourselves to give an intuitive idea which is very close to the formal definition. We can consider a partial recursive function as a computer C which takes a program P (namely a binary string) as an input, performs some computations and gives a string $\sigma = C(P)$, written in the given alphabet $\mathcal{A}$, as an output.

The *AIC* of a string σ is defined as the shortest binary program P which gives σ as its output, namely

$$I_{AIC}(\sigma, C) = \min\{|P| : C(P) = \sigma\}$$

In this case the function $U(\sigma)$ of Definition 2 is just the shortest program which produces σ. We require that our computer is a universal computing machine. Roughly speaking, a computing machine is called *universal* if it can simulate any other machine. In particular every real computer is a universal computing machine, provided that we assume that it has virtually infinite memory. For a precise definition see e.g. [7] or [5]. We have the following theorem due to Kolmogorov.

Theorem 4 *If C and C' are universal computing machine then*

$$\left| I_{AIC}(\sigma, C) - I_{AIC}(\sigma, C') \right| \leq K(C, C')$$

where $K(C, C')$ is a constant which depends only on C and C' but not on σ.

This theorem implies that the *AIC*-information content of σ with respect to C depends only on σ up to a fixed constant and then its asymptotic behavior does not depend on the choice of C. For this reason from now on we will write $I_{AIC}(\sigma)$ instead of $I_{AIC}(\sigma, C)$.

The shortest program which gives a string as its output is a sort of *ideal* encoding of the string. The information which is necessary to reconstruct the string is contained in the program.

Unfortunately this coding procedure cannot be performed by any algorithm (Chaitin Theorem).[2] This is a very deep statement and, in some sense, it is equivalent to the Turing halting problem or to the Gödel Incompleteness Theorem. Then the Algorithmic Information Content is not computable by any algorithm.

This fact has very deep consequences for our discussion of the arrow of time as we will see later. For the moment we can say that the *AIC* cannot be used as a reasonable physical quantity since it cannot be measured nor computed.

3 Computable Information Content

3.1 The Idea of Computable Information Content

Suppose that we have some lossless (reversible) coding procedure $Z : \Sigma(\mathcal{A}) \to \Sigma(\{0, 1\})$ such that from the coded string we can reconstruct the original string (for example the data compression algorithms that are in any personal computer). Since

---

[2]Actually, the Chaitin theorem states a weaker statement: a procedure (computer program) which states that a string σ of length n can be produced by a program shorter than n must be longer than n.

the coded string contains all the information that is necessary to reconstruct the original string, we can consider the length of the coded string as an approximate measure of the quantity of information that is contained in the original string. We can define the information content of the string σ as the length of the compressed string $Z(\sigma)$, namely

$$I_Z(\sigma) = |Z(\sigma)|$$

The advantage of using a Compression Algorithm lies in the fact that, in this way, the information content $I_Z(\sigma)$ turns out to be a computable function and hence it can be used in computer simulations and it can be considered as a measurable physical quantity.

We will list the properties which the notion of computable information content must satisfy for our purposes.

A function

$$I_{CIC} : \Sigma(\mathcal{A}) \to \mathbf{N}$$

is called Computable Information Content if it satisfies the following properties:

- (i) it an information function in the sense of Definition 2.
- (ii) it is computable.
- (iii) $I_{CIC}(\sigma) = M_{CIC}(\sigma) + S_{CIC}(\sigma)$ where $S_{CIC}(\sigma)$ satisfy the following properties:

  - (S1) $S_{CIC}(\sigma) \leq \log W(\sigma)$
  - (S2) $S_{CIC}(\sigma\tau) \leq S_{CIC}(\sigma) + S_{CIC}(\tau)$
  - (S3) $S_{CIC}(\sigma) \geq I_{AIC}(\sigma) - const.$

The properties (i) and (ii) are satisfied by I_Z defined by any reasonable compression algorithm Z. The important peculiarity of the Computable Information Content lies in the possibility of decomposing the global quantity of information in two parts:

- $S_{CIC}(\sigma)$ which we will call computable entropy of σ and represents the *disordered* part of the information.
- $M_{CIC}(\sigma)$ which we will call macroinformation of σ which represent the *regular* part of the information.

The properties (S1), (S2) and (S3) of the entropy are chosen in order to fit our intuitive idea of measure of *disorder*. For example, by (S1), we deduce that a constant string has null entropy: no disorder. (S2) can be interpreted in the following way: the "disorder" of two string is additive unless the two strings are correlated with each other in some way. (S3) gives a lower bound to the quantity of disordered information of a string. Since the best program P which produces σ must be random (in the sense of Chaitin [5]), our string is forced to contain a "quantity of disorder" at least equal to $|P| = I_{AIC}(\sigma)$. The negative constant $-const$ is necessary to make a consistent theory. For example, the entropy of a constant string c is 0, but $I_{AIC}(c) > 0$.

These properties makes the computable entropy come close to the Boltzmann definition of entropy and this fact is very relevant for the interpretation of physical phenomena.

3.2 The Definition of CIC

Functions I_{CIC} which satisfy (i), (ii), and (iii) exist. We will give an example of it.

Suppose to have a string σ and to have computed $\alpha(\sigma)$ and $V(\sigma)$ as in Sect. 2.2.2. If you want to send a message from which a receiver can reconstruct the string σ, a possible strategy is the following one:

- (i) you send to the receiver the dictionary $V(\sigma)$ and the number $n(w,\sigma)$ which specifies the number of times that the word w appears in the parsing $\alpha(\sigma)$.
- (ii) you send another number which select α among all the $S_{com}(\sigma)$ possible strings which have the same dictionary $V(\sigma)$ and the same numbers $n(w,\sigma)$.

In this way, the information content of the full message is divided in two parts: part (i) which specifies the "macroscopic features" of the string and part (ii) which specifies only a number $s \approx S_{com}(\sigma)$ which selects σ among all the strings with the same features.

The above procedure makes possible the following definition of macrostate:

Definition 5 Given two strings σ_1 and σ_2, we say that they belong to the same macrostate if

- $V(\sigma_1) = V(\sigma_2)$
- for every word $w \in V(\sigma_1)$, $n(w,\sigma_1) = n(w,\sigma_2)$

Roughly speaking, the string σ_1 and σ_2 belong to the same macrostate if they can be described in the same way, namely if they have the same dictionary and the same occurrence of each word in the dictionary. So they have the same macroinformation and the same entropy.

4 Information and Dynamics

The various notions of information are useful in many problems. Here we will consider their application to dynamical systems and will investigate the implications relative to the arrows of time which is the main point of this meeting.

We assume to have a dynamical systems consisting of many particles; using the same construction as Sect. 2.2.1 (see table (6)), we can apply the previous results. Our discretized phase space will be given by $\Omega = \mathcal{A}^n$ where $\mathcal{A}$ is the alphabet which corresponds to the graining of the phase space X of a single particle. Notice that the notion of I_{CIC} makes sense also when the number of particles is low, but in this case S_{CIC} will be close to 0 and the statistical behavior is not interesting (unless we decide to study the statistics making the average over long times). For simplicity we assume that time is discrete. The transition map $f : \Omega \to \Omega$ must be considered as the evolution map at time 1. We will consider both Hamiltonian dynamics (reversible dynamics) and dissipative systems. Of course, we may think of dissipative systems as subsystems of an Hamiltonian systems of which the microscopic dynamic variables have been ignored.

Also, we remark that we are not interested in taking the thermodynamic limit. This limit will simplify the equations but will hide some interesting notion such as the notion of macroinformation.

4.1 Physical Systems

If you consider the discretization of a continuous Hamiltonian system of weakly interacting particles, you obtain the usual description of statistical mechanics. In this case it is possible to identify the CIC-entropy with the physical entropy via the Boltzmann equation (1). The concrete computations are the same and any possible difference is of the order of $\log n$ where n is the number of particles.

However, if the particles interact strongly with each other and give a rich structure to the system, our description cannot be reduced to the traditional one, both for the presence of macroinformation (which might become relevant) and for a different notion macrostate. In particular the CIC-entropy of a state is not equal to a probability measure of the macrostate deduced by the Liouville theorem.

At this point it is interesting to stress the differences between this approach and the Brillouin point of view [4]. Also for him, the physical entropy is information, namely the information which the observer does not have; in particular he writes an equation like this

$$I_{\text{tot}} = I_{\text{obs}} + S \tag{12}$$

where I_{tot} is the total information, I_{obs} is the information of the observer, and S is the entropy. In the above equation, he considered I_{tot} constant since the system is reversible and he gets the following equality:

$$\Delta S = -\Delta I_{\text{obs}}$$

which can be interpreted as follows: an increase of entropy ΔS equals the increase of ignorance of the observer. Thus he identifies the entropy as negative information and he can call the information of the observer "negentropy".

In our approach, (12) is replaced by the following one:

$$I_{CIC} = M_{CIC} + S_{CIC} \tag{13}$$

where the macroinformation might be related to the information of the observer, but in no way can be identified with it. In fact in (13) the observer does not play any role. Moreover, in a real system, in general both M_{CIC} and S_{CIC} grow with time. A more detailed description of this scenario is done in next sections.

4.2 Chaotic Reversible Systems

First of all let us consider "chaotic" Hamiltonian system. Most of the states of a system belong to the same macrostate Σ_0 and have the maximum CIC-entropy which is of the order of $\log|\Omega|$ and of course, they have minimum macroinformation.

Thus, for most of the initial condition the system enters the macrostate Σ_0 and it will stay there for a time of the order of Poincaré time. Σ_0 can be considered as the state which corresponds to the thermodynamical equilibrium. If you start with an initial condition with low CIC-entropy and high macroinformation then the CIC-entropy will increase until the maximum entropy while the macroinformation will decrease until the minimum which is a value very small if compared with the value of S_{CIC}.

In this sense time destroys information: namely, the macroinformation of the initial conditions is lost, in the sense that it cannot be recovered by a computable algorithm. In fact, if you have a "disordered" configuration, in general, there is not a computable procedure to know if it is derived by an "ordered" situation or not.

Thus we have obtained the traditional point of view of Boltzmann. The use of CIC makes possible to give a precise sense to the sentences:

information is destroyed

and

the disorder increases.

In fact, in this contest, they simply mean that M_{CIC} decreases and S_{CIC} increases.

4.3 Gradient-Like Systems

A discrete dynamical system (Ω, f) is called gradient-like if it admits a Lyapunov function, namely a function $V : \Omega \to \mathbf{R}$, such that

- $V(f(x)) \leq V(x), x \in \Omega$
- $V(f(x)) = V(x) \Leftrightarrow f(x) = x$

In dissipative physical system the Lyapunov function usually corresponds to the energy. Gradient-like systems evolve until reaching a stable equilibrium configuration x_0. Usually these configurations have low Information Content. Thus the evolution make to decrease both the entropy and the macroinformation; there is an absolute decrease of I_{CIC}. The system loses its memory and any kind of information is destroyed. This is obvious since the transition map f is not injective and different initial conditions lead to the same final configuration. If we embed this system in an invertible system, we will get a chaotic system and the consideration of the previous section apply.

4.4 Self-organizing Systems

If, in a physical dissipative system, there is an input of energy from the outside, in general, stable equilibrium configurations cannot exist. In this case many phenomena may occur; stable periodic orbits, stable tori or even strange attractors. Sometimes, very interesting spatial structures may appear. Analogous phenomena occur in non-reversible cellular automata. The most famous of them is the Conway game of life in which a lot of intriguing shapes appear in spite of the simplicity of the transition map.

From the point of view of Information theory, these are the systems which make the macroinformation to increase. If we start from an initial data with a low macroinformation content, the macroinformation will increase until reaching a limit value. If the system is infinite the macroinformation will increase for ever. For example you may think of the game of life in an infinite grid with initial conditions having only a finite number of black cells (and thus you start with an initial condition which has finite information).

As in the case of gradient-like system, we may embed these systems in a reversible system. In this situation, we will have also an increase of the entropy. Thus the two main arrows of time, described in the introduction, will be present. We believe that a sufficiently large system, in which many nonlinear interactions play a role, is very likely to present such a behavior.

From a general and qualitative point of view, these systems represent a good model for large natural system in which the appearance of complex structures occurs.

5 An Exorcism of the Maxwell Demon

The Maxwell demon acts on a pipe which joins two containers of particles, A and B. This pipe has a gate which can be kept open or closed by a demon. He opens the gate when he sees a particle coming from A and he closes it when he sees a particle coming from B. In this way, at some point all the particles will be in the container B (actually the "original" Maxwell demon made a distinction between slow and fast particles but the argument is the same).

At the end of this operation the entropy of the system of particles will be reduced since all the particles are in one container, namely in the container B.

This seems a violation of the second law of thermodynamics. Where is the catch?

Many different explanations have been proposed to exorcise this demon and to save the second law. We will give a brief sketch of some of them.

5.1 Szilard

The first important contribution to this problem was given by Leo Szilard in 1929. He thought that the measurement performed by the demon cause an increase of entropy in the environment which compensate the decrease of entropy in the containers. He was rather vague about the mechanism responsible of the entropy increase and the questions relative to this point were left open.

5.2 Brillouin and Gabor

The next important contribution came by Léon Brillouin (1956) and Dennis Gabor who saw in the Indetermination Principle of Quantum Mechanics the key point in the exorcism of the demon (see e.g. [4]). When the demon performs its measurement, he needs to send an energetic beam of light on the particle and this fact has an energetic cost which increases the entropy of the environment. We think that this explanation is wrong for at least two reasons: (1) you may imagine that this experiment is performed with big balls and in this case the perturbations of the photons are not relevant, or, to say it in a different way, the explanation should be independent on the scale, while every explanation which includes $\hbar$ depends on the scale; (2) Charles Bennet made a model in which the observation of the demon is independent of the presence of wuantum phenomena. Quantum mechanics has nothing to do with the Maxwell demon.

5.3 Landauer

A big step toward the right answer was made by Rolf Landauer (1961) who studied the constrains imposed to computation by physical laws (see e.g. [6]). He identified some operations which he "called" logically irreversible. These logically irreversible operations are also physically irreversible since they make the entropy of the environment to increase. One of them is the erasing of the memory of the computing machine, whatever its internal nature is. Clearly if you erase the memory you cannot make a time reversal and come back to the initial condition. This fact implies an increase of the entropy of the environment. The entropy balance is easy to calculate if you take in account the distinctions between AIC, CIC and CIC-entropy. If you assume that the computer plus the environment are ruled by reversible equations, then the AIC is preserved. However, this information is not contained in the computer after that its memory has been cleared. Thus, this information has been transferred to the environment, and since we may assume that it is contained in it in a random way, this information makes the CIC-entropy of the environment increase.

5.4 Bennet

The final step was made by Charles Bennet (1982) who gave the following explanation (see e.g. [3]). The demon needs a buffer to store the information that a particle is coming from A or from B in order to keep the door open or closed. Afterwards, he must store the analogous information relative to the next ball in a buffer. Thus he has two choices: or he uses some extra memory or he clears the buffer. According to Bennet, in order to perform a cycle, at some point, the demon needs to clear his memory, and this fact will make the entropy of the environment to increase.

5.5 Our Point of View

I think that this explanation is essentially right even if it presents some weak points which have been pointed out by the conference of David Albert. More or less Albert says the following: first of all, you do not need to consider a cycle; this has nothing to do with the second law which just states that the entropy of an isolated system does not decrease. Now, assume that the containers and the demon D constitute an isolated system and that the demon does not erase its memory. At the end of the process, the system $A + B + D$ will have a lower entropy, at least if you define the entropy as the measure of the final macrostate in the phase space. In fact it is not difficult to imagine a Hamiltonian for which this is true.

However, the Bennet point of view can be easily saved using the notion of CIC-entropy rather than a probability measure in the phase space. In fact, every thing becomes clear if we identify the physical entropy with the CIC-entropy. When the memory of the demon has stored all the past history of this process, it contains a string with a large content of disordered information (CIC-entropy). It is exactly the information which you need to reverse the process. Thus if you make a CIC-entropy balance, you discover that the CIC-entropy is the same. Thus any contradiction disappears.

Moreover, if you assume that our system is not is isolated, this description can say more. When you erase the memory, the AIC contained in the memory of the demon-computer will be discharged in the environment (since the system $A + B + D +$ [environment] is reversible). In this operation the global CIC-entropy will increase, since you cannot find an algorithm which is able to recover the information spread in the environment.

6 Conclusions

We think that the right description of the origin of irreversibility, complexity and the arrow of time lies in a good notion of "information content". A good notion must be independent of the notion of probability for the reasons described in Sect. 2.3.

Moreover, we think that it is very important to distinguish two different meanings of the notion of information:

- the general *abstract* notion of information (such as the AIC) which in reversible system is a constant of motion and exists only in the mind of God (but not in the mind of the demons, at least, if they are submitted to the laws of our universe).

- a computable notion of information (such as the *CIC*) which is related to physical quantities; it changes in time and can be used to describe the observed phenomena.

The distinctions between these two notions is marked by the Turing halting theorem which, we think, is one of the deepest theorems discovered in last century and whose consequences are not yet all completely understood.

Once, we have agreed to consider the *CIC* (or any other "epistemologically" equivalent notion of information) as the relevant physical quantity, it is important to have a mathematical method to separate the *CIC* in two different components:

- the entropy which corresponds to the old idea of "measure of disorder". From a physical point of view this information cannot be used to make exact deterministic predictions. It is the information dispersed in the chaos and it cannot be recovered without a violation of the Turing halting theorem.
- the macroinformation which is related to physical measurable quantities and can be used to make predictions. Moreover, the macroinformation is strictly related to various indicators of complexity.

Thus, in information theory, we have the distinction between macroinformation and *CIC*-entropy. This is similar to the distinction, in thermodynamics, between free energy $F = E - TS$ and bad energy. The *CIC*-entropy cannot be used to make predictions, while TS cannot be used to perform any work. However, it is very important to underline that, in isolated systems, free energy and macroinformation behave in a quite different way: free energy always decreases, while macroinformation might increase. The development of life, in all its forms, determines a decrease of free energy and an increase of macroinformation. Probably there is a deep mathematical relation between the evolution of these two quantities. The study of the interplay between macroinformation, entropy and the other physical quantities is a good way to investigate the origin and the evolution of complex structures.

In Sect. 3.2, we have proposed a mathematical model which makes a distinction between macroinformation and entropy. This is not the only possible model and probably is not the best. However, it seems to me that this is a good direction for investigating this kind of problems.

References

1. Benci, V., Menconi, G.: Some remarks on the definition of Boltzmann, Shannon and Kolmogorov entropy. Milan J. Math. **73**, 187–209 (2005)
2. Benci, V., Bonanno, C., Galatolo, S., Menconi, G., Virgilio, M.: Dynamical systems and computable information. Discrete Contin. Dyn. Syst. **4**, 935–960 (2004)
3. Bennet, C.: Demons, machines, and the second law. Sci. Am. **257**(5) 108–116 (1987)
4. Brillouin, L.: Scientific Uncertainty, and Information. Academic Press, New York (1964)

5. Chaitin, G.J.: Information, Randomness and Incompleteness, Papers on Algoritmic Information Theory. World Scientific, Singapore (1987)
6. Landauer, R.: Fundamental physical limitations of the computational process. Ann. N.Y. Acad. Sci. **426**, 161–170 (1985)
7. Li, M., Vitanyi, P.: An Introduction to Kolmogorov Complexity and Its Applications. Springer, Berlin (1993)

Chapter 20
Two-Way Thermodynamics: Could It Really Happen?

L.S. Schulman

Abstract In previous publications I have suggested that opposite thermodynamic arrows of time could coexist in our universe. This letter responds to the comments of H.D. Zeh (elsewhere in this volume).

Keywords Time's arrows · Two-time boundary conditions · Causality · Cosmology · Quantum measurement theory

1 Context

In 2002 a conference took place in Bielefeld, entitled, "The direction of time: The role of reversibility/irreversibility in the study of nature" [1]. In my presentation I spoke about recent work in which I had demonstrated the compatibility of opposite arrows of time for two subsystems within a larger "universe." As Dieter Zeh explains in his companion article [2], he has reservations about the physical realizability of this phenomenon and our present articles address this issue. However, as a preface to my response to his remarks, I will give a brief review, plus references, to the work that has given rise to this dialog.

When this response was originally written I gave it the title, "The slings and arrows ... whips and scorns of time," [3], not because of the barbs that Zeh was throwing my way, but quite the opposite, because of a review I had written years ago of the first edition of his book, [4], for Science magazine [5]. Instead of giving it its deserved high praise, I looked for faults, some it turns out of my own invention. So this "response" gives me the opportunity to apologize for that review.

In the endnotes there are postscripts added after the original writing of this article.

2 Opposite Arrows

The thesis that the thermodynamic arrow of time follows the cosmological arrow (the universe is *expanding*) was put forth by Gold about 1960 [6]. My contribution,

L.S. Schulman (✉)
Physics Department, Clarkson University, Potsdam, NY 13699-5820, USA
e-mail: schulman@clarkson.edu

S. Albeverio, P. Blanchard (eds.), *Direction of Time*,
DOI 10.1007/978-3-319-02798-2_20,
© Springer International Publishing Switzerland 2014

about 10 years later [7], was to observe that if you wanted to make this case in a logical way, you needed to recognize that the choice of giving *initial* or *final* conditions was itself a choice of the arrow (and that you needed to be careful about this in the sort of arguments Gold was offering). So that to establish Gold's thesis one should argue from time-symmetric boundary conditions. For elaboration and references see my book [8], where I do indeed make such arguments. With this perspective, the thermodynamic arrow of time becomes a *consequence* of the cosmological geometry rather than an independent physical input. This leads to a problem that can be posed purely mathematically. We know that if a low-entropy macroscopic state is given as an initial condition, the entropy will increase. (The definition of "entropy" is discussed below.) That is the Boltzmann H-theorem. By symmetry, if low-entropy data are given as a final condition, then entropy will drop as one approaches T— a bit unintuitive, but it is what you explain to students after they've been exposed to that same theorem and some of its puzzles. But now one can consider a more complicated situation: a compound system, for a part of which initial data are given, and for the other part final data are given. If these portions did not interact, the result is obvious; each acquires its own arrow as if the other were not there. The surprise (perhaps) comes when they *are* permitted to interact. What I showed [9] is that if the interaction is not too strong, the separate portions retain their arrows.

Several questions immediately arise: can sentient beings having opposite arrows communicate? This is perhaps the most entertaining issue [10]. At first I thought they could; now I am not so sure [11]. More important I would say, is whether, if this actually happened, would you notice it? In [9] the effect of one system on the other is an increase of noise, a slight (for small interaction) increase in the rate of entropy increase. At this point I believe that if we were to illuminate such a region with our own light sources and take successive photographs, we would see backward-arrow events, people growing younger, that sort of thing.

Before addressing Zeh's specific criticisms, let me say how I imagine opposite-arrow regions could exist in our actual universe. First, I can think of no way for this to happen except if in our distant future the universe is headed for collapse. Assuming then that the overall geometry of the universe is roughly symmetric, I would expect that the thermodynamic arrow of time would also be symmetric. This is a kind of temporal cosmological principle: our direction of time is not special (with a nod towards Occam's razor as well). Again, these matters are discussed at length in [8]. So sentient creatures in this distant future would also see an expanding universe (this suggestion was made almost immediately after Gold put forward his thesis, although he has said that it was not he who made it). In this distant future, with its time-reversed arrow, one can now imagine that some region and its constituent, highly stable, matter becomes isolated from everything else and is able to avoid equilibration for a very long time (in *its* forward time direction). If this stuff were to show up our neighborhood, the end of its isolation could trigger all sorts of processes that would be visible to us as a decrease in its entropy (in *our* forward direction).

I do not expect to run across this stuff any day now. Not only does it require a roughly symmetric cosmology, but the bang-crunch interval cannot be so long that *everything* has come to equilibrium. And you would need a lot of luck to have a substantial chunk of unequilibrated matter come close enough to observe the destruction of its arrow. (On the other hand, if you were that lucky, you would have found evidence for a time-symmetric universe, exactly because it is so difficult to think of any other way this could happen.)

Besides the material I have already cited, a number of other publications on this subject have appeared[1] [16–18]. In the present article I do not focus on the source of the thermodynamic arrow itself, but recent work on this appears in [19].

3 Dialogue on Opposite Arrows

To keep this response from running to book length, I will focus on what seem to me the most significant points, with bias toward those that can be dealt with most directly.

3.1 Solving the Two-Time Boundary Value Problem

Zeh rightly divides the issue of two-time boundary problems into a number of categories. There is the question of existence, and there is the question of finding the solution. Then there is the important distinction between classical and quantum mechanics.

First, existence: usually I phrase my boundary value problems in terms of macroscopic properties at two widely separated times,[2] so I am asking whether there are paths from one relatively large region of phase space to another. For the classical systems of our usual experience, the answer is yes. I am being cautious in characterizing the dynamics as "our usual experience," since I do not want to address the

---

[1]The idea of having opposite arrows has been taken up by other authors as well. Wiener [12] speculated on this subject, and Creswick [13] has looked into the possibility physically producing systems that in a sense evolve backward in time. Finally there have been recent works of science fiction [14, 15] that explored some of the consequences of these ideas.

[2]The time separation used in any particular two-time problem depends on what is being studied. For looking at Gold's proposal I generally think of the earlier time as being (approximately) the era of recombination, when our present cosmic background radiation was emitted. In a time-symmetric cosmology, I take the other time to be a corresponding time interval before the big crunch (or oscillation minimum). These are times for which matter should be distributed (roughly) uniformly, representing an entropy maximum for a system dominated by short-range forces. As the universe expands and gravitational forces dominate, uniformity becomes an extremely unlikely circumstance, so that what was a maximum becomes a minimum. This justifies two-time low-entropy boundary conditions. For the opposite-arrow boundary value problem, I have in mind a smaller time interval (later and "earlier" than recombination) and regions of space smaller than the entire universe. Finally, for the quantum problems associated with finding "special states". my time range is before and after the operation of a particular apparatus.

question of whether the usual dynamics is ergodic or mixing or whatever. We do know that equipartition is satisfied, that the Fourier heat law is in practice satisfied, so most trajectories do wander quite a bit in phase space, even if the dynamics falls short of certain mathematical idealizations.

A more delicate classical problem occurs when the data specification is set up to give opposite arrows in different regions. The paradoxes that arise in this context may be thought of as assertions on the non-existence of solutions. So there is interest in studying these paradoxes whether or not the two regions can communicate. Zeh refers to my "wet carpet" paradox, which is related to the "grandfather" paradox of time-travel fame. In [10] I take up this existence problem in detail. The simplest resolution, i.e., an existence proof, appeals to continuity and is a takeoff on the treatment of Wheeler and Feynman in their discussions of advanced interactions [20]. But [10] also contains some more down-to-earth mathematical arguments for *microscopic*, mixed boundary conditions. For some boundary value problems, indeed, there is no solution, but for those that most closely resemble the paradox scenario, there are solutions.

What about existence for the quantum problem? I agree with Zeh that this requires more than mere appeal to mixing (or similar) properties of the dynamics. I will argue on several levels.

First I address the last item in Zeh's Sect. 1. How many solutions can one hope for? Even if there are regions of classical phase space that satisfy the two-time boundary conditions, it could happen that their measure is so small that there is not a single quantum state, the point being that quantum states require a minimum volume of $(2\pi\hbar)^N$ (with $2N$ the phase space dimension). I do not have a general answer, but can offer an informative example. Appendix A of Sect. 5.0 in [8] presents the following result: even nonequilibrium *initial* conditions imply a tremendous reduction in available phase space; two-time macroscopic boundary conditions are *also* a tremendous restriction, but not more serious than slightly more demanding initial conditions. Specifically, a cubic centimeter of a monatomic ideal gas of atomic weight 30 at room temperature and atmospheric pressure has about $10^{(10^{20.28})}$ microstates, i.e., lives in a Hilbert space of that dimension. Squeezing them into $1/64$ that volume (in coordinate space) changes the "20.28" to about 20.24. Squeezing them by only $1/8$ and insisting that they reoccupy such a region again at a later time *also* brings the "20.28" to 20.24.[3] There is plenty of room in phase space.

I urge the reader who is troubled by two-time conditioning to reflect on this example. Since we never see all the gas in a cubic cm gather into $1/8$ of its volume spontaneously, we get the impression that this might be impossible rather than

---

[3]The numbers given here differ slightly from those in [8]. Here I use the physically more realistic statistics of indistinguishable particles. These numbers also reflect more detailed state counting. Specifically, for an ideal gas the number of states is $\mathcal{N} = \exp(S/k) = \exp(N \log[(V/N)e^{5/2}/\lambda_{th}^3])$, using the standard expression [21] for the entropy of an ideal gas in three dimensions. N is computed from the pressure using $PV = NkT$, k is the Boltzmann constant, and $\lambda_{th} = h/\sqrt{2\pi mkT}$ is the thermal wavelength. The formula for $\mathcal{N}$ can also be used directly to see the effect of volume changes, as discussed in the text.

merely unlikely. Nevertheless, the reduction of Hilbert space dimension, measured by conventional entropy, is not at all drastic (and corresponds to a slight compression), meaning that Hilbert space has many such states. The reason we do not see this happen is that while such entropy reductions may be small with respect to what macroscopic devices can induce, they are still enormous compared to what will be seen by spontaneous fluctuation.

Now let me give more specific argumentation on the existence of quantum solutions. There is a particular quantum two-time boundary value problem whose solution allows quantum mechanics to retain pure unitary (and deterministic) time evolution (no wave function "collapse") while at the same time does not introduce probability through some back door channel, such as appeal to a collapse-inducing macroscopic world, or many worlds, or degradation of the role of the wave function. This approach involves something I call "special states." These states allow the final condition of the combined apparatus-system to be only a single one of the potential outcomes of the measurement. It would be too much of a distraction to give more detail here; see [8]. In any case, the mathematical problem is formulated as follows. You give projection operators P and Q representing the initial and final subspaces of Hilbert space, $\mathcal{H}$, in which you want your total system (apparatus plus measured system) to be. There is some unitary operator U that evolves this total system between the given times. Then what the two-time boundary value problem seeks is states $\psi(0) \in P\mathcal{H}$ that evolve entirely into $\psi(T) \in Q\mathcal{H}$ (where $[0, T]$ is the time interval for the measurement). This leads you to look at the spectrum of the operator $\widetilde{U}^\dagger \widetilde{U}$, where $\widetilde{U} = QUP$. The operator $\widetilde{U}^\dagger \widetilde{U}$ is Hermitian and has spectrum in the interval $[0, 1]$. Eigenvectors with eigenvalue 1 represent solutions to the boundary value problem. In [8] I report studies of this mathematical problem for several models of apparatus. Generally speaking there are many eigenvalues clustering around 1. (Interestingly, sometimes there can be none, and this gives rise to potential experimental tests of this theory. See [22].) As the size of the apparatus grows there are more and more near-unity eigenvalues as well.[4] Two remarks: (1) For these special states there is no entanglement at the end of a measurement, particularly useful if one wants to think time symmetrically about measurement. (2) A propos Zeh's remarks about trial and error in the finding of solutions, when solutions of this problem were produced, the process consisted of finding the spectrum (including eigenvectors) of the particular operator mentioned above by standard operator techniques (usually numerical).

I also mention that I am not the only one preoccupied with two-time boundary value problems, both classical and quantum. If one wants to split a molecule using a laser pulse, it turns out that simply hitting it with one of its resonance frequencies does not work very well. Instead [23–25] you must shape your pulse and the finding of an appropriate shape involves solving a future-conditioned problem (which can

---

[4]The terms "apparatus" and "system" do not imply that this scheme holds only for laboratory experiments. Any situation that could lead to superpositions of macroscopically different states will have this feature. Again, the present paper is not about quantum measurement theory. and for the many questions that may come to mind please consult [8].

also be considered a form of control theory). Similarly, it is of great interest to people studying tunneling [26, 27] to know the time dependence of a system when it finally does succeed in transiting a barrier. Moreover, the same issues arise in quantum computing, where one wishes to implement a quite specific transformation on a collection of states, at the same time being sure that unwanted entanglement does not arise from amplitudes for exciting other levels [28, 29].

I next turn to the question of how you actually solve two-time boundary value problems. Much of what I have to say can be found in [8], where a chapter is dedicated to this problem. Zeh declares [2] that I "mostly" do it by trial and error. Sometimes I do, sometimes I do not. In one of the articles his critique focuses on, *Causality is an Effect* [18], my calculations are analytic, except for numerical *illustrations* later in the paper. There is nothing wrong with numerical trial and error. Furthermore, for stochastic dynamics explicit analytic results can be attained, and some of my work involves such calculations [30–32].

Actually there is a deeper critique in what Zeh is saying, more than merely complaining about how I go about finding solutions. The important question is whether those relatively simple systems for which you can find solutions are reliable indicators of what happens in more realistic cases. In this regard I point out that [18] deals with recovering, analytically, my results about the flipping of arrows along the way from one low-entropy condition to another. I did not explicitly do the calculation for the simultaneous opposite-arrows case, but it should be an exercise using the same techniques already used for the other result. I mention that in those demonstrations I make fairly strong assumptions on the way the systems go to equilibrium.

However, to judge this last issue, whether the numerical and analytic results may be expected to hold in more general systems, it helps to step back and ask why they hold for the cases that *have* been studied. Consider the case of boundary conditions at $t = 0$ and $t = T$ (>0) with low-entropy macrostates at both ends. I have shown that moving inward—from *both* ends—entropy increases. In the middle, if T is big enough, you have equilibrium. Most significantly, the passage to equilibrium, the entropy increase you get say in going forward from time 0, is macroscopically identical to what you would get moving forward from time 0 with *no* future conditioning. Why is that so? It is because, in a sense, the system *forgets the future*. Suppose there is a relaxation time τ associated with the dynamics. Then the condition for the situation I have described is $T > 2\tau$. Here is why: saying it relaxes in time τ means that starting from low entropy at time 0, the system is likely to reach anywhere in (allowable) phase space by time τ. But then it can also get "back" from wherever it is at time $T/2$ (which by assumption is greater than τ) to the region demanded by the time T final condition. Stated differently and thinking in terms of the backward arrow from T to $T/2$, by time $T/2$ if forgets where it "was" at time T.

This strongly suggests that whatever dynamics one has, if the concepts of relaxation time have relevance, my results on particular models should continue to be valid. Systems for which one cannot assign a relaxation time (or if the time is longer than the T associated with the particular problem) are not expected to give the same results, and indeed they hold independent interest for information such processes might provide about cosmology (again, see [8]).

3.2 Isolation

In Zeh's Sect. 2 he raises the issue of isolating a large system, something that, microscopically, is practically impossible. This observation is useful but not a serious concern. It is useful because it points out an essential feature of any two-time boundary value problem, namely that it is meaningless unless all forces on the system throughout the intermediate interval are included. As such, in their grandest form these boundary value problems should include the entire universe. But there is a second perspective, one that allows a narrower view. Suppose one pretends that one *could* isolate a portion of the universe and then reaches certain conclusions about the solutions of two-time boundary value problems in that context (for example, one might consider opposite-arrow boundary value problems [9] in this way). Then one could consider the same boundary value problem slightly modified, say by the entry of a single photon into the region. Now solve the boundary value problem with the additional force. It will change the microscopic paths, but does it change the qualitative conclusions? Generally I expect not, so that for many purposes perfect isolation is not important. Nevertheless, if problems arise in this formulation (e.g., reaction on the external system), one can go back to talking of the entire universe. These remarks apply for both classical mechanics, as in [8], and quantum mechanics, as in [8] and [33].

3.3 Closed Timelike Curves

As to closed timelike curves, they have nothing to do with my opposite-arrow scenarios. To the extent that I assume any geometrical context, I am happy with Friedmann–Robertson–Walker. So Zeh's criticism of science fiction stories has no relevance.

As an aside, I am not convinced that closed timelike curves that extend over long time spans cannot have both increase and decrease in entropy, hence dissipation and its reverse. Physically there would need to be a reason to single out a low-entropy era, and moving away from that (in both directions) entropy would increase. You also should not have shortcuts, timelike paths of varying lengths for nearby spatial regions. I do not know to what extent these conditions could or could not be met in a Gödel universe.[5] In any case the issue has little to do with my story.

---

[5]Postscript: I've long been suspicious of the alleged paradoxes that would arise in a Gödel universe by virtue of its closed timelike curves. I expect that there could be a reduction in the class of "initial value" problems that have a solution, as for other paradoxes mentioned in this article. ("Initial values" would also be final values and would presumably be on a single spacelike surface. They would involve test particles, not the matter giving rise to the metric itself.) Also, the usual paradoxes are macroscopic, implying the existence, at least locally, of an arrow of time. It's not clear that such could exist. In the summer of 2010 I met another person with similar ideas about this problem, Noam Erez of the Weizmann Institute and my comments here are partly informed by our conversations.

3.4 Entropy Calculation

In his Sect. 3 Zeh declares that the entropy I use is an ensemble entropy and does not
assign an entropy to a microstate. This is not true. As stated in [8], page 32, entropy
is the logarithm of the number of microscopic states consistent with a macroscopic
description. Once you have a coarse graining (i.e., a macroscopic description), you
can take any microstate and use the volume of the coarse grain to which it belongs
to compute its entropy. In cat map studies I implement this as follows: the system
microstate is a point in $\mathbb{I}^{2N}$, with $\mathbb{I}^2$ the unit square and N the number of "atoms"
in the gas. To define coarse grains, $\mathbb{I}^2$ is divided into G regions (usually rectangles)
and the number of points of the projected system point in each region is the coarse
grained description. Thus if a given microstate is $(x_1, y_1, \ldots, x_N, y_N)$, its coarse
grained description is $(n_1, \ldots, n_G)$, where n_k is the number of atoms (x_ℓ, y_ℓ) in
grain k. Following the definition in [18], the entropy is $S = -\sum p_k \log p_k - \log G$,
with $p_k = n_k/N$, if all grains are of equal coordinate space volume. The "$p_k \log p_k$"
as usual arises from the logarithm of $N!/(n_1! \ldots n_G!)$ and represents the missing
information associated with particle identity. The missing information associated
with going from real numbers to finite volumes is the same for all $(n_1, \ldots, n_G)$, and
is dropped, since in this study I am not concerned with comparing coarse grainings.

As remarked in [8] and commented upon by Zeh, the universe is richer than cat
map dynamics. In particular there are fast processes and slow ones. Rather than
a disability, I view this feature as a wonderful opportunity. It is precisely because
the slowest processes may have two-time boundary condition solutions that differ
significantly from their unconstrained counterparts that one might discover indications of a forthcoming big crunch. Specifically you would expect to see impeded or
slowed-down relaxation. This idea is not mine, but was advanced by John Wheeler.
I have elaborated on it in [8], in particular looking for suitable slow processes and
indicators of constrained relaxation.

4 Causality

In his Sect. 4, Zeh discusses causality. Here he addresses the content of my article,
Causality is an Effect, [18], whose main conclusion I will briefly review. This article
is available on the arXiv as cond-mat/0011507. In most of my work I have shown
reversals of the arrow of time by exhibiting the time dependence of the entropy.
Occasionally people ask about other macroscopic quantities, whether they would
show similar behavior. So I decided to deal with the most fundamental such issue,
the appearance of macroscopic causality.

The first problem is defining what you mean by causality. Zeh takes me to task
over this and I entirely agree that defining causality in a fully deterministic world
with fixed boundary conditions at both ends is not easy. Usually what you have in
mind for (macroscopic) causality is that the nail enters the wood *because of* and
subsequent to its being hit by the hammer. It would not go into the wood if the

hammer did not strike. I will not even try to be more precise. The point though is that you'd like to perturb the system and see when the changes take place. But how can one perturb a universe whose dynamics is given and whose past and future are fixed?

My solution was to consider in effect two universes, both with the same macroscopic boundary conditions (at $t = 0$ and $t = T$), but which differ in their dynamics at one particular intermediate time t_0 ($0 < t_0 < T$). So the *microscopic* dynamics will in general differ at *all* times. What I looked at was the *macroscopic* behavior. And indeed, I found that if t_0 was close to 0, all macroscopic changes were confined to $t > t_0$, while if t_0 is close to T, macroscopic changes were confined to $t < t_0$, showing that "causality" follows the same arrow as entropy increase.

As to the weakness of my definition I believe that anyone wanting to define a causality concept close to this will need some such strategy just to keep causality from already being fixed by the use of initial conditions, which as I have often noted is equivalent to fixing an arrow of time. I also remark that my definition was to some extent motivated by discussions of dispersion relations, where *perturbation* is the essential notion, but where there is also implicit a notion of an *un*perturbed system serving as a reference point.

In any case, Zeh would prefer a definition in which it is possible to assign a notion of causality to an individual microstate, which, as far as I can tell, cannot be done with the foregoing definition. Indeed my personal preference is to be able to say things about individual systems, so it would be of interest to first, find such a definition, and second establish that, as for my other definition, this kind of causality is also a *consequence* of other arrow-inducing features in a two-time boundary value problem context.

This article is not the place to carry out the aforementioned program, but I would like to make a suggestion. Again we consider the effect of a "perturbation" on a macroscopic system, but now we have only one system, so the perturbation is only some force that we single out, perhaps because it is large and macroscopically recognizable. Focus on a single microscopic state that satisfies macroscopic two-time boundary conditions (as usual), including the perturbation. The test of causality is the following: if we look at only the macroscopic state of the system on one side of the perturbation and try to calculate what happens on the other side, do we get the right answer?

It is clear that the direction of this kind of causality follows the direction of entropy increase (which I relate to proximity to one or another temporal endpoint). If the system is *not* dissipative,[6] then there is no arrow and predictions will be good from both sides. But if it is dissipative, information is lost in one direction and it is only possible to make reliable predictions in the direction of information loss. It is easy to see how this translates into cat map examples, so I omit detailed illustration. This "causality" is closely tied to arrow-of-time definitions based on the choice of what "initial" means. This is consistent with the main thesis of [18], namely that the notions of macroscopic causality and the arrow of time are essentially the same, and that in particular if one is induced by proximity to low-entropy boundary conditions, so is the other.

---

[6]Bear in mind that this has meaning only with respect to a particular coarse graining.

My second comment on Zeh's Sect. 4 has to do with the oscillator example. The antiquity of this model for studying two-time boundary value problems goes beyond his book [34]; in fact, based on my 1973 work [7] where I used this system for studying two-time boundary value problems, I did not in fact expect to find causality. However, the actual calculation brought a surprise—there *is* causality— and the oscillator example shows interesting subtleties. In Zeh's text he comments that he has used a much larger sample than I did, a remark that puzzled me since my calculation is analytic and only later in my article do I choose a sample for a numerical illustration. If one examines my analytic work it will be seen that there are *two* time scales in this problem. One has to do with the range of frequencies used to smear the oscillators. The other is related to the size of the coarse grains and demands a much larger time interval to see causal effects, as I have defined them. Admittedly my numerical examples are difficult to read, but I would hope the analytic portion would be clear enough. In any case, if you use a long enough total conditioning time, you get causality; if not, you do not.

On the issues raised in Zeh's Sect. 5 on Cosmology and Gravitation I will not comment except to say that everything here rests on much less reliable ground. For example, while some view black holes as the essence of the arrow of time [35, 36], others contemplate their disappearance prior to a big crunch [37]. Moreover, even without total evaporation it is now believed that information (on items that fell in) is returned to the universe through properties of the Hawking radiation. Articles referenced in this section of Zeh's article (but not written by Zeh) concerning absorbing powers of the universe have seemed to me plagued by problems of double counting and incorrect treatment of the boundary value problem that is natural to time-symmetric electrodynamics.

As to Zeh's Sect. 6 on quantum aspects, my views on this are expressed in [8] and, briefly put, are that quantum mechanics, including the measurement process, is fully time symmetric and does not introduce an arrow of time.

Acknowledgements This work was supported by the United States National Science Foundation Grant PHY 00 99471.

References

1. Albeverio, S., Blanchard, P., Drieschner, M., Paycha, S.: The direction of time: the role of reversibility/irreversibility in the study of nature, Bielefeld (2002)
2. Zeh, H.D.: Remarks on the compatibility of opposite arrows of time (2003). physics/0306083
3. Shakespeare, W.: Hamlet, act III, scene 1
4. Zeh, H.D.: The Physical Basis of the Direction of Time, 1st edn. Springer, Berlin (1989)
5. Schulman, L.S.: The problem of time. Book review of "The physical basis of the direction of time" by H. Dieter Zeh. Science **249**, 192 (1990)
6. Gold, T.: The arrow of time. Am. J. Phys. **30**, 403–410 (1962)
7. Schulman, L.S.: Correlating arrows of time. Phys. Rev. D **7**, 2868–2874 (1973)
8. Schulman, L.S.: Time's Arrows and Quantum Measurement. Cambridge University Press, New York (1997)
9. Schulman, L.S.: Opposite thermodynamic arrows of time. Phys. Rev. Lett. **83**, 5419–5422 (1999)

10. Schulman, L.S.: Resolution of causal paradoxes arising from opposing thermodynamic arrows of time. Phys. Lett. A **280**, 239–245 (2001)
11. Schulman, L.S.: Opposite thermodynamic arrows of time. In: Sheehan, D.P. (ed.) Quantum Limits to the Second Law. Am. Inst. of Phys., New York (2002)
12. Wiener, N.: Cybernetics or Control and Communication in the Animal and the Machine, 2nd edn. MIT Press, Cambridge (1961)
13. Creswick, R.J.: Time-reversal and charge echo in an electron gas. Phys. Rev. Lett. **93**, 100601 (2004)
14. Wharton, K.B.: Aloha, Analog Science Fiction and Fact (2003)
15. Wharton, K.: Divine Intervention. Ace Books, New York (2001)
16. Schulman, L.S.: A compromised arrows of time. In: Ancona, V., Vaillant, J. (eds.) Hyperbolic Differential Operators and Related Problems, pp. 355–370. Dekker, New York (2003)
17. Schulman, L.S.: Schulman replies. Phys. Rev. Lett. **85**, 897 (2000)
18. Schulman, L.S.: Causality is an effect. In: Mugnai, D., Ranfagni, A., Schulman, L.S. (eds.) Time's Arrows, Quantum Measurements and Superluminal Behavior, pp. 99–112. Consiglio Nazionale delle Ricerche (CNR), Rome (2001). cond-mat/0011507
19. Schulman, L.S.: We know why coffee cools. Physica E **42**, 269–272 (2009) Postscript
20. Wheeler, J.A., Feynman, R.P.: Classical electrodynamics in terms of direct interparticle action. Rev. Mod. Phys. **21**, 425–433 (1949)
21. Baierlein, R.: Thermal Physics. Cambridge University Press, Cambridge (1999)
22. Schulman, L.S.: Jump time and passage time: the duration of a quantum transition. In: Muga, J.G., Sala Mayato, R., Egusquiza, I.L. (eds.) Time in Quantum Mechanics, pp. 99–120. Springer, Berlin (2002)
23. Vugmeister, B.E., Botina, J., Rabitz, H.: Nonstationary optimal paths and tails of prehistory probability density in multistable stochastic systems. Phys. Rev. E **55**, 5338–5342 (1997)
24. Vugmeister, B.E., Rabitz, H.: Cooperating with nonequilibrium fluctuations through their optimal control. Phys. Rev. E **55**, 2522–2524 (1997)
25. Judson, R.S., Rabitz, H.: Teaching lasers to control molecules. Phys. Rev. Lett. **68**, 1500–1503 (1992)
26. Dykman, M.I., Mori, E., Ross, J., Hunt, P.M.: Large fluctuations and optimal paths in chemical kinetics. J. Chem. Phys. **100**, 5735–5750 (1994)
27. Dykman, M.I., Luchinsky, D.G., McClintock, P.V.E., Smelyanskiy, V.N.: Corrals and critical behavior of the distribution of fluctuational paths. Phys. Rev. Lett. **77**, 5229–5232 (1996)
28. Palao, J.P., Kosloff, R.: Quantum computing by an optimal control algorithm for unitary transformations. Phys. Rev. Lett. **89**, 188301 (2002)
29. Palao, J.P., Kosloff, R.: Optimal control theory for unitary transformations. Phys. Rev. A **68**, 062308 (2003)
30. Schulman, L.S.: Illustration of reversed causality with remarks on experiment. J. Stat. Phys. **16**, 217–231 (1977)
31. Schulman, L.S.: Normal and reversed causality in a model system. Phys. Lett. A **57**, 305–306 (1976)
32. Schulman, L.S.: Models for intermediate time dynamics with two-time boundary conditions. Physica A **177**, 373–380 (1991)
33. Gell-Mann, M., Hartle, J.B.: Time symmetry and asymmetry in quantum mechanics and quantum cosmology. In: Halliwell, J.J., Pérez-Mercader, J., Zurek, W.H. (eds.) Physical Origins of Time Asymmetry, pp. 311–345. Cambridge University Press, New York (1994)
34. Zeh, H.D.: The Physical Basis of the Direction of Time, 4th edn. Springer, Berlin (2001)
35. Penrose, R.: Singularities and time-asymmetry. In: Hawking, S.W., Israel, W. (eds.) General Relativity: An Einstein Centenary Survey. Cambridge University Press, New York (1979)
36. Penrose, R.: Big bangs, black holes and 'time's arrow'. In: Flood, R., Lockwood, M. (eds.) The Nature of Time. Basil Blackwell, Oxford (1986)
37. Steinhardt, P.J., Turok, N.: Cosmic evolution in a cyclic universe. Phys. Rev. D **65**, 126003 (2002)

Chapter 21
Remarks on the Compatibility of Opposite Arrows of Time

H.D. Zeh

Abstract I argue that opposite arrows of time, while being logically possible, cannot realistically be assumed to exist during one and the same epoch of our universe.

Keywords Time arrow · Final conditions · Solvable models · Causality · Retardation · Cosmology · Big crunch · Black holes · Quantum measurements · Wheeler–DeWitt equation

1 Introduction

If, according to the assumptions of statistical physics, the second law is regarded as a "fact" rather than a dynamical law, it could *conceivably* not hold at all, hold only occasionally, or even apply in varying directions of time. Larry Schulman has demonstrated very nicely and convincingly in several publications [1–3] how the latter possibility may occur in principle. The major remaining question then is whether his examples can be regarded as realistic in our universe. This comment was written in response to a review presented by Schulman during the conference on the "Direction of Time", Bielefeld (2002), and originally posted as http://arXiv.org/physics/0306083.

In particular, we may understand from Schulman's examples how a certain time arrow depends on "improbable" (low-entropy) initial or final conditions—regardless of the direction in which we perform our calculation. The latter (apparently trivial) remark may be in place, since many derivations of the second law tacitly assume in a crucial way that the calculation is used to *pre*dict. That is, it is assumed to follow a "physical" direction of time (from an initially given present towards an unknown

This comment was written in reply to Lawrence Schulman's oral conference contribution at Bielefeld, which dealt with models representing opposite arrows of time. Together with the two subsequent Letters, it was first published in the online journal *Entropy* **7**, 199 (2005); **7**, 208 (2005); and **8**, 44 (2006).

H.D. Zeh (✉)
69151 Waldhilsbach, Germany
e-mail: zeh@urz.uni-heidelberg.de
url: http://www.zeh-hd.de

S. Albeverio, P. Blanchard (eds.), *Direction of Time*,
DOI 10.1007/978-3-319-02798-2_21,
© Springer International Publishing Switzerland 2014

future). However, precisely this physical arrow, or the fact that only the past can be remembered and appears "fixed", is a major *explanandum*.

While, for a given dynamical theory, we know in general precisely what freedom of choice remains for initial *or* final conditions, *mixed* ones (such as two-time boundary conditions) are subject to dynamical consistency requirements—similar to an eigenvalue problem with given eigenvalue. This problem remains relevant even for incomplete (for example, macroscopic) initial and final conditions. It is usually difficult to construct an individual solution that is in accord with both of them. In Schulman's examples, individual solutions were mostly found by "trial and error", that is, by exploiting a sufficient number of solutions with given initial conditions and selecting those which happen to fulfill the final ones (or *vice versa*). However, in a realistic situation it would be absolutely hopeless in practice ever to end up with the required low entropy because of the exponential growth of probability with entropy. Only an exponentially small fraction of all solutions satisfies one or the other low-entropy boundary conditions. In the case of complete mixing, it is the square of this very small number that measures the fraction of solutions with two-time boundary conditions.

Being able to find solutions by trial and error thus demonstrates already the unrealistic case. This difficulty in *finding* solutions does not present any problem for their *existence* on a classical continuum of states if mixing is sufficiently complete: any set of solutions with finite measure can be further partitioned at will, since entropy has no lower bound in this classical situation. This conclusion is changed in quantum theory, which would in a classical picture require the existence of elementary phase space cells of Planck size h^{3N}. The product of initial and final probabilities characterizing the required low entropy may then represent a phase space volume smaller than a Planck cell—thus indicating the absence of any solution.

2 Retarded and Advanced Fields

The consistency problem in a classical setting (though without mixing) is discussed in Schulman's "wet carpet" example, intended to prove the compatibility of two interacting systems with different, retarded or advanced, electrodynamics [2]. It is similar to an example studied by Wheeler and Feynman [4], where a charged particle, bound to pass an open trap door, is assumed to shut it *before* passing the door by means of advanced fields which it can create only *after* having passed the door. In both examples, there is but a very narrow band of consistent solutions, in Schulman's example represented by a partly opened window. These narrow bands were found for systems which are far from thermodynamical mixing (cf. the following sections), and they may be consistent only if the model is considered in isolation. In reality, macroscopic objects always interact with their surroundings. In a causal world, this would produce "consistent documents" (not only usable ones) in the thermodynamical future. In this way, information may classically spread without limit, thus leading to inconsistencies with an opposite arrow of other systems. In quantum description, classical concepts even *require* the presence of irreversible decoherence, while microscopic systems would remain in quantum superpositions of all conceivable paths (see Sect. 6).

Philosophers are using the term "over-determination of the past" to characterize this aspect of causality [5, 6]. (Note that the conventional *additive* physical entropy neglects such nonlocal correlations, which would describe the consistency of documents, for being dynamically irrelevant in the future [7].) In a deterministic world, one would thus have to change *all* future effects in a consistent way in order to change the past. For example, classical light would even preserve its usable information content forever in a transparent universe. In classical electrodynamics, there is but *one real* Maxwell field, while the retarded and advanced fields of certain sources are merely auxiliary theoretical concepts. The same real field can be viewed as a sum of incoming and retarded, or of outgoing and advanced fields, for example (see Chap. 2 of [7]). Retarded and advanced fields (of different sources) thus *do not add*. Observing retarded fields (as our sensorium and other registration devices evidently do) means that incoming fields related to unspecified past sources ("noise") are negligible—incompatible with the presence of distinctive advanced radiation.

Problems similar to those with opposite arrows occur with closed time-like curves (CTCs), which are known to exist *mathematically* in certain solutions of Einstein's field equations of general relativity. This existence means that *local* initial and final conditions for the *geometry*, defined with respect to these closed time-like curves, are identical and thus dynamically consistent. However, CTCs are incompatible with an arrow of time for matter, such as an electrodynamic or thermodynamic one. Those clever science fiction stories about time travel, which are constructed to circumvent paradoxes, and thus seem to allow CTCs for human adventurers, simply neglect all irreversible effects which must arise and would destroy dynamical consistency. Since geometry and matter are dynamically coupled, boundary conditions which lead to an arrow of time must also protect chronology (whatever the precise dynamical model). Wet carpet stories belong to the same category as science fiction stories: they do not resolve the *unmentioned* paradoxes that would necessarily arise from opposite arrows.

3 Cat Maps

Borel demonstrated long ago [8] that microscopic states of classically described gases are dynamically strongly coupled even over astronomical distances. This is a time-symmetric consequence of their extremely efficient chaotic behavior, caused by deterministic molecular collisions. Of course, this does not mean that *macroscopic* properties are similarly sensitive to small perturbations, although fluctuations (such as Brownian motion) and their consequences must be affected.

Macroscopic properties characterize the microscopic state of a physical system incompletely, for example by representing a coarse graining in phase space (or, more generally, a *Zwanzig projection*—Chap. 3 of [7]). The deterministic dynamics of initially given coarse grains is often described by measure-preserving dynamical maps.

In contrast to deformations of extended individual objects in space (such as Gibbs' ink drop), and even in contrast to the N discrete points in single-particle phase space which represent a molecular gas, Kac's symbolic "cats" (areas in phase space) [9] represent ensembles, or sets, of *possible* physical states of a given system. Therefore, Schulman's entropy [2] as a function of deformed cats ("cat maps") is an ensemble (or average) entropy—not the entropy of an individual physical state. The entropy of this ensemble is defined to depend on its distribution in phase space, obtained after coarse graining with respect to given and fixed grains as a macroscopic reference system, while the entropy of an individual state (point in phase space) would be given solely by the size of the specific grain that happens to contain it at a certain time.

This is essential (and sufficient) for Schulman's argument that the intersection of two sets representing specific initial or final conditions is not empty *if* mixing is complete. In our universe, however, some variables participate in very strong mixing, while others ("robust" ones, such as electromagnetic waves or atomic nuclei) may remain stable for very long times. They are the ones that may store usable information.

Since cat maps describe sets of states for rather simple dynamical systems, their dynamics is far less sensitive to weak interactions than that of individual Borel type systems. For this reason, two systems described by cat maps with opposite arrows of time may even be consistent for mild interactions [2]. However, these cat maps do *not* form a realistic model appropriate to discuss thermodynamical arrows in our universe.

4 (Anti-)Causality

In order to define causality without presuming a direction of time, one has to refer to the internal structure of the evolving dynamical states. The above-mentioned over-determination of the past (in other words, the existence and consistency of multiple documents) is a typical example. Another one is given by the concentric waves emitted from a local source. In our world, both are *empirically* (not logically) related to a time direction.

While one may expect that all such internal structures can be shown to evolve in time from appropriate initial conditions, they are too complex to be investigated in terms of Schulman's simple models. For example, retarded (concentrically outgoing) waves exist in the presence of sources precisely when incoming fields are negligible. This can be the case "because" of an initial condition for the fields, or because of the presence of thermodynamic absorbers [7].

Instead of these specific structures of physical states, Schulman studied the "effect" (in both directions of time) of "perturbations" defined by small instantaneous changes of the Hamiltonian [3]. This "effect" is not easily defined in a time-symmetric way, since an "unperturbed solution" defined on one temporal side of the perturbation would be exclusively changed on the other one (no matter which is the future or past). If the unperturbed solution obeyed a two-time boundary condition, the perturbed one would in general violate it on this "other" temporal side. In contrast to the above-mentioned internal structures, our conventional concept of perturbations is based on the time direction used in the definition of external operations.

Therefore, in a first step, Schulman considered *sets* of solutions again. The set of *all* solutions obeying the "left" boundary condition (in time) remains unchanged on the left of the perturbation, while the opposite statement is true on the right. However, individual solutions found in the intersection of these two sets (consisting of those ones which fulfill both boundary conditions) in the case of a perturbation are generically different from the unperturbed ones on *both* sides of the perturbation. Now, if mixing is essentially complete on the right of the perturbation (that is, for a sufficiently distant right boundary), the right boundary condition does not affect the solutions which form the intersection (considered as a set) on the left. This means that *mean values of macroscopic variables* in the set of all solutions that are compatible with both boundary conditions may only differ on the right (a consequence regarded as *causality* by Schulman) [3]. This is true, in particular, for the *mean* entropy (if the latter is defined as a function of macroscopic, that is, coarse-grained, variables). Individual solutions can *not* be compared in this way, since there is no individual relation between them. In the case of complete mixing on the right, there is even a small but non-empty subset of solutions of the original two-time boundary value problem which keep obeying the right boundary condition without being changed on the left. However, using them for the argument would mean that only very specific solutions *can* be perturbed in this specific sense.

In a second approach, Schulman studies the "effect" of macroscopic perturbations on *individual* solutions of an integrable system. This system is defined as consisting of a finite number of independent oscillators with different frequencies. Although solutions which fulfill both boundary conditions can be found with and without an appropriate perturbation, they are again not individually related. Therefore, the causal interpretation of the perturbation remains obscure. (For *closed* deterministic systems, any perturbation would itself have to be determined from microscopic boundary conditions, and the consistency problem becomes even more restrictive than for just two boundary conditions.)

Nonetheless, I was pleased to discover that Schulman's model is formally identical with a model of particles freely moving on a periodic interval (a "ring") that I had used in an appendix of [7] for much larger numbers of constituents than used by him (such that finding two-time boundary solutions by trial and error would be hopeless). Particle positions on the ring have merely to be re-interpreted as oscillator amplitudes in order to arrive at Schulman's picture. I used this opportunity to search by trial and error among *analytically* constructed two-time boundary solutions for those ones which happen to possess *slightly* lower entropy than the mean at some given "perturbation time" t_0 (see Fig. 1). (Finding *much* lower entropy values numerically would be too time-consuming for this large number of particles.) Unfortunately, the results do *not* confirm Schulman's claim that these solutions are "affected" by the perturbation only in the direction away from the relevant low-entropy boundary (that is, towards the "physical future") [3]. Evidently, this concept

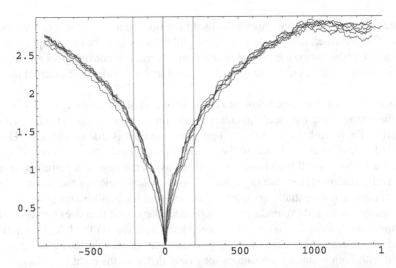

Fig. 1 Four random two-time boundary solutions (forming a narrow bundle in the diagram) are compared with two other ones, selected by trial and error for their slightly lower entropy values at $t_0 = 200$ or $t_0 = -200$. Values for $t < 0$ are identical with those at $t_f - t = 200.000 - t$, although the final condition is actually irrelevant in the range shown. Entropy scattering around $t = 1300$ is accidental (See the appendix of [7] for details of the model and an elementary *Mathematica* program for your convenience)

of causality, defined by means of perturbations, is insufficient. The very concept of a "perturbation" seems to be ill-defined for two-time boundary conditions.

As another example, I calculated the effect on the solution in both directions of time that results from a *microscopic* perturbation *of the state* (in this case simply defined by an interchange of velocities between particles at some time t_0). Both boundary conditions are then violated by the new solution arising from this perturbed state, used as a complete "initial" condition. The results (shown in Fig. 2) are now most dramatic towards the former "past", demonstrating the relevance of fine-grained information (similar to Borel's example) for correctly calculating "backwards in time". Deviations from the original two-time boundary solution close to t_0 also on the right are due to the fact that the coarse graining assumed in this model does not define a very good master equation (as discussed in [7]).

5 Cosmology and Gravitation

It appears evident from the above discussion that opposite arrows of time would in general require almost complete thermalization between initial and final conditions, which is hard to accomplish even in a cosmological setting. In its present stage, this universe is very far from equilibrium. A reversal of the thermodynamical arrow of time together with that of cosmic expansion, as suggested by Gold [10], would therefore require a total life time of the universe vastly larger than its present

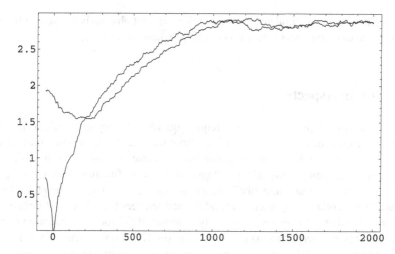

Fig. 2 Time-symmetric "effect" on a solution "caused" by a perturbation of the *microscopic* state at time $t_0 = 200$, defined by an accidental entropy minimum at this time. The perturbed solution drastically violates *both* boundary conditions that were valid for the unperturbed solution

age. Weakly interacting fields may never thermodynamically "mix" with the rest of matter [11].

In particular, Davies and Twamley demonstrated [12] that an expanding and re-contracting universe would remain essentially transparent to electromagnetic waves between the two radiation eras. This means that advanced radiation resulting from all stars which will exist during the recontraction of our universe would be present now, apparently unrelated to any individual future sources because of their distance, but red- *or* blue-shifted, depending on the size of the universe at the time of (time-reversed) emission. According to Craig [13], this radiation would show up as a non-Planckian high-frequency tail of the cosmic background radiation resulting from the past radiation era (where it would be absorbed in time-reverse description). This leads to the consistency problems described in Sect. 2.

While neutrinos from the future would presumably remain unobserved, gravity, despite its weakness, dominates the entropy capacity of this world, and leads to consequences which are the most difficult ones to reverse. Black holes are expected to harbor event horizons which would not be able ever to disappear in classical relativity, while in quantum field theory they are predicted to disappear into Hawking radiation in the distant future in an irreversible manner. However, contraction of gravitating objects, including the formation of black holes, requires that higher multipoles are *radiated away*. This radiation arrow is the basis of the "no hair theorem", which would characterize the asymptotic final states of black holes in an asymptotically flat and time-directed universe. Because of the diverging time dilation close to a horizon, any coherent advanced radiation (with the future black hole as its retarded cause) would be able to arrive in time to prevent the formation of an horizon [14]. This solution of the "information loss paradox" may save a deterministic universe (without leading to inhomogeneous singularities).

While the required initial and final conditions are not obviously consistent in this classical scenario, this problem is relaxed in *quantum* cosmology.

6 Quantum Aspects

Realistic models of physical systems require quantum theory to be taken into account. Since quantum entropy is calculated from the density matrix (that may result from a wave function by means of generalized "coarse graining"), its time dependence has in principle to include a collapse of the wave function during measurements or other "measurement-like" situations (such as fluctuations or phase transitions). If the collapse represents a fundamental irreversible process, it defines an arrow of time that is *never* reversed. Only a universal Schrödinger equation (leading to an Everett interpretation) could be time (or CPT) symmetric. A reversal of the time arrow would then require decoherence to be replaced by recoherence: advanced Everett branches must combine with our world branch in order to produce local coherence. Although being far more complex than a classical model (since relying on those infamous "many worlds") this would still allow us to conceive of a two-time boundary condition for a *global* wave function (see Sect. 4.6 of [7]). Note that Boltzmann's statistical correlations (defined only for ensembles) now become quantum correlations (or entanglement, defined for *individual* quantum states). For example, re-expanding black holes, mentioned in the previous section, would in an essential way require (and possibly be facilitated by) recoherence.

In *quantum gravity* (or any other "reparametrization-invariant" theory), the Schrödinger equation is reduced to the Wheeler–DeWitt equation, $H\Psi = 0$, which does not explicitly depend on time at all. However, because of its hyperbolic form, this equation defines an "intrinsic initial value problem" with respect to the expansion parameter a. In a classical (time-dependent) picture, the initial and final states would have to be identified in order to define *one* boundary condition, while the *formal* final condition (with respect to a) for recontracting universes is reduced to the usual normalizability of the wave function for $a \to \infty$. Big bang and big crunch (distinguished by means of a WKB time, for example) could not even conceivably be different as (complete) quantum states (Chap. 6 of [7]), while forever expanding universes might be said to define an arrow of time that never changes direction during a WKB history. All arrows thus seem to be strongly correlated.

References

1. Schulman, L.S.: Time's Arrows and Quantum Mechanics. Cambridge University Press, Cambridge (1997)
2. Schulman, L.S.: Phys. Rev. Lett. **83**, 5419 (1999)
3. Schulman, L.S.: In: Mugnai, D., Ranfagni, A., Schulman, L.S. (eds.) Time's Arrow, Quantum Measurements and Superluminal Behavior. Consiglio Nazionale delle Ricerche, Roma (2001). cond-mat/0102071

4. Wheeler, J.A., Feynman, R.P.: Rev. Mod. Phys. **21**, 425 (1949)
5. Lewis, D.: Philosophical Papers, vol. II. Oxford University Press, Oxford (1986)
6. Price, H.: Time's Arrow & Archimedes' Point: A View from Nowhen. Oxford University Press, London (1996)
7. Zeh, H.D.: The Physical Basis of the Direction of Time. Springer, Berlin (2001)
8. Borel, E.: Le hasard. Alcan, Paris (1924)
9. Kac, M.: Probability and Related Topics in Physical Sciences. Interscience, New York (1959)
10. Gold, T.: Am. J. Phys. **30**, 403 (1962)
11. Dyson, F.: Phys. Rev. **75**, 1736 (1949)
12. Davies, P.C.W., Twamley, J.: Class. Quantum Gravity **10**, 931 (1993)
13. Craig, D.A.: Ann. Phys. **251**, 384 (1996)
14. Kiefer, C., Zeh, H.D.: Phys. Rev. D **51**, 4145 (1995)

Chapter 22
A Computer's Arrow of Time

L.S. Schulman

Abstract Some researchers believe that the psychological or consciousness arrow of time is a consequence of the thermodynamic arrow. Some do not. As for many issues in this area, the disagreement revolves about fundamental and undebatable assumptions. As a contribution to this standoff I consider the extent to which a computer—presumably governed by nothing more than the thermodynamic arrow—can be said to possess a psychological arrow. My contention is that the parallels are sufficiently strong that little room is left for an independent psychological arrow. Reservations are nevertheless expressed on the complete objectivity of the thermodynamic arrow.

Keywords Time's arrows · Two-time boundary condition · Causality · Cybernetics · Computers

1 Introduction

The manifest asymmetry of past and future was a subject of inquiry long before developments in physical theory enhanced the puzzle through an apparent conflict with the nearly symmetric microscopic laws of physics. This "manifest" asymmetry is sometimes called the psychological arrow of time or the consciousness arrow or the biological arrow [1]. Its characterizations are as diverse as its definition is difficult. One common theme is that the past is *over*, complete, immutable; the future is open to change. Mostly, physicists stick to more clear-cut asymmetries, for example in the Second Law of Thermodynamics (the thermodynamic arrow), the expansion of the universe (the cosmological arrow) or certain microscopic laws of physics (the CP arrow). My own opinion, shared by others, is that the psychological arrow is a consequence of the thermodynamic arrow. I view our psychological processes as an outgrowth of other biological processes and I find no reason to propose an arrow for digestion that is not already covered by that describing other chemical processes, specifically the thermodynamic arrow. See Ref. [2].

L.S. Schulman (✉)
Physics Department, Clarkson University, Potsdam, NY 13699-5820, USA
e-mail: schulman@clarkson.edu

S. Albeverio, P. Blanchard (eds.), *Direction of Time*,
DOI 10.1007/978-3-319-02798-2_22,

Unfortunately, it is difficult to defend this position with anything more than hand-waving. Partly, subtleties in the definition of the thermodynamic arrow get in the way, but more of an obstacle is the intrusion of issues like consciousness, life, free will, and possible indeterminism.

In this article I will show that a contemporary computer has features that parallel the psychological arrow of time. I do not claim that this proves the assertion made in my first paragraph. In fact I expect that no one who does not already agree with me will find my analogies compelling. What I hope to do though is to see how much *can* be said before coming to "undebatable" issues (in the sense that no one ever convinces anyone else) like reductionism.

In Sect. 2 the form of the thermodynamic arrow to be used is presented including its implications for the distinction between prediction and retrodiction (the mathematical details are not essential to the sequel). Next I develop a characterization of a computer. Section 4 discusses computer properties that parallel our own psychological arrow. The final section explicitly states the claimed relationship as well as expresses some reservations.

2 The Thermodynamic Arrow: Causality and Entropy Increase

The thermodynamic arrow is here defined as a kind of causality. Let t be a neutral dynamical time parameter having no a priori thermodynamic directionality. Consider several identically prepared macroscopic systems that are isolated during the time intervals $[0, t_0]$ and $[t_0, t_1]$ ($0 < t_0 < t_1$), and let them be struck by a variety of outside forces at t_0. For a thermodynamic arrow whose direction is in the direction of increasing "t" their behavior in the interval $[0, t_0]$ will be identical, but will be different in the interval $[t_0, t_1]$. This arrow can also be characterized as the use of *initial* conditions for macroscopic problems. The choice of which direction of the parameter t is to be considered "initial' is the arrow. This is essentially equivalent to the usual statements about entropy increase or the forbidding of the conversion of heat to work.

As shown in Ref. [3], both macroscopic causality and entropy increase can be derived within a larger time-symmetric context. An outline of the reasoning follows. For simplicity only a limited range of classical dynamical systems is presented.

The dynamics takes place on a phase space, Ω, with measure μ, and is given by a family of invertible measure-preserving maps, $\phi^{(t)}$, $-\infty < t < \infty$. The coarse graining, necessary to define "macroscopic" as well as entropy, is a partition of Ω, i.e., $\{\Delta_\alpha \subset \Omega\}$, $\alpha = 1, \ldots, G$, with $\bigcup_\alpha \Delta_\alpha = \Omega$, $\Delta_\alpha \cap \Delta_\beta = \emptyset$ for $\alpha \neq \beta$. Let χ_α be the characteristic function of Δ_α and let $\nu_\alpha = \mu(\Delta_\alpha)$ (>0). If f is a function on Ω, its coarse graining is defined to be

$$\hat{f}(\omega) \equiv \sum_\alpha \chi_\alpha(\omega)\langle f \rangle_\alpha, \quad \text{with } \langle f_\alpha \rangle \equiv \frac{1}{\nu_\alpha} \int d\mu \, \chi_\alpha(\omega) f(\omega). \tag{1}$$

Let the system's distribution in Ω be described by a density function $\rho(\omega)$. The primitive entropy is defined as

$$S_{\text{prim}} = -\int_\Omega \rho(\omega) \log(\rho(\omega)) \, d\mu,$$

and is constant in time. The entropy to be used here is defined as

$$S(\rho) \equiv S_{\text{prim}}(\hat{\rho}) = -\int_\Omega \hat{\rho} \log \hat{\rho} \, d\mu, \qquad (2)$$

with $\hat{\rho}$ formed from ρ as in Eq. (1). It is easy to show that

$$S(\rho) = S(\rho_\alpha | v_\alpha),$$

where $\rho_\alpha \equiv \int_{\Delta_\alpha} \rho \, d\mu$, and the function $S(p|q)$ is the *relative entropy* defined by

$$S(p|q) \equiv -\sum_x p(x) \log\left(\frac{p(x)}{q(x)}\right),$$

with p and q probability distributions such that $q(x)$ vanishes only if $p(x)$ does. Of course $\sum \rho_\alpha = \int \rho = 1$. (Note that the sign of our "relative entropy" differs from that of most other authors.)

The selection of coarse grains is itself a question of great interest and elsewhere [4] we have argued that this arises from the dynamics, with dependence on the temporal precision of observers. The physical ideas lying behind Ref. [4] are not new (see Ref. [5]) but as far as I know had not previously submitted to precise implementation.

2.1 Symmetric Behavior of Entropy in the Two-Time Boundary Condition Context

So far everything is time-symmetric. (The transformation $\phi^{(t)}$ is also assumed time-symmetric, with a general definition of this symmetry given in Ref. [6].) To maintain time symmetry one must be careful to set the boundary-value problem symmetric as well. As I have often emphasized, the use of initial conditions can slyly enter a problem, leading occasionally to circular "demonstrations" of an arrow of time. For this reason the dynamical problem for this system is formulated by the demand that the system be found in particular coarse grains at separated times, say ε_0, at time 0 and ε_T at time T. I focus on thermodynamic behavior between these times.[1] For symmetry take $\mu(\varepsilon_0) = \mu(\varepsilon_T)$. The points of Ω satisfying this two-time boundary condition are

$$\varepsilon = \varepsilon_0 \cap \phi^{(-T)}(\varepsilon_T). \qquad (3)$$

---

[1]In discussing the relation of the thermodynamic and cosmological arrows, these times are taken to be cosmologically remote.

To proceed I make the following assumption: the dynamical map, $\phi^{(t)}$ is mixing, and there is a time τ such that for all coarse grains the characteristic decorrelation property holds for $t > \tau$. Specifically, *for $t > \tau$ and for A and B macroscopic* (i.e., unions of coarse grains)

$$\mu\big(A \cap \phi^{(t)}(B)\big) = \mu(A)\mu(B). \tag{4}$$

The usual mixing condition only demands the above factoring, or decorrelation, for $t \to \infty$. The equality in Eq. (4) is shorthand for "equal up negligible quantities" which here correspond to numbers much smaller than the measure of any coarse grain. The time-t image of the set ε is

$$\varepsilon(t) = \phi^{(t)}(\varepsilon_0) \cap \phi^{(t-T)}(\varepsilon_T).$$

To calculate the entropy I need $\rho_\alpha(t)$

$$\rho_\alpha(t) = \frac{\mu(\Delta_\alpha \cap \varepsilon(t))}{\mu(\varepsilon)} = \frac{\mu(\Delta_\alpha \cap \phi^{(t)}(\varepsilon_0) \cap \phi^{(t-T)}(\varepsilon_T))}{\mu(\varepsilon)}.$$

For $T - t > \tau$

$$\mu\big(\Delta_\alpha \cap \phi^{(t)}(\varepsilon_0) \cap \phi^{(t-T)}(\varepsilon_T)\big) = \mu\big(\Delta_\alpha \cap \phi^{(t)}(\varepsilon_0)\big)\mu\big(\phi^{(t-T)}(\varepsilon_T)\big),$$

$$\mu(\varepsilon) = \mu(\varepsilon_0)\mu\big(\phi^{(-T)}(\varepsilon_T)\big).$$

Using the measure-preserving property of $\phi^{(t)}$, a factor $\mu(\varepsilon_T)$ appears in both numerator and denominator leading to

$$\rho_\alpha = \mu\big(\Delta_\alpha \cap \phi^{(t)}(\varepsilon_0)\big)/\mu(\varepsilon_0).$$

This is precisely what one gets *without* future conditioning, so that all macroscopic quantities, and in particular the entropy, are indistinguishable from their unconditioned values.

Working backward from time T one obtains an analogous result. Define $s \equiv T - t$ and set $\tilde{\varepsilon}(s) \equiv \varepsilon(T - s)$. Then

$$\tilde{\varepsilon}(s) = \phi^{(T-s)}(\varepsilon_0) \cap \phi^{(-s)}(\varepsilon_T).$$

If s satisfies $T - s > \tau$, then when the density associated with $\tilde{\varepsilon}(s)$ is calculated its dependence on ε_0 drops out. It follows that

$$\rho_\alpha(s) = \mu\big(\phi^{(-s)}(\varepsilon_T)\big)/\mu(\varepsilon_T).$$

For a time-reversal invariant dynamics this gives the entropy the same time dependence coming back from T as going forward from 0.

The proximity to low entropy boundary conditions thus induces the usual entropically defined thermodynamic arrow, where "proximity" is based on the equilibration time scale, τ. Physical systems typically have more than a single time scale. In fact, as suggested by Ref. [4], the definition of coarse grains generally depends on the existence of a scale shorter than τ, such that on that smaller scale the system relaxes *within* the grain.

2.2 Symmetric Behavior of Causality in the Two-Time Boundary Condition Context

In the same two-time boundary condition context, a perturbation-based notion of macroscopic causality can also be deduced. Using two-time boundary conditions one considers dynamical evolution with unperturbed and perturbed dynamics. "Perturbed" means that at a specified intermediate time an additional force acts. When solving the perturbed and unperturbed boundary-value problems, there will be different microscopic solutions. In principle, the macroscopic solutions could differ at *all* intermediate times. However, in a system with causality they differ on only one side of the perturbation.

Let the time interval for the boundary-value problem be $[0, T]$. Call the unperturbed system A; its boundary conditions and history are as described in the previous section. It evolves under $\phi^{(t)}$, its boundary conditions are ε_0 and ε_T, and its microstates are

$$\varepsilon^{(A)} = \varepsilon_0 \cap \phi^{(-T)}(\varepsilon_T)$$

(formerly called ε). System B, the perturbed case, has an additional transformation act at time t_0. Call this transformation ψ. It should not be dissipative—I do not want an arrow from such an asymmetry [7].[2] ψ is thus invertible and measure preserving and for simplicity is assumed instantaneous. Solutions of the boundary-value problem evolve from ε_0 to ε_T under $\phi^{(T-t_0)}\psi\phi^{(t_0)}$. The microstates for system B are therefore in

$$\varepsilon^{(B)} = \varepsilon_0 \cap \phi^{(-t_0)}\psi^{-1}\phi^{(-T+t_0)}(\varepsilon_T).$$

Clearly, $\varepsilon^{(A)} \neq \varepsilon^{(B)}$. But as I now show, for mixing dynamics and for sufficiently large T, the following hold: (1) for t_0 close to 0, the only macroscopic differences between A and B are for $t > t_0$; (2) for t_0 close to T, the only macroscopic differences are for $t < t_0$. This means that the direction of causality coincides with the direction of entropy increase.

The proof is nearly the same as above. Again use the time τ such that the mixing decorrelation holds for time intervals longer than τ. First consider t_0 close to 0. The observable macroscopic quantities are the densities in grain Δ_α, which are, for $t < t_0$,

$$\rho_\alpha^A(t) = \mu\big(\Delta_\alpha \cap \phi^{(t)}(\varepsilon_0) \cap \phi^{(t-T)}(\varepsilon_T)\big)/\mu\big(\varepsilon^{(A)}\big),$$

$$\rho_\alpha^B(t) = \mu\big(\Delta_\alpha \cap \phi^{(t)}(\varepsilon_0) \cap [\phi^{(t-t_0)}\psi^{-1}\phi^{(t_0-T)}](\varepsilon_T)\big)/\mu\big(\varepsilon^{(B)}\big).$$

As before, the mixing property, for $T - t > \tau$, yields $\rho_\alpha^A(t) = \mu(\Delta_\alpha \cap \phi^{(t)}(\varepsilon_0))/\mu(\varepsilon_0)$, which is the initial-value-only macroscopic time evolution. For ρ_α^B, the only difference is to add a step, ψ^{-1}. Unless ψ^{-1} is diabolically contrived to undo $\phi^{(-u)}$ for large u, this will not affect the argument that showed that the dependence on ε_T disappears. Thus A and B have the same macrostates before t_0.

---

[2]In Ref. [7] an arrow was derived from an asymmetric, dissipative perturbation, rather than from proximity to one or another boundary-value-stipulated low entropy state.

For $t > t_0$, $\rho_\alpha^A(t)$ continues its behavior as before. For $\rho_\alpha^B(t)$ things are different:

$$\rho_\alpha^B(t) = \mu\big(\Delta_\alpha \cap [\phi^{(t-t_0)}\psi\phi^{(t_0)}](\varepsilon_0) \cap \phi^{(t-T)}(\varepsilon_T)\big)/\mu(\varepsilon_B) \quad (t > t_0).$$

Now I require $T - t > \tau$. If this is satisfied the ε_T dependence drops out and

$$\rho_\alpha^B(t) = \mu\big(\Delta_\alpha \cap [\phi^{(t-t_0)}\psi\phi^{(t_0)}](\varepsilon_0)\big)/\mu(\varepsilon_0).$$

This shows that the effect of ψ is the usual initial-conditions-only phenomenon.

If we repeat these arguments for t such that $T - t$ is small, then just as we showed in Sect. 2.1, the effect of ψ will only be at times t *less than* t_0.

2.3 Analysis of a Macroscopic System in This Context

Either based on the above arguments or on other approaches to the thermodynamic arrow, the computer can be treated as a macroscopic system whose underlying microscopic dynamics is reversible, but which nevertheless, when treated macroscopically can have irreversible aspects. Moreover, it will be treated as an *open* system, allowing further introduction of irreversible behavior (Fig. 1). Suppose that a collection of dynamical variables has been identified for the computer. Then it would be reasonable to use the Langevin equation for the motion. The reversible terms in this equation represent the pure underlying dynamics, while the irreversible term plus the noise arise from suppressed degrees of freedom—the usual justification for that equation. Moreover, the sign of the irreversible term would be the expression of the thermodynamic arrow. Finally, considering the density function for the computer's degrees of freedom, it should satisfy a high-dimensional Fokker–Planck equation, as is usual for densities of systems obeying a Langevin equation.

2.4 Prediction and Retrodiction

In Ref. [2] I discussed the equivalence of the arrow of time to the fundamental distinction between prediction and retrodiction for macroscopic states. For prediction one takes equal probability for all microstates consistent with the given macrostate and averages over their subsequent motion. For retrodiction one makes guesses about the earlier microstates and accepts those that arrive in the required macrostate. The guesses are also informed by other considerations so that one is effectively using Bayesian statistics.

It is paradoxical that by this method of knowing the future may be more certain than the past. Take a glass of water with a small piece of ice at 2 p.m. Suppose it to be isolated from 1 p.m. to 2 p.m. and from 2 p.m. to 3 p.m. The 3 p.m. state is not in doubt: a colder glass of water. But what is the 1 p.m. precursor? Two pieces of ice, one of them small? One big cube? One big sphere? There is no way to know.

But the paradox is only that: when we—or a computer—"knows" the past we do not attain this knowledge by retrodicting (but see Sect. 4). If someone is an eyewitness (seeing an ice *cube*, say) that observation is transmitted and stored in the brain.

Fig. 1 The computer as an
open system

Without worrying about the exact storage mechanism, what has been done is the creation of a *record* of the past. The states maintaining this record have the property that they do not change—if memory is good—so for them the retrodiction problem is trivial. (This property can also be stated as the possession of few predecessors [2].) It is this record that is the past. Indeed we distinguish between "knowing" the past in this way and "knowing" it by retrodictive calculation: "I saw it was a cube of 2 cm," versus "I suppose it was a 2 cm cube because the local source of ice is a freezer that only makes cubes and it would have had to be about 2 cm to reach the present size in this environment." (Note too the Bayes-like use of outside information.)

3 A Computer

However abstract this discussion may become, the computer is to be thought of as a *physical* system, like a steam engine or a cuckoo clock. It is attached to a power source, usually thought of as supplying energy, but more significantly characterized as a source of negentropy. (The total energy in the machine is secondary; in fact effort must be expended to keep it cool. Similar energy balance issues exist for the planet: the role of the computer power supply thus resembles the role of the sun vis a vis the earth.)

Each bit of data or program is held in a "two-state" physical subsystem. Ideally this is pictured as a double-well potential with a high barrier. Actual computers have far more internal degrees of freedom for each bit; so many that for example one can generally assign a temperature to the storage unit. The characteristic features though are the high barrier when the system is left alone and the existence of a mechanism that easily moves the bit from one "state" (which is really a collection of microstates) to another. The high barrier, preventing spontaneous transitions, ensures the reliability of retrodiction. Call the state and system, $\omega \in \Omega$. The function of the CPU (central processing unit) is to move the system from one point to another within Ω. For humans the states are more subtle with actual storage mechanisms far from understood [8–14].

There is also a clock. Although asynchronous computers exist, most machines march to a definite beat. For humans there is no overall synchronization, although locally (as in pacemakers) it can be crucial. I mention this because in the context of psychological arrows there is often discussion of the meaning of the "present." For the computer I do not believe the ticks of its clock define the duration of "present," so that one need not be concerned with the presence or absence of mental synchronous processing. Rather I expect the computer's present to be the interval between writes to the record file, as will be described below.

Computations are accompanied by dissipation, so much so that one of the principal issues for Intel's Itanium chip is its power consumption.[3] More fundamentally, Landauer [15, 16] has shown that computation requires irreversible processes and heat generation. From the standpoint of our two-state systems (where those "two" states are *macro*states), the system will typically enter a new macrostate in a microstate with relatively high energy. Dropping to a lower energy of the same macrostate produces heat energy and allows the system to "forget" its recent arrival and be indistinguishable from a system that had been in this state indefinitely. This represents a loss of information.

There is also an evolutionary process that applies to computers. It is not Darwinian survival of the best software and hardware, as is evident to anyone who has had an effective tool made obsolete by the ongoing march of commercial interests. Nevertheless, consumers do have a vote, and what pleases them and fulfills their needs tends to survive.

Both computers and animals find it convenient to have (at least) two kinds of memory, long term and short term. In view of the difficulty of finding a full physiological basis for *any* memory in the brain, one does not expect there to be much resemblance in the physical mechanisms of the two systems. Nevertheless, the usefulness of maintaining both sorts of memory appears to be common. One might also construe the overall architecture of humans or machines to be a kind of memory, in the sense that a good deal of the underlying programming of both machines and people is built into the structures of the respective entities. Thus the genome is a kind of memory as is the wiring diagram, evolved and extended from earlier versions, of a chip.

4 Arrow-Like Features in Computer States

I consider a computer whose job it is to record, predict (and perhaps influence, as explained below) the weather. It is an open system and interacts with the external world in three principal ways: (1) acquisition of needed resources (electricity, air for cooling, etc.), (2) input via "sensory" channels (keyboard, mouse, updates on

---

[3]See for example the *New York Times* article, "Intel's Huge Bet Turns Iffy," by J. Markoff and S. Lohr (Sep. 29, 2002) or the more recent, "Intel Takes The Heat Off Its Chips," *Information Week*, Feb. 7, 2005, by A. Ricadela.

current weather, both through links to raw data and through connections to other computers that get and process similar data) and (3) output to monitor, disks, links to other machines. It was created, hooked up and turned on by a human at some time in the past. "Past" here is in accordance with the thermodynamic arrow, which is given. The operation of the computer is that of a physical device (transistors, motors, cables, etc.) in the context of this thermodynamic arrow. The objective is to see how many properties of the consciousness or psychological arrow may be attributed to the computer, given this thermodynamic arrow.

I assume that the programming of the machine is such that at any given moment the weather information in the computer is of several kinds:

1. In long term memory: records of actual weather patterns (including collected data).
2. In long term memory: records of weather patterns computed by the machine, for times at which the machine also has actual weather patterns (Item 1).
3. In long term memory: records of weather patterns computed by the machine for times at which the machine does not have actual weather patterns. These can be both for times before and after the current external present.
4. In short term memory: records of weather patterns that are already computed but not yet stored in long term memory. (Relation to external time as in Item 3.) Also external weather patterns currently being input.
5. In short term memory: temporarily stored numbers involved in computing the next weather pattern.

In addition a considerable portion of the machine memory may hold computer programs, which themselves span a hierarchy of types: programs written for this task, software that implements these programs (e.g., codes for Fortran), low level utility programs as well as the operating system. The physical device takes the machine from one "state" to another, where "state" is a list of all bits in the foregoing inventory.

For a well-written program the way the machine handles these different kinds of information parallels the way we do. The past will include all patterns in Item 1. A separate part of the past will be the memory in Items 2 and 3. The computer will distinguish these as its own "opinions," its guesses, some of which have been checked against authority (Item 1). If the computer needs to check the information in Item 4 (perhaps while moving forward with the next calculation), this too will be considered past.

What then is future? In practice this is what the machine *will* compute or will receive from external links. But in the machine itself there isn't any. It is prepared to accept new data to add to Items 1, 2 and 4, and to this extent shows awareness that there *is* a future. Moreover, provisions for the future can go beyond the programming necessary for the computer to be ready to accept new data. There can also be an ability to act, to influence the future. For example, in response to inadequate data it may automatically launch a weather balloon, or inform a human of the need to do so. It might even institute cloud seeding operations in an attempt to increase rainfall (presumably having been programmed to do this and linked to appropriate external devices). What I call "awareness" of the future is thus the fact that built into the program is the ability to accept new data, continue the computation while receiving new data or transferring data between short and long term memory, as well as provisions in the program to respond to certain states by issuing particular kinds of output, such as seeding a cloud. This is not different from our relation to the future, except that our "programming" developed through a process of biological evolution (which certain computer programs emulate [17]).

And the concept of "present"? Comparison with our own experiences suggests that the interval between writing calculated patterns to short term memory is the appropriate analogue. If the machine multitasks by accepting external weather input simultaneously with its calculations, the human analogy is less clear although we too are fitted for multitasking: most of us *can* chew gum and walk at the same time.[4]

The distinctions above regarding past, present and future apply to what I called a well-written program. This recalls the existence of humans who do not possess a "normal" sense of time. Saniga [18–21] has collected many examples of this from the literature of psychopathology and interpreted these unusual perspectives in terms of projective spaces.

For a computer, as for a person, a check of memory is technically speaking a retrodiction. (This check may be part of its continuing program, perhaps to improve performance by looking for sources of error in previous work, or it may be introspection due to encountering unanticipatable events, such as a query from a human using the machine.) When pulling up "old" records it equates the stored 0 s and 1 s in its memory as a weather pattern, effectively retrodicting by "believing" those bits to be the same as it earlier wrote. Here is where my earlier remarks on the characteristics of good memory registers plays a role. They should be states with few precursors, in fact one precursor, the same state. Here too "state" should be interpreted as a coarse grain in which (e.g.) only the magnetic configuration is relevant, temperature and small variations in magnetization being ignored. For these systems retrodicting is reliable.

There is also poor memory. Files may be corrupted (including by viruses) and the computer may or may not be aware of this, where "aware" means the bad data are flagged, perhaps having failed some check-digit test. A computer may also have false or implanted memories, the skullduggery being different from that in the human phenomenon, but the result analogous.

5 Conclusions

The computer has a past that is in many ways as rich as our own, complete with memories of actual events, of its impressions of those events, of its calculated predictions for future events. For the example given above, it also maintains an image

---

[4]Lyndon Johnson is said to have unkindly suggested that Gerald Ford was incapable of this bit of multitasking. See the *Columbia World of Quotations*, no. 22545, Columbia Univ. Press, 1996.

of the world. Further, it has a present delineated by intervals between the creation of new memories, probably a bit more well defined than our own present. It is prepared for the future and may act to affect that future. It does all this without an independent arrow of time, retaining the past/future distinction by virtue of its being part of a mechanistic world with a thermodynamic arrow in a particular direction. For the computer, as for us, the past is over, complete. In a well-written program, files in the enumerated categories of Sect. 4 are not tampered with. Similarly, the future is open, in the sense that it is nowhere contained in a memory file. It has an existence in that the machine is programmed to deal with certain kinds of input ("contingencies") as well as the results of its own calculations.

The point of this article is that in view of all this parallel structure there is no reason to postulate an independent psychological arrow. This is a reductionist view that may not be acceptable to some.

I close with a disclaimer. It is assumed throughout this article that the Second Law of Thermodynamics is an objective statement about the world, like Einstein's General Relativity, whereas the psychological arrow is lacking in full definition because of its subjective nature. But the Second Law has subjective elements as well. At its core is an essential distinction between the microscopic and the macroscopic, equivalent to the distinction between work and heat, equivalent in turn to the selection of coarse grains in phase space or Hilbert space. The choice of coarse grains has important aspects of subjectivity, so that the superior position of the Second Law vis à vis the definition of a psychological arrow may be questioned. Recent work [4] has addressed this question in a precise way and implementation of the physical idea that coarse grains correspond to objectively slow variables has begun. My belief in the ultimate success of this program leads me back to the conclusion that the psychological arrow is the dependent concept, but one should not be too dogmatic.

Acknowledgements   I thank E. Mihóková for helpful discussions. This work was supported by the United States National Science Foundation Grant PHY 00 99471.

References

1. Savitt, S.F.: Time's Arrows Today. Cambridge University Press, Cambridge (1995)
2. Schulman, L.S.: Time's Arrows and Quantum Measurement. Cambridge University Press, New York (1997)
3. Schulman, L.S.: Causality is an effect. In: Mugnai, D., Ranfagni, A., Schulman, L.S. (eds.) Time's Arrows, Quantum Measurements and Superluminal Behavior, pp. 99–112. Consiglio Nazionale delle Ricerche (CNR), Rome (2001)
4. Schulman, L.S., Gaveau, B.: Coarse grains: the emergence of space and order. Found. Phys. **31**, 713 (2001)
5. Landau, L.D., Lifshitz, E.M.: Statistical Physics. Pergamon Press, Oxford (1980)
6. Schulman, L.S.: Time reversal for unstable particles. Ann. Phys. **72**, 489 (1972)
7. Schulman, L.S., Shtokhamer, R.: Thermodynamic arrow for a mixing system. Int. J. Theor. Phys. **16**, 287 (1977)

8. Milton, J.G., Mackey, M.C.: Neural ensemble coding and statistical periodicity: speculations on the operation of the mind's eye. J. Physiol. **94**, 489 (2000)
9. Freeman, W.J.: A physiological hypothesis of perception. Perspect. Biol. Med. **24**, 561 (1981). Reprinted in [13]
10. Alkon, D.L.: Learning in a marine snail. Sci. Am. **249**, 70 (1983)
11. Lynch, G., Baudry, M.: The biochemistry of memory: a new and specific hypothesis. Science **224**, 1057 (1984)
12. Mishkin, M., Appenzeller, T.: The anatomy of memory. Sci. Am. **256**, 80 (1987)
13. Shaw, G.L., Palm, G.: Brain Theory. World Scientific, Singapore (1988)
14. Marinaro, M., Tagliaferri, R.: Neural Nets: WIRN VIETRI-98. Springer, Berlin (1999)
15. Landauer, R.: Irreversibility and heat generation in the computing process. IBM J. Res. Dev. **5**, 183 (1961), Reprinted in [16]
16. Leff, H.S., Rex, A.F.: Maxwell's Demon: Entropy, Information, Computing. Princeton University Press, Princeton (1990)
17. Gould, H., Tobochnik, J.: An Introduction to Computer Simulation Methods: Applications to Physical Systems. Addison-Wesley, Reading (1996)
18. Saniga, M.: Pencil of conics: a means towards a deeper understanding of the arrow of time? Chaos Solitons Fractals **9**, 1071 (1998)
19. Saniga, M.: On a remarkable relation between future and past over quadratic galois fields. Chaos Solitons Fractals **9**, 1769 (1998)
20. Saniga, M.: In: Buccheri, R. et al. (eds.) Studies on the Structure of Time: From Physics to Psycho(patho)logy. Kluwer Academic/Plenum, Norwell/New York (2000)
21. Saniga, M.: Quadro-quartic Cremona transformations and four-dimensional pencil-space-times with the reverse signature. Chaos Solitons Fractals **13**, 797 (2002)

Chapter 23
Remarks on the Compatibility of Opposite Arrows of Time II

H.D. Zeh

Abstract In a series of papers (Schulman in Time's Arrows and Quantum Mechanics, 1997; Phys. Rev. Lett. 83:5419, 1999; Time's Arrow, Quantum Measurements and Superluminal Behavior, 2001), Lawrence Schulman presented examples which demonstrate the compatibility of opposite arrows of time in various situations. In a previous letter to Entropy (Zeh in Entropy 7(4):199, 2005)—in this volume reproduced in Chap. 21, I questioned some of them for not being realistic in spite of being logically correct. Schulman replied (Entropy 7(4):208, 2005) to these objections in a letter directly succeeding my one (Chap. 20). I am here trying to clarify some aspects of the dispute, thereby further explaining and supporting my previous conclusion that simultaneous opposite arrows are incompatible in practice.

Keywords Time arrow · Final conditions · Solvable models · Causality · Retardation · Quantum measurements · Gravitational entropy · Cosmology

1 Introduction

It is always an intellectual pleasure arguing with Larry Schulman, in particular on fundamental problems such as the arrow of time. Let me therefore first emphasize that this discussion can be meaningful (and hopefully useful for the reader) only because we already agree on many basic assumptions—for example, that an arrow of time is not *intrinsic* to the concept of time (as traditionally assumed by philosophers), but the consequence of boundary conditions which seem to characterize this specific universe. Although even this may ultimately be a matter of definition (of time), the boundary value approach is the appropriate one in a setting that is based on physical laws which are deterministic and, up to certain "compensating symmetry transformations" [6], even symmetric under time reversal. The physical

First published in *Entropy* **8**, 44 (2006).

H.D. Zeh (✉)
69151 Waldhilsbach, Germany
e-mail: zeh@urz.uni-heidelberg.de
url: http://www.zeh-hd.de

S. Albeverio, P. Blanchard (eds.), *Direction of Time*,
DOI 10.1007/978-3-319-02798-2_23,
© Springer International Publishing Switzerland 2014

287

meaning of the determinism of the Schrödinger equation is, of course, itself a fundamental issue (see Sect. 4).

For example, we seem to agree that the *global* arrow of time could as well point in the opposite formal direction (although this would not make an observable difference), that there might be no arrow at all, or that there could be opposite arrows in dynamically isolated systems (such as causally disconnected regions of spacetime). We also seem to agree that the arrow of time characterizing most physical and physiological processes may in principle change direction if and when the universe starts recontracting, while the nature of phenomena occurring in the transition region (which may have to last quite a while) would then probably have to be quite unusual, and presumably exclude anything resembling memory or information.

So our remaining problem may be described as whether or not this transition can be drastically inhomogeneous, such that local "pockets" of an arrow may persist beyond the turning point of cosmic expansion, and remain consistent with the then globally dominating opposite arrow. Since two non-interacting universes with opposite arrows could trivially co-exist, the answer to this problem must depend on the strength of the interaction between the pockets and the rest—and here we still disagree. For example, it would be particularly fascinating to observe from outside "time going backward" in such a pocket. However, we are no Laplacean demons with their unrestricted observation capacities and independent arrows, but participators of this universe. Since sufficiently isolated pockets can hardly be expected to exist here on earth or within the solar system, it appears technically difficult, for example, to "illuminate them with our own light" for this purpose. While *their* light would be advanced from our point of view, we should nonetheless be able to see it, since in a classical description there is but one (the "real") electromagnetic field that interacts with our retina—although the latter would then react in *our* causal direction.

Even if we *could* use our own light for this purpose, it would slightly disturb matter in the pocket. So the question arises, what this would mean in a situation with varying direction of "causality". This problem is illustrated in Fig. 2 of [4] by means of the exact microscopic evolution of a system consisting of many degrees of freedom. This evolution is assumed to start from a state of low entropy and, for statistical reasons, then to evolve into a high-entropy state. If a very small change of the microscopic state is now assumed at a certain time, one may re-calculate the trajectory from this new boundary condition in both directions of time. While the evolution would in general only microscopically be changed in what used to be the direction of growing entropy (the physical future), it becomes macroscopically different in the opposite direction—as demonstrated in the figure by means of the entropy. Hence, a small change of state in the "time pocket" must be expected to drastically change its history (defined as its own past—that is, in the direction of our future), thereby destroying its previously assumed opposite arrow.[1] Even if we

---

[1]It would be illustrative to generalize this model to a set of *local* degrees of freedom with local interactions in order to study the *propagation* of the distortion in space, and the resulting "causal structures".

assume that this distortion, caused by us, induced changes in the pocket only in *its* future (in the direction of our past), it would then also somewhere have to affect our past, and, by the same argument, destroy the low entropy condition at the big bang that our existence relies on.

This consequence, which seems to demonstrate the inconsistence of arbitrary boundary conditions at different times, is a *probable* result, obtained for a "typical" (arbitrarily chosen) solution taken from the pocket. However, Schulman's argument is more subtle. His question is whether solutions which obey boundary conditions at both ends *exist* (even though they may be very improbable). Indeed, for the same reason that any global solution which characterizes an arrow is improbable when compared with the overwhelming majority of quasi-equilibrium states, but nonetheless readily accepted as describing reality, we should similarly be ready to accept even less probable solutions which obey two low entropy boundary conditions if there are reasons to consider such conditions as given. One may even speculate whether an appropriate assumption of this kind could give rise to a unique solution for cosmic evolution—given the kinematical concepts and dynamical laws. Since anti-causality is counterintuitive, one cannot refer to causal intuition in order to disprove its possibility.

As we discussed various consequences of this situation in our previous letters and former publications, I will now briefly address some of Schulman's arguments in his reply to my objections, essentially in the order he presented them.

2 Wet Carpets and Detective Stories

In his *wet carpet* example [2], Schulman explained how a warning from the future by means of retarded radiation to close the window and save one's carpet from getting wet by a sudden rain shower may be consistent with determinism. Similarly, Wheeler and Feynman [7] had discussed the case of a bullet passing a trap door and causing this door to close before the passage by means of advanced radiation sent from the space behind the door. Would the door or the window close because of the warning, or not, since there would be no warning if it was closed? In both cases, a consistent solution of the problem requires a continuum of possible states in between a closed and an open door, that is, the possibility of a half-open door. This continuum is related to the infinite information capacity of classical phase space, which will be quantitatively discussed and compared with the existence of phase space cells in the next section.

However, while these examples are correct and helpful for an understanding, they are also unrealistic in neglecting uncontrollable but nonetheless important degrees of freedom, which are responsible for the "overdetermination of the past" that characterizes a causal world and lets the past appear fixed. In quantum theory, this unavoidable information spreading leads to the phenomenon of decoherence. For this reason I compared these examples, which employ isolated systems, with certain clever science fiction stories that describe time travel into the past in an apparently consistent manner—namely by neglecting all uncontrollable phenomena which are related to irreversibility. A similar strategy is known for good detective stories, where the murderer must come from a given set of persons (in contrast to crime in reality), and is expected to be "logically determined" by only those events which were explicitly mentioned in the story.

Of course, I did *not* want to say that Schulman's or Wheeler and Feynman's important examples represent science fiction. Quite the contrary: they are very informative, even though they neglect an important aspect that must be added in order to describe the real world.

3 Phase Space Volume

In order to comply with quantum theory, an effective classical phase space must be assumed to consist of finite cells of size h^{3N}, where $3N$ is the number of degrees of freedom. In my arguments I had raised the question if a two-times boundary condition of low entropy would possibly reduce phase space to less than one cell—thus signaling dynamical inconsistence in the case that the conditions can be regarded as statistically independent. Schulman replied that a typical phase space volume for a gas would contain something like $10^{10^{20.28}}$ phase space cells. A reduction of the spatial volume by a factor of 64, say, would reduce the second exponent only in the second figure after the decimal point, such that this reduction could easily be applied twice.

This is correct, but arguments with double exponentials may easily be deceiving: a small entropy difference may require a huge probability ratio. For simplicity, let me consider a gas consisting of $N = 10^{20}$ particles. Under customary conditions here on earth, the phase space for each particle in a gas is of the order $10^{10} h^3$. The resulting N-particle phase space $(10^{10})^N h^{3N} = 10^{10^{21}} h^{3N}$ is slightly larger than Schulman's choice. Reducing the single-particle phase space by a factor of 64 would indeed lead to a very "small" change, given by $(10^{10}/64)^N = 10^{10^{20.916}}$, although it represents a reduction of phase space by a factor of $(1/64)^{10^{20}}$. Even forgetting Gibbs' paradox and enlarging phase space by the enormous factor of $N! \approx N^N$ would lead to "no more" than $(10^{20} 10^{10})^N = 10^{10^{21.477}}$. The entropy is therefore almost exclusively determined by the particle number N.

On the other hand, entropy differences in the solar system are governed by density and temperature ratios such as those between the sun and interstellar space. So, a change of single-particle phase space by a factor of 10^5 (that is, a reduction from 10^{10} to 10^5) in an irreversible cosmic process appears quite conservative, and would be far from requiring degenerate matter. Applied once, it reduces total phase space to $(10^5)^N = 10^{10^{20.699}}$ —apparently not drastically different from the numbers given above, but applied twice (at two sufficiently distant times) it leads to $(10^0)^N = 1$ (independent of N)! A similar consistency problem would arise in Wheeler and Feynman's time-symmetric absorber theory (see Chap. 2 of [6]).

In cosmological context, entropy is dominated by gravity and black holes (see Sect. 5). A two-times low entropy condition would then lead to severe consistency problems even in the case of a simultaneous transition between opposite arrows. They may perhaps be overcome in time-less quantum gravity [8].

4 Quantum Measurement

Schulman claims furthermore that "special quantum states" (obeying appropriate final conditions) would be able to solve the quantum measurement problem without requiring deviations from unitary dynamics. This would be a dramatic achievement, but it could not be explained just by assuming separate "pockets" with an opposite arrow. It would instead require special final conditions whenever and wherever measurements or measurement-like events occur. Special *initial* conditions of low entropy are in general readily accepted, since their causal origin can be confirmed "historically" by means of consistent documents. No such documents exist about a low entropy future. Simply assuming "special states" in this way, just as they are required, would therefore replace science by arbitrary wishful thinking.

However, there exist arguments which seem to strictly rule out such possibilities by investigating the dynamics of *individual* quantum measurements [9, 10]. They are based on von Neumann's interaction, and valid regardless of all individual complexity in the device, such as the presence of many particles or metastable states. No freedom for a selection of final states then remains—except for the addition of irrelevant degrees of freedom.

In fact, the example defined in Schulmann's Sect. 6.2 of [1] is not concerned with a superposition of different measurement results, but with a superposition of just two states: one with a droplet *somewhere* in the Wilson chamber (itself a superposition of many locations), and one with no droplet at all (no measurement). Moreover, the system is treated as closed, hence neglecting any decoherence, which would necessarily characterize a visible droplet or any other "pointer position". The model, which uses a spin lattice, leads to a final state that oscillates in time between "droplet" (here represented by many correlated spin flips) and "no droplet", and so allows one to assume *ad hoc* that the coupling between these two states happens to cease at "special times", namely when one of the two oscillating amplitudes vanishes. However, such oscillations are known to occur only for isolated "pathological" systems, such as harmonic oscillators, while realistic complex systems behave irreversibly. As mentioned above, this model would *not* solve the problem of superpositions of *different* pointer positions, while the superposition considered by Schulman could easily be avoided by an irreversible formation of droplets (with 100 % efficiency rather than oscillation). Superpositions of different pointer positions would in fact immediately and irreversibly be decohered by the environment, and thereafter be dynamically robust (with coefficients remaining essentially fixed), but continue to exist in different Everett worlds if the Schrödinger equation were universally valid. Schulman's proposal to solve the measurement problem does not even seem to be possible as 'wishful thinking'.

5 Gravity

It has been known at least since Bekenstein's discovery of black hole entropy and Penrose's subsequent estimates [11] that cosmic entropy is overwhelmingly dominated by gravity. The major low entropy property of the universe is its homogeneity. The formation of inhomogeneities in the form of stars, galaxies and ultimately black holes defines the most important process of entropy production—regardless of what happens inside the event horizon of a black hole, or what kind of new physics may apply there or close to the horizon [12]. It is often overlooked that solar (or even larger) black holes would *lose* mass by radiation only in the very distant future of an ever expanding universe: at present their accumulation of 3 K background radiation by far outweighs their 10^{-7} K (or weaker) Hawking radiation. In order to get rid of gravitating objects for a cosmic "time reversal" in the not extremely distant future, one would need advanced (incoming coherent) radiation to reverse their gravitational contraction [8]. These objects are, therefore, strongly coupled to the general arrow of time, and a further indication that the arrow cannot vary from place to place.

To conclude, let me emphasize again that situations with opposite arrows may be very useful for pedagogical purposes, but cannot be expected to "really happen". Therefore, they appear particularly misleading when presented as possibilities to solve the quantum measurement problem.

References

1. Schulman, L.S.: Time's Arrows and Quantum Mechanics. Cambridge University Press, Cambridge (1997)
2. Schulman, L.S.: Phys. Rev. Lett. **83**, 5419 (1999)
3. Schulman, L.S.: In: Mugnai, D., Ranfagni, A., Schulman, L.S. (eds.) Time's Arrow, Quantum Measurements and Superluminal Behavior. Consiglio Nazionale delle Ricerche, Roma (2001). cond-mat/0102071
4. Zeh, H.D.: Entropy **7**(4), 199 (2005). http://www.mdpi.org/entropy/list05.htm#issue4. Reprinted in Chap. 21 of this volume
5. Schulman, L.S.: Entropy **7**(4), 208 (2005). http://www.mdpi.org/entropy/list05.htm#issue4. Reprinted in Chap. 20 of this volume
6. Zeh, H.D.: The Physical Basis of the Direction of Time. Springer, Berlin (2001)
7. Wheeler, J.A., Feynman, R.P.: Rev. Mod. Phys. **21**, 425 (1949)
8. Kiefer, C., Zeh, H.D.: Phys. Rev. D **51**, 4145 (1995)
9. Wigner, E.P.: Am. J. Phys. **31**, 6 (1963). Reprinted in Wheeler, J.A., Zurek, W.H.: Quantum Theory and Measurement. Princeton University Press, Princeton (1983)
10. Zeh, H.D.: Found. Phys. **1**, 69 (1970). Reprinted in Wheeler, J.A., Zurek, W.H.: Quantum Theory and Measurement. Princeton University Press, Princeton (1983)
11. Penrose, R.: In: Isham, C.J., Penrose, R., Sciama, D.W. (eds.) Quantum Gravity II. Clarendon, Oxford (1981)
12. Zeh, H.D.: Phys. Lett. A **347**, 1 (2005)

Index

S. Albeverio, P. Blanchard (eds.), *Direction of Time*,
DOI 10.1007/978-3-319-02798-2,
© Springer International Publishing Switzerland 2014

Printed in the United States
By Bookmasters